Wild Germplasm for Genetic Improvement in Crop Plants

Wild Germplasm for Genetic Improvement in Crop Plants

Edited by

Muhammad Tehseen Azhar
Institute of Molecular Biology and Biotechnology, Bahauddin Zakariya University, Multan, Pakistan

Shabir Hussain Wani
Mountain Research Centre for Field Crops, Sher-e-Kashmir University of Agricultural Sciences and Technology of Kashmir, Jammu and Kashmir, India

ELSEVIER

ACADEMIC PRESS
An imprint of Elsevier

Academic Press is an imprint of Elsevier
125 London Wall, London EC2Y 5AS, United Kingdom
525 B Street, Suite 1650, San Diego, CA 92101, United States
50 Hampshire Street, 5th Floor, Cambridge, MA 02139, United States
The Boulevard, Langford Lane, Kidlington, Oxford OX5 1GB, United Kingdom

Library of Congress Cataloging-in-Publication Data
A catalog record for this book is available from the Library of Congress

British Library Cataloguing-in-Publication Data
A catalogue record for this book is available from the British Library

ISBN: 978-0-12-822137-2

For information on all Academic Press publications
visit our website at https://www.elsevier.com/books-and-journals

Publisher: Charlotte Cockle
Acquisitions Editor: Nancy Maragioglio
Editorial Project Manager: Cole Newman
Production Project Manager: Kumar Anbazhagan
Designer: Mark Rogers

Typeset by Thomson Digital

Contents

Contributors

Muhammad Qadir Ahmad, Xuzhou Institute of Agricultural Sciences in Jiangsu Xuhuai District/Key Laboratory of Biology and Genetic Breeding of Sweetpotato, Ministry of Agriculture and Rural Affairs/ Sweetpotato Research Institute, CAAS, Xuzhou, Jiangsu, China; Department of Plant Breeding and Genetics, Faculty of Agricultural Sciences and Technology, Bahauddin Zakariya University, Multan, Pakistan

Muhammad Saeed Ahmad, Department of Plant Breeding and Genetics, University of Agriculture, Faisalabad, Pakistan

Emre Aksoy, Niğde Ömer Halisdemir University, Faculty of Agricultural Sciences and Technologies, Department of Agricultural Genetic Engineering, Niğde, Turkey

Zulfiqar Ali, Institute of Plant Breeding and Biotechnology, MNS University of Agriculture, Multan, Pakistan

Mohamed Hamed Arisha, Xuzhou Institute of Agricultural Sciences in Jiangsu Xuhuai District/Key Laboratory of Biology and Genetic Breeding of Sweetpotato, Ministry of Agriculture and Rural Affairs/ Sweetpotato Research Institute, CAAS, Xuzhou, Jiangsu, China; Department of Horticulture, Faculty of Agriculture, Zagazig University, Sharkia, Egypt

Sue J. Armstrong, School of Bioscience, University of Birmingham, Birmingham, United Kingdom

Muhammad Jawad Asghar, Nuclear Institute for Agriculture and Biology, Faisalabad, Pakistan

Rana Muhammad Atif, Centre for Advanced Studies in Agriculture and Food Security, University of Agriculture, Faisalabad, Pakistan; Department of Plant Breeding and Genetics, University of Agriculture, Faisalabad, Pakistan

Muhammad Tariq Awan, Department of Biotechnology, Mirpur University of Science and Technology (MUST), Mirpur, Pakistan

Farrukh Azeem, Department of Bioinformatics and Biotechnology, Government College University, Faisalabad, Pakistan

Muhammad Tehseen Azhar, Institute of Molecular Biology and Biotechnology, Bahauddin Zakariya University, Multan, Pakistan; School of Agricultural Sciences, Zhengzhou University, Zhengzhou, Henan, China

Allah Bakhsh, Niğde Ömer Halisdemir University, Faculty of Agricultural Sciences and Technologies, Department of Agricultural Genetic Engineering, Niğde, Turkey

Melike Bakir, Department of Agricultural Biotechnology, Faculty of Agriculture, Erciyes University, Kayseri, Turkey

Ikram Bashir, Postgraduate Program in Agronomy, Federal University of Pelotas, Rio Grande do Sul, Brazil

Muhammad Tanees Chaudhary, Department of Plant Breeding and Genetics, University of Agriculture, Faisalabad, Pakistan

Xiongming Du, State Key Laboratory of Cotton Biology, Institute of Cotton Research of the Chinese Academy of Agricultural Sciences, Anyang, China

Noreen Fatima, Department of Plant Breeding and Genetics, University of Agriculture, Faisalabad, Pakistan

Claudia Teixeira Guimarães, Maize Germplasm Bank at Embrapa Maize and Sorghum, Sete Lagoas, Minas Gerais, Brazil

Amjad Hameed, Nuclear Institute for Agriculture and Biology, Faisalabad, Pakistan

Muhammad Hammad Nadeem Tahir, Institute of Plant Breeding and Biotechnology, Muhammad Nawaz Shareef University of Agriculture, Multan, Pakistan

Gustavo Heiden, Postgraduate Program in Agronomy, Federal University of Pelotas, Rio Grande do Sul, Brazil; Embrapa Temperate Agriculture, Pelotas, Rio Grande do Sul, Brazil

Amanda M. Hulse-Kemp, US Department of Agriculture-Agricultural Research Service, Genomics and Bioinformatics Research Unit, Raleigh, NC, United States; Department of Crop and Soil Sciences, North Carolina State University, Raleigh, NC, United States

Tehreem Jabbar, Centre for Agricultural Biochemistry and Biotechnology, University of Agriculture, Faisalabad, Pakistan

Muhammad Kashif, Department of Plant Breeding and Genetics, University of Agriculture, Faisalabad, Pakistan

Yalcin Kaya, Trakya University Engineering Faculty, Genetics and Bioengineering Department, Edirne, Turkey

Shahzad Khalid, Forestry Research Institute, Gatwala, Faisalabad, Pakistan

Asif Ali Khan, Institute of Plant Breeding and Biotechnology, MNS University of Agriculture, Multan, Pakistan

Shoaib Liaqat, Institute of Plant Breeding and Biotechnology, MNS University of Agriculture, Multan, Pakistan

Sajid Majeed, Department of Plant Breeding and Genetics, University of Agriculture, Faisalabad, Pakistan

Amir Maqbool, Niğde Ömer Halisdemir University, Faculty of Agricultural Sciences and Technologies, Department of Agricultural Genetic Engineering, Niğde, Turkey

Rodrigo Nicolao, Postgraduate Program in Agronomy, Federal University of Pelotas, Rio Grande do Sul, Brazil

Qandeel-e-Arsh, Centre for Agricultural Biochemistry and Biotechnology, University of Agriculture, Faisalabad, Pakistan

Qiang Li, Xuzhou Institute of Agricultural Sciences in Jiangsu Xuhuai District/Key Laboratory of Biology and Genetic Breeding of Sweetpotato, Ministry of Agriculture and Rural Affairs/ Sweetpotato Research Institute, CAAS, Xuzhou, Jiangsu, China

Iqrar Ahmad Rana, Centre for Agricultural Biochemistry and Biotechnology, University of Agriculture, Faisalabad, Pakistan; Centre for Advanced Studies in Agriculture and Food Security, University of Agriculture, Faisalabad, Pakistan

Humera Razzaq, Department of Plant Breeding and Genetics, University of Agriculture, Faisalabad, Pakistan

Hafiz Mamoon Rehman, Centre for Soybean Research of the State Key Laboratory of Agrobiotechnology and School of Life Sciences, The Chinese University of Hong Kong, Shatin, Hong Kong, China

Muhammad Rizwan, Nuclear Institute of Agriculture, Tandojam, Sindh, Pakistan

Muhammad Abu Bakar Saddique, Institute of Plant Breeding and Biotechnology, MNS University of Agriculture, Multan, Pakistan

Asif Saeed, Department of Plant Breeding and Genetics, University of Agriculture, Faisalabad, Pakistan

Hina Saleem, Department of Plant Breeding and Genetics, University of Agriculture, Faisalabad, Pakistan

Huma Saleem, Department of Plant Breeding and Genetics, University of Agriculture, Faisalabad, Pakistan

Duygu Sari, Department of Field Crops, Faculty of Agriculture, Akdeniz University, Antalya, Turkey

Hatice Sari, Department of Field Crops, Faculty of Agriculture, Akdeniz University, Antalya, Turkey

Muhammad Naveed Shahid, Department of Botany, Division of Science and Technology, University of Education, Lahore, Pakistan

Muhammad Shahid, Nuclear Institute for Agriculture and Biology, Faisalabad, Pakistan

Amir Shakeel, Department of Plant Breeding and Genetics, University of Agriculture, Faisalabad, Pakistan

Zeeshan Shamim, Department of Biotechnology, Mirpur University of Science and Technology (MUST), Mirpur, Pakistan

Shadab Shaukat, Department of Plant Breeding and Genetics, University of Agriculture, Faisalabad, Pakistan

Muhammad Ali Sher, Institute of Plant Breeding and Biotechnology, MNS University of Agriculture, Multan, Pakistan

Flavia França Teixeira, Maize Germplasm Bank at Embrapa Maize and Sorghum, Sete Lagoas, Minas Gerais, Brazil

Roumiana Vasilevska-Ivanova, Institute of Plant Physiology and Genetics, Sofia, Bulgarian Academy of Sciences, Sofia, Bulgaria

Shabir Hussain Wani, Mountain Research Centre for Field Crops, Sher-e-Kashmir University of Agricultural Sciences and Technology of Kashmir, Jammu and Kashmir, India

Muhammad Waqas, Department of Plant Breeding and Genetics, University of Agriculture, Faisalabad, Pakistan

About the editors

Dr Muhammad Tehseen Azhar is working as Associate Professor at the Institute of Molecular Biology and Biotechnology, Bahauddin Zakariya University Multan-Pakistan, his primary responsibility is teaching to graduate and post-graduate classes. Besides, his focus is screening and development of segregating population resulting in the development of several bulks of cotton with more yield and fibre traits. Dr Azhar has specific interest in the development of cotton germplasm having tolerance to biotic and abiotic stresses. Being a Borlaug Alumnus, he worked with various cotton geneticists namely, Dr Richard Percy, Dr Lori L. Hinze and Dr Jame Frelichowsk during a Norman E. Borlaug Fellowship at Texas A&M University and USDA-ARS, USA. Besides, Dr Azhar is Endeavour Alumnus and appointed as Adjunct Lecturer at the School of Plant Biology, UWA. Recently he is appointed as Adjunct Associate Professor, School of Agriculture Sciences, Zhengzhou University, Zhengzhou, China. In addition, Dr Azhar is appointed as a Chair of Germplam and Genetic Stock in International Cotton Genome Initiative (ICGI).

Shabir Hussain Wani is a scientist and assistant professor at the Sher-e-Kashmir University of Agricultural Sciences and Technology of Kashmir, where his research focuses on plant breeding and genetics/plant biotechnology. Included among his awards are Fellow of the Linnean Society of London (2017); the Young Scientist Award (2017) from the Indian Ecological Society, Punjab Agricultural University, Ludhiana; Young Scientist of the Year (2016) by the International Foundation for Environment and Ecology, Kolkata, India; the Young Achiever Award (2016) from the Society for Advancement of Human and Nature (SADHNA), Dr. YS Parmar University of Horticulture and Forestry; the Young Scientist Award (2015) from the Society for Plant Research, Meerut, India; the Young Scientist Award (2009) from the Society for Promotion of Plant Science Research, Jaipur National University, Jaipur; and Fellow of Society for Plant Research Award (2011–12).

Foreword

I have known Dr. Azhar for the past 15 years, first as a PhD student in my lab, and then as a faculty member at the Department of Plant Breeding and Genetics, University of Agriculture, Faisalabad, Pakistan. Because this book is an edited collection of literature on wild resources of crop plants, I want to add that Dr. Azhar has also studied wild relatives of the genus *Gossypium* as a part of his PhD research. Breeding for innovative cultivars adapted to biotic and abiotic stresses is a key to guaranteeing an increase in the productivity and profitability of crop plants. Crop wild relatives (CWRs) are related to contemporary crops, including their ancestors. CWRs have escaped the process of human selection and domestication and survived various challenges in their ecosystems; they provide a gigantic array of valuable genes/alleles and can be a source for enhancing genetic diversity if used in well-planned breeding programs. Diverse studies on important agricultural crops, such as wheat, cotton, rice, maize, oilseeds, vegetables, barley, and pulses, offer numerous examples of bringing resistance for biotic and abiotic stresses, such as heat and drought, from CWRs. With the recent advances in plant omics, it has now become routine to efficiently transfer and utilize resistance from diverse sources. Advances in next-generation sequencing (NGS) have revolutionized genome sequencing and identifying genomic variants that are related to important agronomic traits. Lessons learned by better understanding the genomes of CWRs are likely to help in the transfer of important traits.

Editors of this volume have done exceptional work by including various chapters from international as well as national authors in various research fields. These well-written chapters come from scientists and researchers working in various developed and developing nations. This book describes recent developments with CWRs; their efficient transfer and further utilization through modern genetic, biotechnological, and genomic approaches; and their utilization with important agricultural crops, such as rice, wheat, maize, barley, and pulses. I congratulate the editors for compiling this volume and hope that it will be a valuable reference for researchers, students, and policy makers.

Dr. Shahid Mansoor
Sitara-e-Imtiaz
Chief Scientist and National Distinguished Professor
National Institute for Biotechnology and
Genetic Engineering (NIBGE), Faisalabad, Pakistan

Chapter 1

Introduction: Crop Wild Relatives in Plant Breeding

Sajid Majeed[a], Muhammad Tanees Chaudhary[a], Amanda M. Hulse-Kemp[b,c] and Muhammad Tehseen Azhar[d,e]

[a]Department of Plant Breeding and Genetics, University of Agriculture, Faisalabad, Pakistan; [b]US Department of Agriculture-Agricultural Research Service, Genomics and Bioinformatics Research Unit, Raleigh, NC, United States; [c]Department of Crop and Soil Sciences, North Carolina State University, Raleigh, NC, United States; [d]Institute of Molecular Biology and Biotechnology, Bahauddin Zakariya University, Multan, Pakistan; [e]School of Agricultural Sciences, Zhengzhou University, Zhengzhou, Henan, China

1.1 Background

Crop wild relatives (CWRs) are the ancestors of cultivated crops as well as other closely related wild species that have been integrated to recent agriculture and have potential to provide useful genetic resources to plant breeders [1]. CWRs are utilized to meet the challenges of global food security. These species are found in a wide range of habitat but are always at greater risk of disappearance because genetic reservoirs are continued to be degraded due to fluctuations in global climate and human interventions. A lot of efforts have been made in the past few decades to conserve CWRs to ensure the availability of genetic diversity for crop breeding programs [2]. As a result, various gene banks and agricultural institutions are now working across the globe for their collection, conservation, and utilization [3]. Since plant breeding preliminary depends upon the existing diversity of field crops and their wild relatives, thus CWRs provides an opportunity to make use of their untapped variability to develop high yielding and climate resilient crops. Wild species could be utilized in modern plant breeding through detection and transfer of useful wild alleles via interspecific and wide hybridization as well as genetic engineering approaches for the improvement of beneficial attributes of the crops [4]. Despite the poor quality and low yield, these resources are still the building blocks from which all modern varieties are developed. Therefore, this chapter is written on a broader prospective of conservation background and utilization strategies of wild germplasm in plant breeding through conventional and molecular approaches to uplift the yield and quality barriers in cultivated crops under ever

Wild Germplasm for Genetic Improvement in Crop Plants. http://dx.doi.org/10.1016/B978-0-12-822137-2.00001-1

evolving demands and climate scenarios. The detail for each crop is provided in further chapters of this book.

1.2 Potential of CWRs to meet food security challenges

The population of humans has increased from 1 billion to 7.7 billion in 200 years and this number is predicted to reach 10.9 billion by the end of 21st century [5]. To feed the additional 3.2 billion mouths is one of the biggest challenges to crop scientists. To achieve this goal, agriculturists around the globe are making continuous efforts to introduce new crops, enhance quality of the food, increase crop yield per unit area and expansion in agricultural land. Diversity in gene pools for wild relatives can potentially be exploited under such challenging circumstances to attain food security [6]. Wild germplasm have co-evolved with the changing climate and is therefore an immense source of genes that can be mined to introduce novel traits in existing cultivars for adaptability, yield, and quality [7]. The additional diversity in wild germplasm is attributable to the fact that it has generally escaped the genetic bottleneck events which transpired throughout the domestication of modern crop cultivars [8,9]. It is more acceptable to introduce allelic diversity of wild relatives into existing crop varieties through prebreeding, since genetic modification of crops is still contentious [10,11].

In spite of the fact that CWRs are typically poor for yield and agronomic features, they have been used in breeding programs to increase yield. Still not many examples are reported for direct deployment of yield related genes from wild germplasm to cultivars [12]. In most of the instances, yield increase was associated with stress tolerance features of CWRs (will be discussed in later sections). It is predicted that more than 10,000 wild relatives have direct potential to combat food insecurity [13]. For example, heterotic loci are associated with the number of spikelets and quantitative trait loci (QTLs) related to abscisic acid responsiveness has been identified and mapped in *Oryza rufipogon* [14,15]. High brix percentage and harvested weight per unit area in cultivated tomatoes can be gained by utilizing reported QTLs derived from *Solanum pennellii* [16]. Availability of wild tomatoes provided the opportunity to bypass negative epistasis and to counteract enigmatic variants causing breeding hurdles to attain better yield [17,18]. Fabrications in growth habit and flower colors in cultivated peanut are achieved through the utilization of mapped QTLs from chromosome segment substitution lines of wild peanut species [19]. Higher pod yield with an early maturing feature under water deficit conditions has been attained by introgression of alleles from wild to cultivated species of peanut [20]. Two wild relatives (*Cajanus cinereus* and *Cajanus mollis*) have been discovered with high number of seeds per pod in pigeon pea [21].

Improving quality along with yield by employing natural variations from the crop gene pool is increasingly indispensable to "feed more from less" [22,23].

The quality features are predominantly depending upon the category and usage of the crop such as quality measures for grain cereals will not be the same as for oil seed crops and vice versa. Even within cereals, quality standards for wheat are different from rice, and within rice, bran has different nutritional standards as compared to the white grain portion, as both can be used for an edible purpose [24–26]. Quality of a crop also depends upon geographic regions, food propensity and genetic background of the inhabitants. For example, a higher proportion of the population in Europe and America are sensitive to gluten related disorders in comparison to the Asia pacific population [27]. Consequently, the majority of wheat breeding programs in Asian countries are based on enhancing gluten content in contrast to Europe and United States of America, where development of gluten-free cereal products is also the primary objective of numerous research projects [28–31]. Therefore, crop breeding goals must be formulated by contemplating the aforementioned aspects and then the wild crop germplasm can be exploited correspondingly.

Depletion of mineral and protein content of cereals in parallel to yield improvement on account of immense breeding efforts over the years has been reported [32–35]. Hence, wild relatives are the reservoir of diversity of imperative nutrients for biofortification of staple food crops [36,37]. It was reported that wild germplasm of barley contained approximately 50% more zinc, iron, and protein than cultivated accessions [38]. Diversity array technology (DArT) derived single nucleotide polymorphism (SNPs) markers and *HvNAM* genes related to grain protein content have been mapped in wild barley for future use [39–41]. The *Aegilops* Genus is proved to be highly diversified in contemporary alleles for micronutrients, glutenin and gliadin proteins for biofortified wheat development [42]. Wild relatives of cassava (*Manihot esculenta*) are recognized as substantial source of untapped diversity for protein, starch, provitamin A, zinc, and iron contents [43,44]. Even though transgenic cassava with revamped nutritional values has been developed to address the hidden hunger of Sub-Saharan Africa, plant breeding generated biofortified cassava would have a vast scope by virtue of the great acceptability among scientific and social communities [45–53]. In addition, wild relatives had a remarkable role in fixing heterosis in F_1 generation either as a source of cytoplasmic male sterility or fertility restorer lines for hybrid seed production at commercial scale [54–56]. This peculiarity of CWRs have been effectively used and is still under refinement in cotton, rice, sunflower, pearl millet, and several other crops [57–64].

1.3 Utilization of CWRs under climate changes scenarios

Utilization of CWRs in plant domestication is an evolutionary aspect to develop new crop varieties by altering plant features to meet the food demand [65]. Historically, few CWRs or progenitors have been utilized and most of the selected traits were only related to improvement in yield and adaptability [66].

The rigorous selection for particular traits and from particular set of genotypes have narrowed the genetic bases of crop plants with the passage of time [67]. It was reported that about half of the genetic diversity of soybean has been wiped out as a result of domestication [68]. Domestication reduced the diversity at both inter and intra genomic levels and genetic distances between the wild relatives and cultivated crop plants have become wider. Thus, the transfer of genes from wild to cultivated has become difficult and is now the major drawback for the exploitation of genetic variations of CWRs [69]. Wild relatives are continuously evolving and have survived under harsh climatic conditions by developing novel traits which can be further utilized in practical plant breeding. Therefore, these novel genes and traits can be taken into account under the changing climatic conditions and ever evolving environmental stresses. As described earlier, most of the time yield increase through prebreeding is associated with the stress tolerance features of CWRs as described in the following sub-sections with a few suitable examples.

1.3.1 Drought

As the global fresh water resources are depleting as a result of rise in global temperature, water scarcity is prevailing in agricultural zones, and drought stress is emerging as a major crop production threat [70]. Drought or water deficit conditions limit crop productivity by disturbing the normal functioning of plants and induce certain metabolic changes in the plant cells [71]. It may alter the physiological and agronomic attributes of crop plants and cause significant reduction in the yield and quality [72]. Introgression approaches, which involves the transferring of selective stress tolerant genes from wild relatives to cultivated plants is an effective strategy to enhance drought tolerance ability [73]. For instance, wild semmer wheat (*Triticum dicoccoides*) has been identified as a beneficial source of candidate genes related to water deficit conditions and can be exploited as a donor parent in wheat breeding programs to combat water shortage problems [74]. Similarly, several potential drought tolerance conferring QTLs have been mapped in wild progenitors of tomatoes [75] and barley [76]. These identified QTLs and genes are still not fully applicable and are limited in use primarily due to complex polygenic inheritance where each QTL has minor effect [77].

1.3.2 Salinity

Salt stress inhibits various metabolic pathways of crop plants causing alterations in biochemical mechanisms. More than 45 million hectares of agricultural land has been effected by salinity and 1.5 million hectares are adding to this number annually [78]. Therefore, improving the plant productivity under a saline environment is essential to meet the increasing food demand [79]. Salinity tolerance ability of crops vary between species of the same genera, for example, wild relatives of rice (*O. rufipogon* and *O. nivara*) are more tolerant to high salt concentration as compared to Asian cultivated rice (*Oryza sativa*). Efforts have been made to introduce genes from wild to cultivated rice such as a deployment of $PcINO_1$ gene (encodes L-myoinositol 1-phosphate synthase enzyme) from *O. coarctata* to *O. sativa* which resulted into significant improvement in salt tolerance ability [80]. Likewise, a soybean wild relative *Glycine soja* is more tolerant to salt stress as compared to cultivated soybean (*G. max*). Recently, two salt tolerant genes *GmCHX1 and GmSALT3* have been identified in *G. soja* and *G. max*, respectively. The existence of *GmSALT3* gene in few accessions of cultivated soybean suggests that salt tolerant alleles might have been lost in most of the genotypes during domestication [81]. It further highlights the importance of conserving cultivated accessions along with wild relatives. Both of these salt tolerant genes are involved in regulating ion homeostasis and can offer great potential for the development of commercial soybean varieties with improved salt tolerance [82,83]. Another achievement is the development of a salt-tolerant commercial durum wheat accession by introducing an allele from *T. monococcum via* cross breeding which produce approximately 25% more yield in high saline fields as compared to its recurrent parent [84].

1.3.3 High temperature

The world ambient temperature is rising as a result of global warming which is highly affecting the productivity of the crops. Reproductive stages of most of the crops are more vulnerable to heat stress, thus reproductive failures are getting common due to increased level of heat waves [85,86]. High reproductive success rate in crops is vital for seed settling, therefore breeding for thermotolerant cultivars is requisite to ensure food security in the current situation [87]. Wild accessions could be effective source to hunt for heat tolerant genetic resources as they have co-evolved with changing climate. For example, *Oryza meridionalis* (a wild rice) possessed higher photosynthetic rate and accumulation of protective proteins, *i.e.*, *Hsp70*, *Hsp90*, and *Cpn60* under high temperature conditions as compared to cultivated rice [88]. Further screening of wild rice accessions (*O. barthii, O. australiensis* and *O. meridionalis*) revealed the considerable variability for tillering under thermal and drought stress which can be exploited [89]. The study of wild tomatoes revealed *S. pimpinellifolium* (a wild tomato) as a substantial source to attain high pollen viability under heat stress [90,91]. Therefore, an ample opportunity is available to utilize wild species as a source of heat tolerance in crop plants.

1.3.4 Insect pests and pathogens

Biotic stress involves the invasion by living organisms such as pathogens (fungi, viruses, and bacteria), nematodes, and insect pests. They may cause damage to vegetative and reproductive parts of the plant at every stage of its life cycle [92]. Although several resistant crop varieties have been developed in the past, pathogens are also co-evolving with the plants, and are therefore continuously breaking the resistance. Most of the pathogens have a high rate of mutation and recombination which makes it more difficult to control their populations. Scientists are always seeking novel resistance genes for pest and pathogens to overcome this genetic hindrance. Under such circumstances, CWRs serve as useful genetic material to find resistance genes [93]. Utilization of various molecular and omics approaches further facilitates the researchers to characterize these resistance genes for successful pyramiding to provide broad spectrum resistant plants [81]. One example is Southern Corn Leaf Blight in the United States which damaged the vegetative and reproductive parts of the corn and caused up to 55% yield losses. Resistant genes for this pathogen have identified in a wild ancestor *Tripsacum dactyloides* and it has been transferred into modern cultivars [94,95]. In tomatoes, five *Ty* genes exhibiting resistance to tomato yellow leaf curl virus (TYLCY) have been successfully introgressed into cultivated varieties from wild species. Pyramiding of these *Ty* genes has provided durable horizontal resistance to TYLCY [96]. Insect resistance can also be enhanced by understanding the chemical compositions and genetic architecture of wild relatives. The morphological and physiological attributes of wild relatives make them more resistant to insect-pest attack. For example, teosinte leaves have more toxic secondary metabolites for insect resistance as compared to maize [97]. Similarly, antixenosis levels of a wild strawberry (*Fragaria vesca*) for strawberry leaf beetle are found to be higher as compared to cultivated [98]. Wild cotton species also possess certain traits which make them resistant to insect pests attack such as hairiness, presence of gossypol, redness of stems, and okra type leaves. These features of wild cotton have been successfully exploited in past are still under investigation [97].

1.4 Crop wild relative derived genetic markers

Most crop species have undergone some form of genetic bottleneck event, at least corresponding to the time of domestication, but many species have even undergone multiple bottleneck events, including polyploidization, domestication, and continued selection of advanced types. These genetic bottleneck events have greatly reduced the available diversity in the germplasm pool to develop elite cultivated lines. In several cases this brings breeders back to the necessity of utilizing the crop wild relatives as a necessary source of genetic diversity. Utilizing the diversity in CWR is uniquely dependent upon the biological system of each individual crop species, but in each system the crop wild relatives can be broken down into tiers of materials based on the ease of feasibility of utilization.

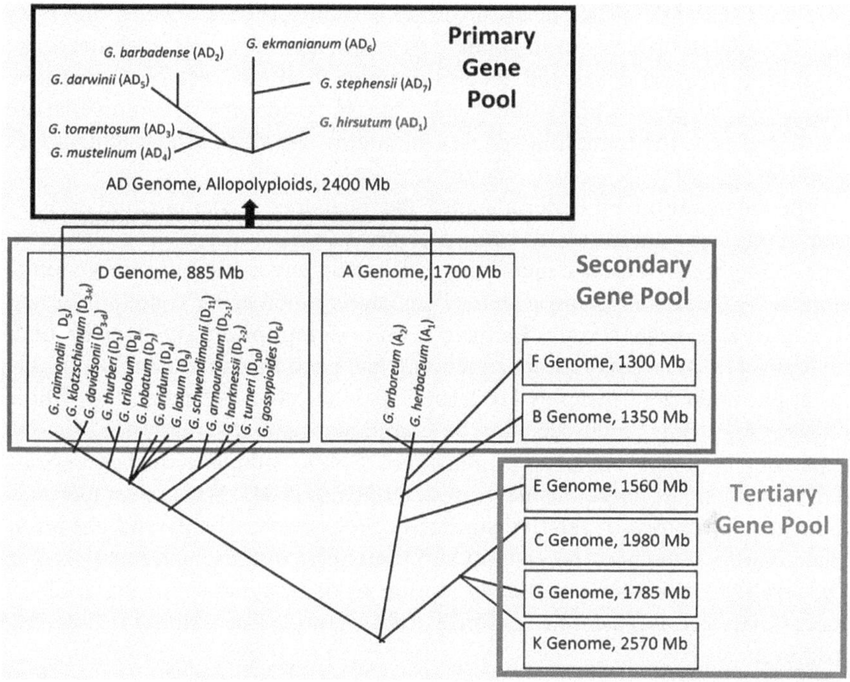

FIGURE 1.1 Depiction of the primary, secondary and tertiary gene pools of the *Gossypium* genus. Gray text (red in web version) indicates the primary cultivated type and therefore the primary species to which the comparisons and designations for gene pool classifications are depicted. Letters in parenthesis indicate the genome assignment for each species. *(Adapted from Wendel et al., 2003 [101] and Shim et al., 2018 [100]).*

The *primary gene pool* (GP-1) of CWR consists of typically the most closely genetically related individuals that contain equivalent genome structure and chromosome number, making crosses with these individuals and the elite germplasm relatively "easy" [99]. In the case of the polyploid *Gossypium* genus (**Fig. 1.1**), this would be the set of seven (AD) tetraploid species (*G. hirsutum*, *G. barbadense*, *G. mustelinum*, *G. tomentosum*, *G. darwinii*, *G. stephensii*, and *G. eckmanianum*) which can be crossed with relative ease with the main cultivated species *G. hirsutum* [100]. The *secondary gene pool* (GP-2) of CWR consists of materials which are more distantly related and thus harder to integrate, they may potentially be of differing ploidy level than cultivated species. For *Gossypium* the secondary gene pool consists of diploid species containing genome compliments that are highly related to the subgenome content of the tetraploid species, including the A, B, D and F genomes of cotton. However, utilization of this level of material relies on bridging the ploidy gap in cotton as well as other genome structure issues. The *tertiary* *gene pool* (GP-3) consists of even further distantly related species that typically have large differences in genome structure and of ploidy level that make integration into the cultivated species quite difficult. In case of *Gossypium*, these materials consist of diploid species that have genome structures that are not shared with the tetraploid species, including the C, E, G and K genomes of cotton.

The designation of materials into the primary, secondary and tertiary gene pools have traditionally been determined utilizing breeding and cytogenetic methods. With the increasing utilization and availability of genomic sequencing, these designations have been also refined using genetic information. Efforts to sequence CWRs have been ongoing, primarily since second-generation or next-generation sequencing has become economically feasible on large numbers of samples [102,103]. This allowed for generating whole genome sequencing that can then be compared against a reference genome sequence to detect variants, primarily of SNPs, simple sequence repeats (SSRs), and small insertions/deletions (INDELs) [104]. While this approach is limited to regions of relative similarity between the CWRs and the reference genome sequence, the amount and variability of these variants will give a generalized idea how genetically similar or diverse the set of materials of interest may be; and thus the feasibility of utilizing a line within the breeding program.

Next-generation sequencing technologies have allowed most communities to assess general levels of SNPs, SSRs and small INDELs, this technology limits the ability to detect large structural differences at the genome level as all analyses rely on comparison to typically a single reference genome. With the increasing accuracy and cost-effectiveness of third-generation sequencing technologies such as Pacific BioSciences and Oxford Nanopore [105,106] sequencing of long reads, crop communities are able to better assess structural differences between species and individuals within a genus. These efforts are becoming defined generally as "pan genome" efforts, which will allow for capturing the landscape of true differences between the set of materials of interest including large structural differences, copy number variants, and large INDELs within breeding materials [107–110]. These efforts will help breeders better define the feasibility for utilizing specific germplasm resources as well as the targeted gene regions and how they correspond between the cultivated types and the CWR. Development and understanding of these differences will ultimately help allow breeding programs to intelligently utilize CWRs and to track regions of introgression using genetic information.

1.5 Conservation background of plant genetic resources

Urgency of conserving wild plants was first highlighted by Russian scientist Nikolai Vavilov in 1930s [111]. Afterward, The Rockefeller Foundation and Ford Foundation planted the seeds of Green Revolution by establishing the International Rice Research Institute (IRRI) and International Maize and Wheat Improvement Center (CIMMYT) which drew the worldwide attention of plant scientists and policy makers to preserve the "hidden treasure" of field crops [112]. As a result, the Consultative Group of International Agriculture Research (CGIAR) was established in 1971, initially four centers including CIMMYT, IRRI, International Center for Tropical Agriculture (CIAT) and International Institute of Tropical Agriculture (IITA) were affiliated with CGIAR. By that time, the natural diversity of CWRs was under serious threat as the result of genetic erosion caused by urbanization, deforestation, industrialization, desertification, genetic vulnerability, overgrazing, and certain climate change aspects. It was reported that 75% of the natural diversity in global crops was lost during the 20th century [113]. Therefore, it was crucial to conserve the plant genetic resources at the global scale for their sustainable use in crop breeding projects to tackle food security challenges. Another eminent event in the history of crop diversity conservation was the establishment of International Board for Plant Genetic Resources (IBPGR) supported by CGIAR in 1974, and later on upgraded to International Plant Genetic Resources Institute (IPGRI) in 1991 and Biodiversity International in 2006. Currently, about 15 non-profit research centers/organizations are working across the globe as independent organizations under the leadership of CGIAR. In addition to the aforementioned, the Africa Rice Center, International potato center (CIP), International Center for Agricultural Research in Dry Areas (ICARDA), and International Crops Research Institute for the Semi-Arid Tropics (ICRISAT) are now also part a of CGIAR.

Currently there are ~1750 gene banks with the deposit of 7.4 million plant accessions stored. These gene banks are engaged in collection, testing, multiplication, characterization, storage, and distribution of germplasm for research purpose. Interestingly, CGIAR based gene bank represents10% of this total collection but accounts for 94% of germplasm distribution throughout of the world. On an average, CGIAR gene banks provide 100,000 accessions to more than 100 nations every year. Although there are many gene banks in different countries with advanced facilities, but the most interesting one to discuss here is the Svalbard Global Seed Vault. This world's largest and most secure seed bank is based at the North Pole in the permafrost region of Norway which is 1300 km beyond the Arctic Circle. It has a storage capacity of more than 4.5 million seed samples, but currently has more than 1 million samples. It serves as a backup storage facility of seed samples in case of natural catastrophes and man-made disasters as other seeds bank are quite vulnerable to such events [114,115]. Therefore, it is also known as "Doomsday seed bank." Since, the existence of natural diversity in crop plants is essential for the survival of human, considerable achievements have been made to preserve the crop germplasm. Scientist are making continuous efforts to ensure conservation of biodiversity [116,117]. Some of the collection information and web-links of renowned seed banks are provided in Table 1.1.

TABLE 1.1 Germplasm collection information of some renowned seed banks of the world.

S. No.	Institutes/seed bank	Headquarter	Crops	Total accessions	References
1	International Maize and Wheat Improvement Center (CIMMYT)	Mexico	Wheat	153,000	https://www.cimmyt.org
			Maize	28,000	
2	International Crops Research Institute for the Semi-Arid Tropics (ICRISAT)	India	Sorghum	39000	https://www.icrisat.org/
			Millets	23,000	
			Chickpea	17,000	
			Pigeon pea	18,000	
			Groundnut	15,000	https://ciat.cgiar.org/
3	International Center for Tropical Agriculture (CIAT)	.Colombia	Beans	38,000	
			Cassava	6,600	
			Alfalfa	17,476	
4	International Institute of Tropical Agriculture (IITA)	Nigeria	Banana	228	https://www.iita.org/
			Cassava	2,035	
			Cowpea	16,005	
			Yam	1,433	
5	International Potato Center (CIP)	Peru	Potato	4,415	https://cipotato.org/
6	International Center for Agricultural Research in the Dry Areas (ICARDA)	Syria	Barley	25,113	https://www.icarda.org/
			Chickpea	10,599	
			Lentil	11,940	
			Wheat	33,459	
			Fava bean	25,83	
7	International Rice Research Institute (IRRI)	Philippine	Rice	128,000	https://www.irri.org/
8	Africa Rice Center (ARC)	Ivory Coast	Rice	20,000	http://www.africarice.org/
9	International Livestock Research Institute (ILRI)	Kenya	Napier grass	2,300	https://www.ilri.org/

10	Brazilian Agricultural Research Corporation (EMBRAPA)	Brazil	Soybean	27,050	https://www.embrapa.br/en/international
			Wheat	15,118	
			Sorghum	7,215	
			Beans	16,447	
11	N.I.Vavilov Institute of Plant Genetic Resources (VIR)	Russia	Wheat	54,000	http://www.vir.nw.ru/unu-kollektsiya-vir/
			Barley	20,000	
			Oat	13,000	
			Potato	10,000	
			Rye	3,000	
12	Commonwealth Scientific and Industrial Research Organisation (CSIRO)	Australia	Signal grass	800	https://www.csiro.au/
13	Biodiversity International	Italy	Banana	15,000	https://www.bioversityinternational.org/
14	United State Department of Agriculture (USDA)	US	Rice	19,000	https://www.usda.gov/
			Soybean	22,497	
			Cotton	9,812	
			Maize	8,506	
			Barley	2,000	
15	National Bureau of Plant Genetic Resources (NBPGR)	India	Chickpea	14,704	https://www.devex.com/organizations/national-bureau-of-plant-genetics-resources-nbpgr-124737
			Barley	9,161	
			Wheat	35,889	
			Maize	6,909	
			Sorghum	17,466	
			Pearl millet	5,772	
			Groundnut	13,144	

Acknowledgments

The authors are thankful to Higher Education Commission (HEC), Islamabad-Pakistan for provision of research grant and Department of Crop and Soil Sciences, North Carolina State University, Raleigh, NC, USA for establishment of research collaboration.

Web references

https://www.cimmyt.org
https://www.icrisat.org/
https://ciat.cgiar.org/
https://www.iita.org/
https://cipotato.org/
https://www.icarda.org/
https://www.irri.org/
http://www.africarice.org/
https://www.ilri.org/
https://www.embrapa.br/en/international
http://www.vir.nw.ru/unu-kollektsiya-vir/
https://www.csiro.au/
https://www.bioversityinternational.org/
https://www.usda.gov/
https://www.devex.com/organizations/national-bureau-of-plant-genetics-resources-nbpgr-124737

References

[1] Castañeda NP, Khoury CK, Achicanoy HA, Bernau V, Dempewolf H, Eastwood RJ, et al. Global conservation priorities for crop wild relatives. Nat Plants 2016;2(4):1–6.
[2] Van ZM, Rakha M, yee Tan S, Chou Y-Y, Chang C-H, Yen J-Y, et al. Mapping patterns of abiotic and biotic stress resilience uncovers conservation gaps and breeding potential of Vigna wild relatives. Sci Rep 2020;10(1):1–11.
[3] Contreras-Toledo AR, Cortés-Cruz M, Costich DE, de Lourdes Rico-Arce M, Brehm JM, Maxted N. Diversity and conservation priorities of crop wild relatives in Mexico. Plant Genet Resour 2019;17(2):140–50.
[4] Dulloo ME, Maxted N. Plant genetic resources conservation and utilization–crop wild relatives. Plant Genet Resour 2019;17(2):101–2.
[5] Roser M, Ritchie H, Ortiz-Ospina E. World population growth. Our World in Data 2019. Available from: https://ourworldindata.org/world-population-growth.
[6] Dempewolf H, Baute G, Anderson J, Kilian B, Smith C, Guarino L. Past and future use of wild relatives in crop breeding. Crop Sci 2017;57(3):1070–82.
[7] Zhang H, Mittal N, Leamy LJ, Barazani O, Song BH. Back into the wild-Apply untapped genetic diversity of wild relatives for crop improvement. Evol Appl 2017;10(1):5–24.
[8] Reynolds M, Dreccer F, Trethowan R. Drought-adaptive traits derived from wheat wild relatives and landraces. J Exp Bot 2007;58(2):177–86.

[9] Fielder H, Brotherton P, Hosking J, Hopkins JJ, Ford-Lloyd B, Maxted N. Enhancing the conservation of crop wild relatives in England. PLoS One 2015;10(6).
[10] Muñoz N, Liu A, Kan L, Li M-W, Lam H-M. Potential uses of wild germplasms of grain legumes for crop improvement. Int J Mol Sci 2017;18(2):328–35.
[11] Zimny T, Sowa S, Tyczewska A, Twardowski T. Certain new plant breeding techniques and their marketability in the context of EU GMO legislation–recent developments. N Biotechnol 2019;51:49–56.
[12] Hajjar R, Hodgkin T. The use of wild relatives in crop improvement: a survey of developments over the last 20 years. Euphytica 2007;156(1–2):1–13.
[13] Maxted N, Kell S. Establishment of a global network for the in situ conservation of crop wild relatives: status and needs. Rome, Italy: FAO Commission on Genetic Resources for Food and Agriculture; 2009. 266.
[14] Luo X, Wu S, Tian F, Xin X, Zha X, Dong X, et al. Identification of heterotic loci associated with yield-related traits in Chinese common wild rice (*Oryza rufipogon Griff*). Plant Sci 2011;181(1):14–22.
[15] Cal WY, Tan LB, Liu FX, Sun CQ. Identification of quantitative trait loci and candidate genes associated with ABA sensitivity in common wild rice (*Oryza rufipogon Griff*). J Integr Agric 2017;16(11):2375–85.
[16] Gur A, Zamir D. Mendelizing all components of a pyramid of three yield QTL in tomato. Front Plant Sci 2015;6:1096–103.
[17] Soyk S, Lemmon ZH, Oved M, Fisher J, Liberatore KL, Park SJ, et al. Bypassing negative epistasis on yield in tomato imposed by a domestication gene. Cell 2017;169(6):1142–55.
[18] Soyk S, Lemmon ZH, Sedlazeck FJ, Jiménez-Gómez JM, Alonge M, Hutton SF, et al. Duplication of a domestication locus neutralized a cryptic variant that caused a breeding barrier in tomato. Nat Plants 2019;5(5):471–86.
[19] Fonceka D, Tossim H-A, Rivallan R, Vignes H, Lacut E, De Bellis F, et al. Construction of chromosome segment substitution lines in peanut (*Arachis hypogaea* L.) using a wild synthetic and QTL mapping for plant morphology. PLoS One 2012;7(11):65–78.
[20] Dutra WF, Guerra YL, Ramos JP, Fernandes PD, Silva CR, Bertioli DJ, et al. Introgression of wild alleles into the tetraploid peanut crop to improve water use efficiency, earliness and yield. PloS One 2018;13(6):123–36.
[21] Sharma S, Upadhyaya HD. Pre-breeding to expand primary genepool through introgression of genes from wild Cajanus species for pigeonpea improvement. Legume Res 2016;(11):17–20.
[22] Pretty J. Can ecological agriculture feed nine billion people? Monthly Rev 2009;61(6):46.
[23] Fanzo J, Hunter D, Borelli T, Mattei F. . Diversifying food and diets: using agricultural biodiversity to improve nutrition and health: Routledge 2013;61(6):46–55.
[24] Liu J, He Z, Zhao Z, Pena R, Rajaram S. Wheat quality traits and quality parameters of cooked dry white Chinese noodles. Euphytica 2003;131(2):147–54.
[25] Fang C, Hu X, Sun C, Duan B, Xie L, Zhou P. Simultaneous determination of multi rice quality parameters using image analysis method. Food Anal Methods 2015;8(1):70–8.
[26] Bagchi TB, Sharma S, Chattopadhyay K. Development of NIRS models to predict protein and amylose content of brown rice and proximate compositions of rice bran. Food Chem 2016;191:21–7.
[27] Ashtari S, Pourhoseingholi MA, Rostami K, Aghdaei HA, Rostami-Nejad M, Busani L, et al. Prevalence of gluten-related disorders in Asia-Pacific region: a systematic review. J Gastrointestin Liver Dis 2019;28(1):95–105.
[28] Arendt E, Dal Bello F. Gluten-free cereal products and beverages, 191. Elsevier; 2011. 21-27.

[29] Durham S. Defatted corn protein produces palatable gluten-free bread. Agric Res 2010;58(10):28–32.
[30] Ahmed M. Response of spring wheat (*Triticum aestivum* L.) quality traits and yield to sowing date. PloS One 2015;10(4):e0126097.
[31] García-Molina MD, Giménez MJ, Sánchez-León S, Barro F. Gluten free wheat: Are we there? Nutrients 2019;11(3):487.
[32] Simmonds NW. The relation between yield and protein in cereal grain. J Sci Food Agric 1995;67(3):309–15.
[33] Oury F-X, Berard P, Brancourt-Hulmel M, Heumez E, Pluchard P, Rousset M, et al. Yield and grain protein concentration in bread wheat: a review and a study of multi-annual data from a French breeding program [*Triticum aestivum* L]. J Gen Breed 2003;22(4):315–24.
[34] Shewry PR. Improving the protein content and composition of cereal grain. J Cereal Sci 2007;46(3):239–50.
[35] Fan M-S, Zhao F-J, Fairweather-Tait SJ, Poulton PR, Dunham SJ, McGrath SP. Evidence of decreasing mineral density in wheat grain over the last 160 years. J Trace Elem Med Biol 2008;22(4):315–24.
[36] Gómez-Galera S, Rojas E, Sudhakar D, Zhu C, Pelacho AM, Capell T, et al. Critical evaluation of strategies for mineral fortification of staple food crops. Transgenic Res 2010;19(2):165–80.
[37] Andersson MS, Saltzman A, Virk PS, Pfeiffer WH. Progress update: crop development of biofortified staple food crops under HarvestPlus. Afric J Food Agric Nutr Dev 2017;17(2):11905–35.
[38] Wiegmann M, Thomas WT, Bull HJ, Flavell AJ, Zeyner A, Peiter E, et al. Wild barley serves as a source for biofortification of barley grains. Plant Sci 2019;283:83–94.
[39] Wenzl P, Carling J, Kudrna D, Jaccoud D, Huttner E, Kleinhofs A, et al. Diversity arrays technology (DArT) for whole-genome profiling of barley. Proc Nat Acad Sci 2004;101(26):9915–20.
[40] Rodríguez-Suárez C, Giménez M, Gutiérrez N, Ávila C, Machado A, Huttner E, et al. Development of wild barley (*Hordeum chilense*)-derived DArT markers and their use into genetic and physical mapping. Theor Appl Genet 2012;124(4):713–22.
[41] Cai S, Yu G, Chen X, Huang Y, Jiang X, Zhang G, et al. Grain protein content variation and its association analysis in barley. BMC Plant Biol 2013;13(1):35–47.
[42] Kumar A, Kapoor P, Chunduri V, Sharma S, Garg M. Potential of Aegilops sp. for improvement of grain processing and nutritional quality in wheat (*Triticum aestivum*). Front Plant Sci 2019;10:308–17.
[43] Bredeson JV, Lyons JB, Prochnik SE, Wu GA, Ha CM, Edsinger-Gonzales E, et al. Sequencing wild and cultivated cassava and related species reveals extensive interspecific hybridization and genetic diversity. Nat Biotechnol 2016;34(5):562–7.
[44] An F, Chen T, Stéphanie DMA, Li K, Li QX, Carvalho LJ, et al. Domestication syndrome is investigated by proteomic analysis between cultivated cassava (*Manihot esculenta* Crantz) and its wild relatives. PloS One 2016;11(3):e0152154.
[45] Zhang P, Gruissem W. Production of transgenic cassava (*Manihot esculenta* Crantz). Transgenic crops of the world. Springer; 2004. 301-319.
[46] Adenle AA. Response to issues on GM agriculture in Africa: Are transgenic crops safe? BMC Res Notes 2011;4(1):388–9.
[47] Maghari BM, Ardekani AM. Genetically modified foods and social concerns. Avicenna J Med Biotechnol 2011;3(3):109–19.
[48] Adenle AA, Aworh OC, Akromah R, Parayil G, Developing GM. super cassava for improved health and food security: future challenges in Africa. Agric Food Sec 2012;1(1):11–8.

[49] Gomes P, Nassar N. Cassava interspecific hybrids with increased protein content and improved amino acid profiles. Genet Mol Biol 2013;12(2):1214–22.
[50] Telengech P, Maling'a J, Nyende A, Gichuki S, Wanjala B. Gene expression of beta carotene genes in transgenic biofortified cassava. Biotech 2015;5(4):465–72.
[51] National Academies of Sciences, Engineering, and Medicine, Division on Earth and Life Studies; Board on Agriculture and Natural Resources. Social and economic effects of genetically engineered crops. Genetically engineered crops: Experiences and prospects. United States: National Academies Press (US); 2016.
[52] Oparinde A, Abdoulaye T, Mignouna DB, Bamire AS. Will farmers intend to cultivate Provitamin A genetically modified (GM) cassava in Nigeria? Evidence from a k-means segmentation analysis of beliefs and attitudes. PloS One 2017;12(7):345–55.
[53] Wolfe MD, Bauchet GJ, Chan AW, Lozano R, Ramu P, Egesi C, et al. Historical introgressions from a wild relative of modern cassava improved important traits and may be under balancing selection. Genet 2019;213(4):1237–53.
[54] Li S, Yang G, Li S, Li Y, Chen Z, Zhu Y. Distribution of fertility-restorer genes for wild-abortive and Honglian CMS lines of rice in the AA genome species of genus *Oryza*. Ann Bot 2005;96(3):461–6.
[55] Schröder R, Prasse R. Do cultivated varieties of native plants have the ability to outperform their wild relatives? PLoS One 2013;8(8):e71066.
[56] Montenegro de Wit M. Stealing into the wild: conservation science, plant breeding and the makings of new seed enclosures. J Peasant Stud 2017;44(1):169–212.
[57] Shih-Cheng L, Loung Ping Y, editors. Hybrid rice breeding in China. Innovative approaches to rice breeding: Selected papers form the 1979 International Rice Research Conference International Rice Research Institute, Los Banos, Philippines; 1980.
[58] Seiler GJ. Utilization of wild sunflower species for the improvement of cultivated sunflower. Field Crops Res 1992;30(3-4):195–230.
[59] Rai K. A new cytoplasmic-nuclear male sterility system in pearl millet. Plant Breed 1995;114(5):445–7.
[60] Mackenzie S. Male sterility and hybrid seed production. Plant Biotechnology and Agriculture. Elsevier; 2012. 185-194.
[61] Luo D, Xu H, Liu Z, Guo J, Li H, Chen L, et al. A detrimental mitochondrial-nuclear interaction causes cytoplasmic male sterility in rice. Nat Genet 2013;45(5):573–6.
[62] Liu Z, Wang D, Feng J, Seiler GJ, Cai X, Jan C-C. Diversifying sunflower germplasm by integration and mapping of a novel male fertility restoration gene. Genet 2013;193(3):727–37.
[63] Pattanashetti SK, Upadhyaya HD, Dwivedi SL, Vetriventhan M, Reddy KN. Pearl millet. Genetic and genomic resources for grain cereals improvement:. Elsevier; 2016. 253-89.
[64] Zhang B, Zhang X, Liu G, Guo L, Qi T, Zhang M, et al. A combined small RNA and transcriptome sequencing analysis reveal regulatory roles of miRNAs during anther development of upland cotton carrying cytoplasmic male sterile *Gossypium harknessii* (D2) cytoplasm. BMC Plant Biol 2018;18(1):242–9.
[65] Anderson JE, Campbell A, Kantar MB. Crop wild relatives of root vegetables in North America. North American crop wild relatives, Vol. 2. Springer; 2019. 243-279.
[66] Cruz-Cárdenas CI, Cortés-Cruz M, Gardner CA, Costich DE. Wild relatives of maize. North American crop wild relatives, Vol. 2. Springer; 2019. 3-39.
[67] Riordan EC, Nabhan GP. Trans situ conservation of crop wild relatives. Crop Sci 2019;59(6):2387–403.

[68] Bruce RW, Torkamaneh D, Grainger C, Belzile F, Eskandari M, Rajcan I. Genome-wide genetic diversity is maintained through decades of soybean breeding in Canada. Theor Appl Genet 2019;132(11):3089–100.
[69] Kioukis A, Michalopoulou VA, Briers L, Pirintsos S, Studholme DJ, Pavlidis P, et al. Intra-specific diversification of the crop wild relative *Brassica cretica* Lam. using demographic model selection. BMC Genom 2020;21(1):48–56.
[70] Hussain S, Hussain S, Qadir T, Khaliq A, Ashraf U, Parveen A, et al. Drought stress in plants: An overview on implications, tolerance mechanisms and agronomic mitigation strategies. Plant Sci Today 2019;6(4):389–402.
[71] Laxa M, Liebthal M, Telman W, Chibani K, Dietz K-J. The role of the plant antioxidant system in drought tolerance. Antioxidants 2019;8(4):94–9.
[72] Deepak SB, Thakur A, Singh S, Bakshi M, Bansal S. Changes in crop physiology under drought stress: A review. J Pharm Phytochem 2019;8(4):1251–3.
[73] Geda C, Repalli S, Dash G, Swain P, Rao G. Enhancement of drought tolerance in rice through introgression of arabidopsis DREB1A through transgenic approach. J Rice Res 2019;7(208):2–12.
[74] Pour-Aboughadareh A, Omidi M, Naghavi MR, Etminan A, Mehrabi AA, Poczai P, et al. Effect of water deficit stress on seedling biomass and physio-chemical characteristics in different species of wheat possessing the D genome. Agron 2019;9(9):522–9.
[75] Gao L, Gonda I, Sun H, Ma Q, Bao K, Tieman DM, et al. The tomato pan-genome uncovers new genes and a rare allele regulating fruit flavor. Nat Genet 2019;51(6):1044–51.
[76] Jia Z, Liu Y, Gruber BD, Neumann K, Kilian B, Graner A, et al. Genetic dissection of root system architectural traits in spring barley. Front Plant Sci 2019;10:400–10.
[77] Sharma M, Gupta BB. Role of wild relatives for development of climate-resilient varieties. Rediscovery of genetic and genomic resources for future food security. Springer; 2020. 303-14.
[78] Carillo P, Annunziata MG, Pontecorvo G, Fuggi A, Woodrow P. Salinity stress and salt tolerance. Abiotic stress in plants–Mechanisms and adaptations, vol. 1. IntechOpen; 2011. 21-38.
[79] Tyerman SD, Munns R, Fricke W, Arsova B, Barkla BJ, Bose J, et al. Energy costs of salinity tolerance in crop plants. New Phytol 2019;221(1):25–9.
[80] Grene R, Provart NJ, Pardo JM. Resistance to salinity and water scarcity in higher plants. Insights from extremophiles and stress-adapted plants: Tools, discoveries and future prospects. Front Plant Sci 2019;10:373–83.
[81] Pandotra P, Bakshi P, Singh AK, Gupta S. Exploring genetic resources for identification of potential novel genes for crop improvement. Rediscovery of genetic and genomic resources for future food security. Springer; 2020. 225-237.
[82] Qi X, Li M-W, Xie M, Liu X, Ni M, Shao G, et al. Identification of a novel salt tolerance gene in wild soybean by whole-genome sequencing. Nat Commun 2014;5:4340.
[83] Liu Y, Yu L, Qu Y, Chen J, Liu X, Hong H, et al. GmSALT3, which confers improved soybean salt tolerance in the field, increases leaf Cl-exclusion prior to Na+ exclusion but does not improve early vigor under salinity. Front Plant Sci 2016;7:1485–92.
[84] Genc Y, Taylor J, Lyons GH, Li Y, Cheong J, Appelbee M, et al. Bread wheat with high salinity and sodicity tolerance. Front Plant Sci 2019;10:1280–7.
[85] Cheabu S, Moung-Ngam P, Arikit S, Vanavichit A, Malumpong C. Effects of heat stress at vegetative and reproductive stages on spikelet fertility. Rice Sci 2018;25(4):218–26.
[86] Nadeem M, Li J, Wang M, Shah L, Lu S, Wang X, et al. Unraveling field crops sensitivity to heat stress: Mechanisms, approaches, and future prospects. Agron 2018;8(7):128–37.

[87] Waqas MA, Kaya C, Riaz A, Li Ye. Potential mechanisms of abiotic stress tolerance in crop plants induced by Thiourea. Front Plant Sci 2019;10:1336–43.
[88] Bierschenk B, Melle Tilahun Tagele BA, d Ashrafuzzaman M, Wu L-B, Becker M, Frei M. Evaluation of rice wild relatives as a source of traits for adaptation to iron toxicity and enhanced grain quality. PloS One 2020;15(1):e0223086.
[89] Fujino K, Hirayama Y, Obara M, Ikegaya T. Introgression of the chromosomal region with the Pi-cd locus from *Oryza meridionalis* into *O. sativa* L. during rice domestication. Theor Appl Genet 2019;132(7):1981–90.
[90] Driedonks N, Rieu I, Vriezen WH. Breeding for plant heat tolerance at vegetative and reproductive stages. Plant Reprod 2016;29(1–2):67–79.
[91] Driedonks N, Wolters-Arts M, Huber H, de Boer G-J, Vriezen W, Mariani C, et al. Exploring the natural variation for reproductive thermotolerance in wild tomato species. Euphytica 2018;214(4):67–76.
[92] Madani B, Mirshekari A, Imahori Y. Physiological responses to stress. Postharvest physiology and biochemistry of fruits and vegetables. Elsevier; 2019. 405-23.
[93] van Zonneveld M, Rakha M, Tan S-y, Chou Y-y, Chang C-h, Yen J, et al. Understanding patterns of abiotic and biotic stress resilience to unleash the potential of crop wild relatives for climate-smart legume breeding. bioRxiv 2019;15:596072.
[94] Murray SC, Mayfield K, Pekar J, Brown P, Lorenz A, Isakeit T, et al. Tx741, Tx777, Tx779, Tx780, and Tx782 inbred maize lines for yield and southern United States stress adaptation. J Plant Regist 2019;13(2):258–69.
[95] Jovovic Z, Andjelkovic V, Przulj N, Mandic D. Untapped genetic diversity of wild relatives for crop improvement. Rediscovery of genetic and genomic resources for future food security. Springer; 2020. 25-65.
[96] Bai Y, Verlaan MG, Hutton S. Tomato yellow leaf curl virus resistance. Google Patents; 2019.
[97] Mammadov J, Buyyarapu R, Guttikonda SK, Parliament K, Abdurakhmonov IY, Kumpatla SP. Wild relatives of maize, rice, cotton, and soybean: treasure troves for tolerance to biotic and abiotic stresses. Front Plant Sci 2018;9:886–96.
[98] Nellist CF. Disease resistance in polyploid strawberry. The genomes of Rosaceous berries and their wild relatives. Springer; 2018. 79-94.
[99] Stelly DM, Lacape J-M, Dessauw D, Kohel RJ, Mergeai G, Sanamyan M, editors. International genetic, cytogenetic and germplasm resources for cotton genomics and genetic improvement. Omnipress; 2008.
[100] Shim J, Mangat P, Angeles-Shim R. Natural variation in wild *Gossypium* species as a tool to broaden the genetic base of cultivated cotton. J Plant Sci Curr Res 2018;2(005):45–56.
[101] Wendel JF, Cronn RC. Polyploidy and the evolutionary history of cotton. Adv Agron 2003;78:139–45.
[102] Brozynska M, Furtado A, Henry RJ. Genomics of crop wild relatives: expanding the gene pool for crop improvement. Plant Biotechnol J 2016;14(4):1070–85.
[103] Hulse-Kemp AM, Ashrafi H, Stoffel K, Zheng X, Saski CA, Scheffler BE, et al. BAC-end sequence-based SNP mining in allotetraploid cotton (*Gossypium hirsutum*) utilizing resequencing data, phylogenetic inferences, and perspectives for genetic mapping. G3: Genes Genom Genet 2015;5(6):1095–105.
[104] Migicovsky Z, Myles S. Exploiting wild relatives for genomics-assisted breeding of perennial crops. Front Plant Sci 2017;8:460–7.
[105] Roberts RJ, Carneiro MO, Schatz MC. The advantages of SMRT sequencing. Genom Biol 2013;14(7):405.

[106] Jain M, Olsen HE, Paten B, Akeson M. The Oxford Nanopore MinION: delivery of nanopore sequencing to the genomics community. Genom Biol 2016;17(1):239–43.
[107] Tao Y, Zhao X, Mace E, Henry R, Jordan D. Exploring and exploiting pan-genomics for crop improvement. Mol Plant 2019;12(2):156–69.
[108] Khan AW, Garg V, Roorkiwal M, Golicz AA, Edwards D, Varshney RK. Super-pangenome by integrating the wild side of a species for accelerated crop improvement. Trend Plant Sci 2019;12(2):156–69.
[109] Song J-M, Guan Z, Hu J, Guo C, Yang Z, Wang S, et al. Eight high-quality genomes reveal pan-genome architecture and ecotype differentiation of *Brassica napus*. Nat Plant 2020;6:34–45.
[110] Lyu J. Tomato pan-genome. Nat Plants 2019;5(6):558–64.
[111] Vavilov NI, Vavylov MI, Vavílov NÍ, Dorofeev VF. Origin and geography of cultivated plants. Cambridge University Press; 1992.
[112] Hamilton AC, Karamura D, Kakudidi E. History and conservation of wild and cultivated plant diversity in Uganda: forest species and banana varieties as case studies. Plant Divers 2016;38(1):23–44.
[113] Wang S, Zhang Z. The state of the world's plant genetic resources for food and agriculture. J Plant Genet Resour 2011;12(3):325–38.
[114] Brodal G, Asdal Å, editors. The Svalbard Global Seed Vault and the ongoing 100 years seed storage experiment. Seventh international symposium on seed, transplant and stand establishment of horticultural crops-SEST2016 1204; 2016.
[115] Qvenild M. Svalbard Global Seed Vault: a 'Noah's Ark' for the world's seeds. Dev Prac 2008;18(1):110–6.
[116] Meilleur BA, Hodgkin T. In situ conservation of crop wild relatives: status and trends. Biodiver Conserv 2004;13(4):663–84.
[117] Hawkes JG, Maxted N, Ford-Lloyd BV. The ex situ conservation of plant genetic resources. Springer Science & Business Media; 2012.

Chapter 2

Wild Cotton Genepool: An Unopened Treasure

Huma Saleem[a], Muhammad Tanees Chaudhary[a], Amir Shakeel[a], Shabir Hussain Wani[b], Xiongming Du[c] and Muhammad Tehseen Azhar[d,e]
[a]*Department of Plant Breeding and Genetics, University of Agriculture, Faisalabad, Pakistan;* [b]*Mountain Research Center for Field Crops, Sher-e-Kashmir University of Agricultural Sciences and Technology of Kashmir, Anantnag, J&K, India;* [c]*State Key Laboratory of Cotton Biology, Institute of Cotton Research of the Chinese Academy of Agricultural Sciences, Anyang, China;* [d]*Institute of Molecular Biology and Biotechnology, Bahauddin Zakariya University, Multan, Pakistan;* [e]*School of Agricultural Sciences, Zhengzhou University, Zhengzhou, Henan, China*

2.1 Introduction

The global agricultural trends are changing with the increase in food demand which is expected to double by 2050. Besides this, natural resources are also in continuous state of depletion. In current scenario, modern agriculture can only achieve sustainable level by developing crop varieties with high nutritional value, good yielding potential, tolerance against various stresses, and adaptable to environmental adversities. Cultivation and utilization of limited genotypes as parental source in breeding programs is one of the major reasons of diminishing the genetic potential of varieties. As a result, modern day cultivars have became vulnerable to insect-pest and diseases because pathogens are evolving more rapidly with every passing year as compared to crop genotypes [1]. Mostly, the field crops are the outcomes of evolutionary processes and intensive selection over the years from breeding materials to enhance yield and quality. Therefore, it shows clear differences among the traits when compared between cultivated germplasm and their wild relatives [2,3]. These crop improvement efforts have narrowed the genetic bases of modern cultivars and created a bottleneck for further improvement. For instance, 98% of cotton that is currently being grown in the United States is *Gossypium hirsutum* [4] and 85% of that is modified for pest resistance and herbicide. The uniformity at genetic level in cultivated varieties has put the US cotton industry at a high risk of collapse. Thus, the use of crop wild relatives (CWRs) is a promising approach to expand the variation and diversity of cultivated crops [5]. Moreover, high selection pressure reduced the adaptability and fitness of existing crop cultivars and advanced breeding lines. Under these circumstances, wild

Wild Germplasm for Genetic Improvement in Crop Plants. http://dx.doi.org/10.1016/B978-0-12-822137-2.00002-3

progenitors are the best candidates for genetic improvement. Wild relatives are always a rescue source for breeders to improve disease resistance and to overcome the uniformity and vulnerability issues of the crops. In most of the instances, yield and stress tolerance are negatively correlated, but it is reported that tolerance can be achieved without compromising the yield and quality [6]. Plant germplasm resources are reservoirs of desirable characteristics/genes, and breeder can pyramid them in a single plant. It provides a foundation stone to develop and protect the future of agriculture from various natural calamities.

2.2 Evolution and domestication of cultivated *Gossypium*

The lineage of *Gossypium* is assumed to be originated 10–20 million years ago (mya) [4]. The wild species of *Gossypium* spread across America, Africa, Australia, and Asia. The word cotton is an English name that is derived from Arabic language "al qutn" [7]. Africa is considered to be center of origin of first and foremost cotton species. Old World cottons are diploid in nature, whereas this genus is divided into eight cytogenetically categorized genome groups that have African/Asian, American, and Australian origin. Interspecific hybridization and subsequent polyploidization between Asian cottons (diploid and A-genome) and American cottons (diploid and D-genome) formed the five allotetraploid (AD) lineages that include Pima or Egyptian long staple silk fibere cotton and Upland or Acala/Mocó, short staple coarse fibere cotton [8].

The diploid species are classified into eight genomic groups, A–G and K [9]. The A, B, E, and F genomes belongs to African clade but has connections in Asia too; while the D genome refers to America; and C, G, and K genomes represents the Australian clade [10] [11]. *Gossypium* genus includes more than 50 well established perennial species that are distributed globally, and out of which 46 are diploid ($2n= 2x = 26$) and the remaining are tetraploid ($2n= 4x = 52$) [12]. The earliest cotton cultivar belong to *G. herbaceum* and *G. arboreum* both diploid progenitor of cultivated cotton belong to A-genome evolved in Africa and southern Asia, respectively. Another diploid species *G. raimondii* with D-genome belongs to new world and evolved over 1-2 mya [4]. An interspecific hybridization event has changed the ploidy level of currently cultivated cotton in the best interest of farmers and fabric industry. According to cytogenetic research [13], African origin species *G. herbaceum* (A1) and the American origin species *G. raimondii* (D5) are parents of *G. hirsutum* and *G. barbadense* [10]. The wild cotton relatives of *Gossypium* are considered relatively inferior in terms of fiber quality and production as compared to field grown species. Despite this fact, wild genepool of *Gossypium* is considered as a rich treasure of novel genes that can be incorporated to enhance the performance of other traits in cotton varieties [4,14]. Interspecific hybridization is always considered as an important and beneficial step to improve the genetic and morphological architecture of existing cultivars [15]. Several wild species have been identified to harbor genes conferring biotic and abiotic resistance, in addition to fiber quality traits.

The first polyploid hybrids were exceptionally vigorous in their ability for fiber quality and production. *G. hirsutum* which is also called American/Maxican cotton cultivated for their spinnable fibers, contribute over 90% of the global cotton cultivation. *G. barbadense* which is also called Egyptian cotton adds its share in 8% of the world cotton production. While diploids *G. arboreum* and *G. herbaceum* cultivated in some areas of Pakistan, India, Bangladesh, Iran, and China by providing ~2% share in cultivated cotton of the world [16,17]. The wild relatives are important source of genetic diversity. These are also source of important traits responsible for resistance to biotic and abiotic stress, male sterility and superior fiber properties (Table 2.1). For the improvement of cultivated cotton, molecular and classical plant breeders have exploited several gene resources of important traits from the wild relatives [12].

Cotton was first domesticated in old world about 7,000 years ago. In 6th millennium BC, earliest evidence for the cotton use was found in kachi plain in

TABLE 2.1 Origin, genome assignment, and useful agronomic traits of *Gossypium* species.

Wild species	Useful agronomic traits	Origin	Genome
G. hirsutum	Widely cultivated	Central America	AD_1
G. barbadense	Widely cultivated, long and high quality lint, resistance to *Verticillium* wilt	South America	AD_2
G. tomentosum	Tolerance to heat, source of the nectariless trait for resistance against tarnished plant bug, fleahoppers, boll rot, and bollworm, resistance to jassids and thrips, high fiber quality, fiber length, and fiber fineness	Hawaiian Islands	AD_3
G. mustelinum	Long fibers	NE Brazil	AD_4
G. darwinii	Fine fibers, drought tolerance, resistance to *Fusarium* and *Verticillium* wilt	Galapagos Islands	AD_5
G. ekmanianum	Salt tolerance	Dominican Republic	AD_6
G. stephensii	Sprawling growth habit, light reddish brown fiber, dense leaf pubescence	Wake Atoll	AD_7
G. africanum	High fiber strength	Southern Africa	A
G. herbaceum L.	Resistance to hoppers, white flies, thrips, aphids, and leaf curl virus	Southern Africa	A_1
G. arboreum L.	Resistance to hoppers, white flies, aphids, leaf curl virus, thrips, drought tolerance, resistance to black root rot, reniform nematodes, and spider-mites	Indus valley, Madagascar	A_2
G. anomalum	Resistance to cotton wilt, angular leaf spot, springtails and aphids, drought tolerance, high fiber quality, cytoplasmic male sterility	Africa (Angola, Namibia)	B_1

(Continued)

TABLE 2.1 Origin, genome assignment, and useful agronomic traits of *Gossypium* species. (*Cont.*)

Wild species	Useful agronomic traits	Origin	Genome
G. triphyllum	Flower color range from blue to purple	Namibia in Africa	B_2
G. capitis-viridis	High fiber quality, resistance to *Verticillium* and *Fusarium* wilt	Cape Verde Islands	B_3
G. trifurcatum	Salinity tolerance	Somalia	B
G. sturtianum	Glandless-seed and glanded-plant, resistance to *Fusarium* wilt	Australia	C_1
G. nandewarense	*Verticillum* wilt resistance	Australia	$^{C}1N$
G. robinsonii	Cotton leaf curl virus resistance	WA, Australia	C_2
G. thurberi	Tolerance to mild frost via defoliation, high resistance to *Verticillium dahlia*	Mexico	D_1
G. armourianum	Resistance to white flies, bacterial blight, and jassids	Mexico	$D_{2\text{-}1}$
G. harknessii	Cytoplasmic male sterility and fertility restorer	Mexico	$D_{2\text{-}2}$
G. davidsonii	Salinity tolerance	Mexico	$D_{3\text{-}d}$
G. klotzschianum	Salinity tolerance	Galapagos Islands	$D_{3\text{-}k}$
G. aridum	Salinity tolerance, resistance to reniform nematode	Mexico	D_4
G. raimondii	Resistance to jassids	Peru	D_5
G. gossypioides	Resistance to cotton leaf curl disease	Mexico	D_6
G. lobatum	Resistance to *Verticillium* wilt	Mexico	D_7
G. trilobum	Cytoplasmic male sterility and restorer factor. Drought tolerance, resistance to bollworm, pink worm, boll rot, *Verticillium* and *Fusarium* wilt	Western Mexico	D_8
G. laxum	Arborescent growth habits, pest and disease resistance	Mexico	D_9
G. turneri	Caduceus involucels	Mexico	D_{10}
G. schwendimanii	Cherry petal color, cordate leaf shape	Mexico	D_{11}
G. stocksii	Strong fibers, resistance to leaf curl virus, resistance to reniform nematode	East Africa, Arabia	E_1
G. somalense	Resistance to reniform nematode, extra fiber strength, resistance to Egyptian bollworm and pink bollworm, arid tolerance	Northeastern Africa	E_2

TABLE 2.1 Origin, genome assignment, and useful agronomic traits of *Gossypium* species. (*Cont.*)

Wild species	Useful agronomic traits	Origin	Genome
G. areysianum	Epicalyx bracts lanceolate, main leaf veins anastomosing	Arabia	E_3
G. incanum	Corolla gland dotted, calyx with distinct triangular teeth	Arabia	E_4
G.benadirense	Epicalyx bracts entire or subentire with all the veins anastomosing	Somalia, Ethiopia, Kenya	E
G. bricchettii	Drought tolerance	Somalia	E
G. vollesenii	Resistance to jassids	Somalia	E
G. longicalyx	Resistance to reniform nematode	Africa	F_1
G. bickii	Glandless-seed and glanded-plant	Central Australia	G_1
G. australe	Glandless-seed and glanded-plant, resistance to aphids and spider, mites, resistance to *Fusarium* and *Verticillium* wilts, drought tolerance	Australia	G_2
G. nelsonii	Drought tolerance	Australia	G_3
G. costulatum	Resistance to aphids	Australia	K_1
G. populifolium	Bacterial blight resistance	WA, Australia	K_2
G. cunninghamii	Cytoplasmic male sterility	Northern NT, Australia	K_3
G. pulchellum	Jassid resistance	WA, Australia	K_4
G. pilosum	Heat tolerance	WA, Australia	K_5
G. anapoides	Compact plant architecture, sessile leaves	Australia	K_6

Source: Shim et al. (2018).

Balochistan, Pakistan from Neolithic occupation of Mehrgarh. In Indus-valley of Pakistan and India, cultivation of *G. arboreum* began and then spread over Asia and Africa. Various cotton scientists have consensus that African species were the wild progenitor of *G. herbaceum,* while ancesstor of *G. arboreum* is still unknown [4]. In the Indus- valley, earliest agriculture village holds evidence of use and domestication of *G. arboreum* occurs 5000 years ago. Finished cloth and raw material were exported from South Asia in Jordan. Cotton fabric has also been found in Greece, Iraq, and Iran in 5th–8th century BC.

G. arboreum is a sub-tropical and tropical plant, and this species was introduced into China as an ornamental plant, about thousand years ago. During Islamic agricultural revolution, cotton was adapted to grow in temperate climates [18]. *G. herbaceum* is known to grow in grasslands and open forests of Africa.

Wild species have taller plants, smaller fruits, and thicker seeds as compared to domesticated plants. Closest wild progenitors of *G. herbaceum* suggested northward distribution toward near East and North Africa.

Recent taxonomic classification revealed that genus *Gossypium* included 14 species from Arabia and Africa. *G. trifurcatum* belongs to desert area of eastern Somalia. It was considered that it does not include in genus *Gossypium* due to the presence of dentate leaves but recent molecular work showed that it is included in cotton species. Arabian and African species contain considerable diversity [4]. These account four of the eight genomes. These species of cotton are placed in three genepool (GP) keeping in view of their genome type (Fig. 2.1).

2.3 Wild cotton species in Pakistan

The indigenous cultivated cotton (*G. arboreum* L) is locally named as "Desi cotton", holding the A-genome and this species is evolved from primitive *G. herbaceum* L. Six distinct races of *G. arboreum* have been reported: *indicum, burmanicum, soudanense, sinense, bengalense, and cernuum.* In Pakistan, Desi cotton belongs to *bengalense* race of *G. arboreum*. Characterization of *G. arboreum* has been carried out on the basis of agronomical, morphological, and physiological features and it suggested that this species is tolerant to biotic and abiotic stresses [19]. Through breeding, two varieties of *G. arboreum*, namley 278-Mollisoni and Z.Mollisoni were developed. These two varieties gradually replaced the older types. In 1930, 15-Mollisoni was approved for cultivation due to its high ginning outturn (GOT) of 35%. As compared to 15-Mollisoni, a new variety 39-Mollisoni exhibited 36%–37% GOT [6]. Another variety, 119-Sanguineum (119-S), exhibited 36.4% GOT that was approved for cultivation during 1941. This variety was developed through selection from mixture of *G. arboreum* biotypes. Another elite line 231-R was also approved for general cultivation during 1959.

Four elite lines were identified based on flower color and leaf morphology at Cotton Research Institute, Multan during 1952. Another line 73/3 showed 42% GOT as compared to 37%–38% GOT, of already cultivated varieties. However the yield was same as that of already cultivated Desi cotton [20]. At Cotton Research Station, Faisalabad, crosses were made between Chinese varieties Million Dollar and 39-Mollisoni. These efforts resulted in the development of Jubilee strains. These have higher GOT and improved staple length but lower yield as compared to Mollisoni (control variety). In 1944, Jubilee strains were crossed with *G. anomalum* from Nigeria. Resulted genotypes showed improved staple length but none of these were accepted by famers. At Multan, interspecific crosses were made between *G. thurberi* and *G. arboreum* but these were lower in yield as compared to existing genotypes [21].

There are seven land races of *G. hirsutum*, out of which *latifolium* was mostly used for improvement of cotton in Pakistan. It covers >98% area but less than 2% is under cultivation of *G. arboreum* whereas *G. barbadense* L. is not grown in Pakistan due to low yield and high photo-sensitivity [22]. There are numerous institutes in Pakistan that collect and store wild and cultivated germplasm. Plant

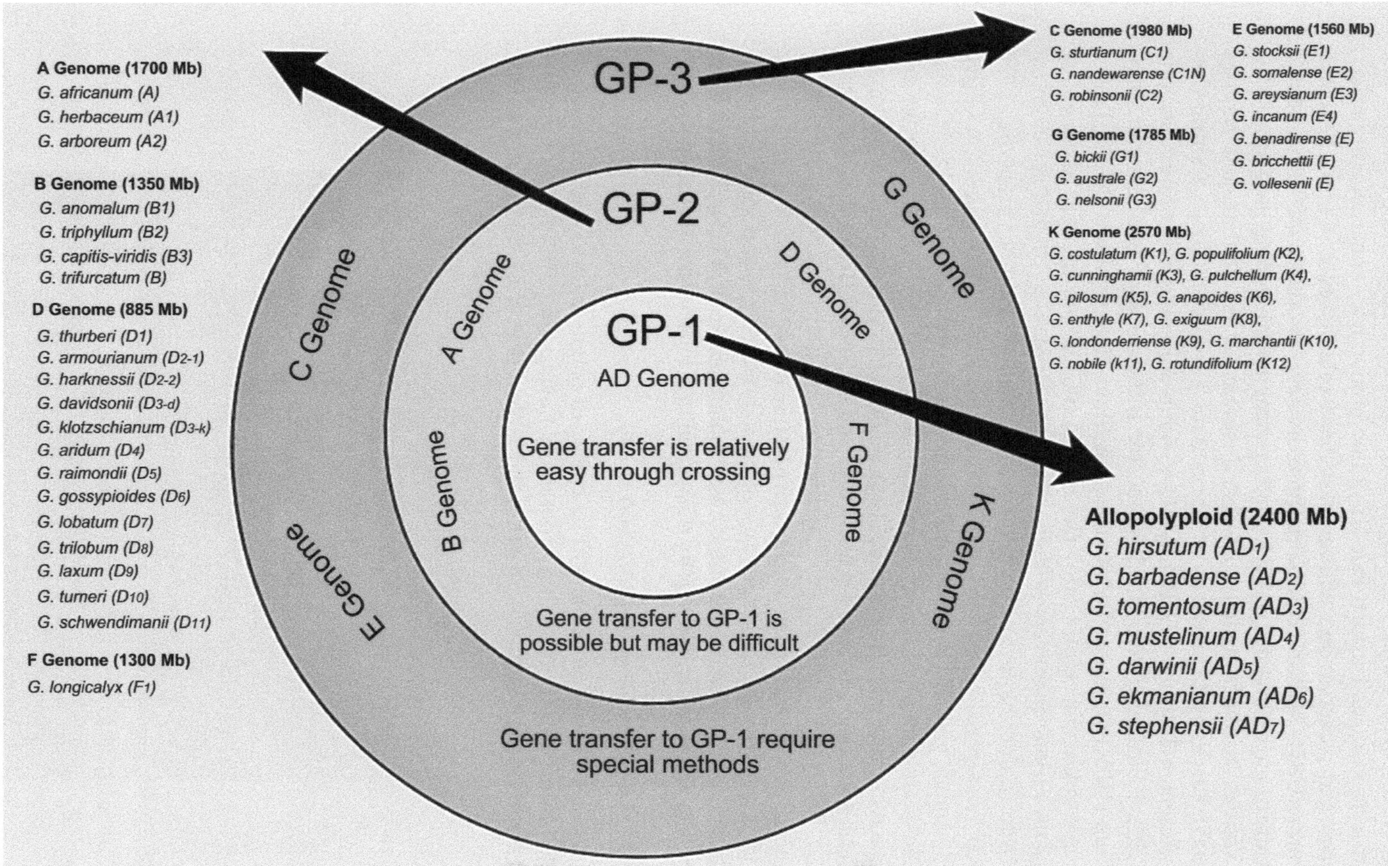

FIGURE 2.1 **Placement of genomes of *Gossypium* in various genepool and the opportunities for utilization in breeding program for inter-specific hybridization for improving most cultivated *G. hirsutum*.**

Genetic Resources Institute (PGRI), NARC at Islamabad provides medium- and short-term storage. Through Pak-US cotton productivity enhancement project, storage and preservation facilities have been established at Cotton Research Institute (CRI), Multan (Pakistan) approximately 28 cotton species (wild and cultivated) are maintained at CRI, Multan useable for breeding programs [23].

Various institutes namely, Nuclear Institute for Agriculture and Biology (NIAB), Faisalabad; University of Agriculture, Faisalabad (UAF); Central Cotton Research Institute (CCRI), Multan; Ayub Agriculture Research Institute (AARI), Faisalabad; National Institute for Biotechnology and Genetic Engineering (NIBGE), Faisalabad; Nuclear Institute of Agriculture (NIA), Tandojam and Pakistan Central Cotton Committee (PCCC) are engaged in collecting notes on various characters of cotton plant for morphological and agronomical characterization, biotic and abiotic stress evaluation, and fiber properties [22]. NIAB Faisalabad is also working on biochemical properties of cotton. CCRI, CRS Multan, and NIAB Faisalabad are involved in cytogenetic studies. Studies on the quantification of Bt toxin are being conducted at CEMB (Center for Excellence in Molecular Biology, Lahore), NIBGE Faisalabad, AARI Faisalabad, NIGAB (National Institute for Genomics and Advanced Biotechnology, Islamabad). FSC&RD (Federal Seed Certification and Registration Department, Islamabad) is working on the seed quality testing of cotton [23].

Description of wild species of cotton is presented in the following paragraphs. The herbarium of these species is maintained in the Department of Plant Breeding and Genetic, University of Agriculture, Faisalabad (Pakistan).

2.3.1 *Gossypium arboreum* (2n = 26)

This species has A_2 genome and its center of origin is India and Pakistan. It is also called Desi cotton, that is perennial having long trailing thin branches and stems. Plant height is between 6 and 10 feet (Fig. 2.2). Leaves are mostly

FIGURE 2.2 Growth habit of *Gossypium arboreum* L.

leathery with dotted glands and sub-glabrous (abortively short stellate hairs on the blade). Flowers are yellow with purplish red spots on petals. Capsules are round to slight tapered with 3–4 locules (Fig. 2.3A–D). Disease, insects, and drought resistance are the important breeding characteristics of this species.

FIGURE 2.3 Leaf (A), Flower (B), Bud (C), and Boll (D) of *G. arboreum*.

2.3.2 *Gossypium herbaceum* (2n = 26)

This cotton species belongs to A_1 genome and center of origin is Pakistan, India, and Iran. It might have been introduced before or after domestication to Near East including Arab, Baluchistan, and perhaps Sindh via trade route between Arab and Africa. It is annual in growth habit having shrub like stems and branches, slightly bent at the joints, sparsely hairy (Fig. 2.4). Leaves leathery and deeply cordate (heart shaped) (Fig. 2.5). Flowers are medium sized, yellow or red with purple center. Calyx is comparatively large, broader and its upper margin serrated into 6-8 teeth. Capsules are small, round with 3-4 locules (Fig. 2.6A–D). Disease, insect, and drought resistance are its key breeding characters.

FIGURE 2.4 Growth habit of *G. herbaceum*.

FIGURE 2.5 Leaf of *G. herbaceum*.

2.3.3 *Gossypium somalense* (2n = 26)

G. somalense belongs to E_2 genome and its center of origin is Pakistan. It is perennial shrub with tall, upright growth habit. Plant is usually 3 m tall (Fig. 2.7). Stem is tomentose, green in color, and have hairs on its tips. Leaves are petiolate and its midrib has nectars beneath the surface bracts. Calyx has five lobes with black glands. Corolla is yellow with noticeable dark red spots at the base. Anther and pollens are yellow and non-glandular with protruding stigma. Staminal column is glabrous, non-glandular. Capsules are glanded, 3–4 celled and ovoid in shape with densely pubescent seeds. Fiber color is brownish (Fig. 2.8A–C). Resistance to bollworm and drought are its key characters for breeding.

FIGURE 2.6 Flowers (A,B), Bud (C), and Boll (D) of *G. herbaceum*.

2.3.4 *Gossypium klotzschianum* (2n = 26)

G. klotzschianum belongs to $D_{3\text{-}k}$ genome and Central America is considered the center of origin. It is upright shrub or tree having perennial growth habit. It has spreading branches which are usually reddish brown when young. Stem tips are glabrous (Fig. 2.9). Leaf length is generally of 2.5 inches with unclear dotted glands (Fig. 2.10). Inflorescence is axillary and shoots are lateral. Petiole size is 0.6 cm but peduncle is double the length of it. Flowers are large sized and 1.5 inches broad. Corolla pale yellow but petal spots are unnoticeable. Fruits round to ovate with four cells having large flattened free seeds.

FIGURE 2.7 Growth habit of *G. somalense*.

FIGURE 2.8 Leaf (A), Flower (B), and Boll (C) of *G. somalense*.

FIGURE 2.9 Growth habit of *G. klotzschianum*.

FIGURE 2.10 Leaf of *G. klotzschianum*.

2.3.5 *Gossypium thurberii* (2n = 26)

This species belongs to D_1 genome and its center of origin is Central America. It is upright shrub have perennial growth habit with 2 m height (Fig. 2.11). Leaves are glabrous with 3–5 lobes but basally constricted central lobe (Fig. 2.12). Petals have cream color spots at the base and bracteole size is 0.8 cm long and 0.4 cm wide. Pollen color is bright yellow. Style is a bit larger than androecium. Capsules are glabrous and have 3 locules with black naked seeds. Bolls have smooth surface and roundish oval in shape with pointed tip. Fiber quality is moderate. It possesses resistance to frost, fusarium wilt, and bollworms. High GOT, plentiful boll bearing, and good spinning are its prominent breeding traits.

FIGURE 2.11 Growth habit of *G. thurberii.*

FIGURE 2.12 Leaf of *G. thurberii.*

2.3.6 *Gossypium australe* (2n = 26)

This species belongs to G genome and it is originated in N. Central Australia. It has shrub type growth habit which is 2–3 m tall (Fig. 2.13). Stem is tomentose with glands on it. Leaves are ovate to elliptical, hairy, and reddish. Nectar is elongate with three bracts. Calyx constricted at the base, petals pink in color with dark red spot non-glanded, glabrous staminal column with pinkish anthers. Densely glanded style, which is exceeding the androecium (Fig. 2.14A,B). Capsule has 3–4 cells and fiber color is brownish. It is resistant to bacterial blight and drought and have high GOT. Gossypol glands are absent in seeds and this is the prominent feature for breeding perspective.

2.3.7 *Gossypium hirsutum* (4n = 52)

G. hirsutum belongs to $(AD)_1$ genome and its center of origin is Central America. This species has annual growth habit. Stem is green but can also be pigmented to densely hairy (Fig. 2.15). Leaves can be broad as well as narrow. Pedicel are variable, bracteole size large but variable. Corolla is of yellow color without spots on petals. Anthers are greenish yellow in color. Stigma is non-protruding type. Bolls are ovate, roundish, or elongated. Locules are smooth, 3–5 in number with blunt tip. Fiber color has different shades *i.e.*, white, cream, brown and green (Fig. 2.16A–C). Resistance to verticillium wilt, fiber yield, and quality are dominant characteristics to exploit in breeding programs.

FIGURE 2.13 **Growth habit of *G. austral.***

FIGURE 2.14 Leaf (A) and Bud (B) of *G. australe.*

FIGURE 2.15 Growth habit of *G. hirsutum.*

2.3.8 *Gossypium barbadense* (2n = 52)

This species belongs to $(AD)_2$ genome and originated in South America. It is annual in growth habit. It is shrub with a height of 1.5–2 m (Fig. 2.17). Presence of spots and deeply pitted points on bolls are key characters for identification of this species. Stem is slightly hairy, corolla is spotted; yellow in color

FIGURE 2.16 Leaf (A), Boll (B), and Flower (C) of *G. hirsutum*.

and its size ranges from 4.5–5.5 cm. Anther and pollen are yellow in color (Fig. 2.17). Stigma is protruding type. Bolls shape is elongated with pointed tip and 3 locules. Fiber fineness is a key character for identification of Pima cotton (Fig. 2.18).

FIGURE 2.17 **Growth habit of *G. barbadense*.**

FIGURE 2.18 **Leaf of *G. barbadense*.**

2.3.9 *Gossypium capes-viridis* (2n = 26)

G. capes-viridis belongs to genome B_3 and Cape Verde Island is center of origin of this species. It is shrub in nature of 1.5–2 m tall (Fig. 2.19). Leaves are petiolate, lobed, and pubescent. Stem also have pubescence on it. Flowers have bracts, glands, nectars, and yellowish in color with dark red spot at base. Long petiolate and bifurcated stigma are characters of identification (Fig. 2.20). It is resistant to bacterial blight.

FIGURE 2.19 **Growth habit of *G. capes-viridis*.**

FIGURE 2.20 **Leaf of *G. capes-viridis*.**

2.3.10 *Gossypium nelsonii* (2n = 26)

G. nelsonii belongs to G genome and center of origin is N. Central Australia. It is distributed in Australia. Plant is shrub in nature which can be of 2–2.5 m in height (Fig. 2.21). Leaves are oblong without lobes. Flowers are showy and beautiful with pink in color with reddish base (Fig. 2.22A,B).

FIGURE 2.21 **Growth habit of *G. nelsonii.***

FIGURE 2.22 Leaf (A) and flower (B) of *G. nelsonii.*

2.3.11 *Gossypium longicalyx* (2n = 26)

This wild cotton species belongs to F_1 genome and center of origin and distribution is Africa. It is perennial shrub with trailing habit as well. It has slender stem with pubescence as well (Fig. 2.23). Leaves are ovate with glands. Nectars are absent in flower. Petals are yellowish in color. Stigma is non-protruding. Pollens are yellow in color and capsules have 3 locules, whereas bolls are elongated in nature. Elongated leaves and flower without spot are its key characters for identification (Fig. 2.24). It can be used in breeding programs due to fiber fineness and source of cytoplasm male sterility.

FIGURE 2.23 **Growth habit of *G. longicalyx*.**

FIGURE 2.24 **Leaves of *G. longicalyx*.**

2.3.12 *Gossypium stocksii* (2n = 26)

This species belongs to E_1 and center of origin is Pakistan. It is wild diploid species distributed in America. It is perennial in growth habit but small in size (Fig. 2.25). Stem is woody, and leaves are simple or lobed in nature. Flowers are produced in axillary shoots, and flowers are large, bractete, tapering, yellow, glanded, and have red spot. Bolls are black dotted with dark blue surface, and boll have 3 celled with 2–3 seeds in it. Seed is smooth, blackish with brownish lint on it. Bushy Habit, petals rotating toward right, newly appeared flower is yellow in color whereas old flower turn into pink (Fig. 2.26A,B). It possesses drought tolerance related genes.

FIGURE 2.25 **Growth habit of *G. stocksii*.**

FIGURE 2.26 New flower color (yellow) (A) and Old flower color (pink) (B) of *G. stocksii*.

2.3.13 *Gossypium lobatum* (2n = 26)

This species belongs to genome D_7 and center of origin is Central America. It is also American diploid species. It is tree type that grows very tall. Stem have hair like pubescent (Fig. 2.27). Leaves are petiolate and have glands on them. Flower and fruits are produced in leaf axils. Calyx is yellow in color. Petals have purple spot at its base. Anthers are purplish but pollens are yellowish orange in color (Fig. 2.28A,B). This species has resistant genes against bollworms.

FIGURE 2.27 Growth habit of *G. lobatum*.

FIGURE 2.28 Leaf (A) and flower (B) of *G. lobatum*.

2.3.14 *Gossypium laxum* (2n = 26)

G. laxum genome is D_9 and center of origin is Central America. It is perennial in growth habit (Fig. 2.29). Plant has more than 3–4 m height. Leaves are single lobed and of heart shape (Fig. 2.30).

FIGURE 2.29 Growth habit of *G. laxum.*

FIGURE 2.30 Leaves of *G. laxum.*

2.3.15 *Gossypium gossypioides* (2n = 26)

This species belongs to genome D_6 and center of origin is Central America. It is also distributed in America. Plant has shrub growth habit. Stem has very small dense hair on it. Leaves are of three lobed and glanded without nectaries. Flower color is pink with purplish spot at its base. Calyx has dotted glands on it. Capsule have 3 celled, seed has sparse fiber (Fig. 2.31A–C). Resistance against Jassid is an importance feature of this species.

FIGURE 2.31 Growth habit (A), flower (B), and leaves (C) of *G. gossypiodes*.

FIGURE 2.32 Growth habit (A), leaf (B), and flower (C) of *G. harknessii*.

2.3.16 *Gossypium herkensii* (2n = 26)

It belongs to genome $D_{2\text{-}2}$ and center of origin is Central America. It is also distributed in America. It is perennial in growth habit. Plant spread like shrubs. Leaves are small, thick, glabrous, three lobed, and with prominent veins. Flowers are of yellowish color, petals have purple spot on them (Fig. 2.32A–C). It is resistance to fusarium wilt, tolerant to drought, and can be used as CMS line in hybrid development program.

2.3.17 *Gossypium aridum* (2n = 26)

This species belongs to genome D_4 and center of origin is Central America. It is also tall diploid species of cotton which can be grow up to 4–8 m. It has glands and hair on stem (Fig. 2.33). Leaves are ovate with petiole (Fig. 2.34). Flowers are solitary, bracteate, necteries present, and yellowish color. Anthers are purplish with yellowish pollen. Seeds are with dense brown fiber. It is drought tolerant, high in seed number, and can be used as CMS line in breeding.

FIGURE 2.33 **Growth habit of *G. aridum.***

FIGURE 2.34 **Leaf of *G. aridum.***

FIGURE 2.35 Growth habit (A), leaf (B), and flower (C) of *G. lanceolatum*.

2.3.18 *Gossypium lanceolatum* (4n = 52)

This wild cotton species belongs to genome AD and center of origin is Southern Mexico. It is tetraploid and distributed in America. It has sparse growth habit that can grow up to 3–4 m. Leaves are tri-lobed with clear venation. Flowers are of yellow color with reddish spot at central position near base. Anther color is yellow (Fig. 2.35A–C).

2.3.19 *Gossypium tomentosum* (4n = 52)

G. tomentosum belongs to genome AD_3 and center of origin and distribution is Hawaiian Island. It is tetraploid and has trailing type growth habit (Fig. 2.36). Leaves are hairy and has three clear cut lobes with dense reticulate veins (Fig. 2.37). This species have drought tolerance related genes.

2.3.20 *Gossypium bickii* (2n = 26)

This species belongs to genome G_1 and its center of origin and distribution is N. Central Australia. It is diploid and spreading type shrub which is 1.5 m tall.

FIGURE 2.36 **Growth habit of *G. tomentosum*.**

FIGURE 2.37 Leaves of *G. tomentosum*.

Leaves are elliptical, sometimes ovate in shape. Flowers are pink with red spot at its base (Fig. 2.38A,B). Style is glanded and seeds are appressed from both sides but do not have glands.

FIGURE 2.38 Growth habit (A) and leaf (B) of *G. bickii*.

2.4 Wild cotton species in China

China has long history of growing cotton and native people of South China used Jibei accession abundantly for cotton to weave colorful cloth. *G. arboreum* and *G. herbaceum* were two primary species of cotton in China. *G. herbaceum* was grown in West China and *G. arboreum* in South China. In south *G. arboreum* kept growing and then brought and spread to North China but across the country *G. herbaceum* was not popularized. *G. barbadense* and *G. hirsutum* was imported to China during1892–93. Now China is the largest consumer and producer of cotton in the world [24]. Yangtze River, the Yellow River area, and Xinjiang area are three main cotton growing areas in China. Chinese Academy of Agricultural Sciences (CAAS) was established during 1958 for collection of wild and cultivated germplasm for long and medium term storage. At Hainan Island, in vivo collection of wild cotton species is also conserved. From 1958, in vivo garden and germplasm bank is under the supervision of six Professors for development of cotton germplasm. Major work of these Professors include: development of early system of cotton preservation, collection, exploration, and evaluation of *G. arboreum* in South-West China, evaluation of these cultivars in germplasm bank, development of molecular methods for evaluation of genetic variation and diversity in cotton germplasm, genes identification linked with important traits, in vivo preservation and evaluation of germplasm with available molecular methods [24]. In 1950s, delta-pine line of upland cotton was imported from United States for the enhancement of new cultivars, in addition Stonville and Acala were also introduced in China. In last 100 years, about 2236 accessions of cotton were introduced into China that includes: 198 *G. barbadense,* 14 *G. arboreum*, and 2024 *G. hirsutum* accessions. While, Egypt, Pakistan, United States, Russia, and Australia are major countries with a great number of introductions [24]. Total germplasm collection of China is 8,868 accessions that include: 32 wild species and four field grown species *G. hirsutum*, *G. barbadense*, *G. arboreum*, and *G. herbaceum.* Most of the accessions originated from all over the China and, remaining was imported from 52 countries around the globe [24].

These germplasm accessions are evaluated for different agro-economic, morphological, and fiber quality related characteristics. The major characters are different maturity stages, strong and long fiber, high oil, protein and gossypol content, dwarfness, pubescence, resistance against aphids, bollworms, verticillium wilt, salinity, tolerance against drought, humidity and cold, genetic markers, color of fiber, and male sterility [25]. Germplasm are divided into three categories, that is, temperate, sub-tropical, and tropical type. The imported germplasm from United States, that is, DPL-4, DPL-15, Stoneville Empire, and TX 53 were the primary material for the promotion of breeding program in China. From these basic lines, thousands of new lines were developed [26]. One of the typical cultivar is Guannong No.1 that was utilized as basic line. New basic lines, that is, Zhongmiansuo 4, Zhongmiansuo 7, Zhongmiansuo 2,

Xuzhou 181, Ganmian 1, Jingmian 1, Liaomian 1, Shanmian 5, Mcnair 210, GL-5, KK1543, Jiangsumian 3, Jiangsumian 1, and Yishuhong were grown and commercialized on large area. Crossing between Xingtai 6871 and Uganda 3 resulted in Zhongmiansuo 12 cultivar which was resistant to *Fusarium* wilt and high yielding. During 1992–96, it was grown on approximately 50% cotton area of China. Okra leaf cultivar TX 8207 was the oldest cultivar that showed reduced vegetative growth and Syllepte derogata Fabricius resistance [26].

Scientists from China have started to utilize and study wild species, and Shiyuan 321 was developed by hybridization using *G. hirsutum*, *G. barbadense*, and *G. thurberi*. Through hybridization of *G. thurberi*, *G. anomalum*, and *G. hirsutum,* a second commercialized variety Jinmian 2 was developed which was high yielding, early maturing with good fiber quality, and resistant to some diseases. HST2 was developed from *G. hirsutum* and *G. sturtianum* which showed resistance to Fusarium wilt, Verticillium wilt, and drought. Line BZ201 from *G. hirsutum* and *G. thurberi*, line JSCG7235 from *G. anomalum* and *G. hirsutum* were developed for more strong and long fiber [27]. It is suggested that for enhancement of cotton breeding program in China, more wild species and landraces should be collected and preserved. Likewise, exchange of germplasm between different countries should be continued in order to enhance the genetic diversity [27].

2.5 Wild cotton species in India

Cotton has played an integral and pivotal role in uplifting the economic state of India. In the Indian subcontinent evidence of growing cotton has been found since the 4th millennium BC [19]. Fabric recovered from Mohenjo-Daro was found to be originated from the plants of cotton and scientists suggested that those old fabrics are closely related to species of *G. arboreum* [28]. Out of four lint bearing species of cotton, *G. arboreum* L. and *G. herbaceum* L. are diploid and native to Africa and Asia. New World cottons *G. barbadense* and *G. hirsutum* that are tetraploid are introduced in India in 18th century AD [29]. India was a leading country in term of largest cotton growing area and growing diverse cultivars by the end of 20th century. During 2002–03, area of cotton in India has increased due to introduction of transgenic cotton. *G. hirsutum* × *G. hirsutum* hybrids were grown more on than 90% area of cotton in the country [19]. Two “AD” genome allotetraploid species and two “A” genome diploid are cultivated for their fiber lint. Breeding lines and germplasm genetic stock is developed by crossing between wild and cultivated species. In India the gene banks of Central Institute for Cotton Research (CICR), Nagpur and its regional station and certain other agricultural centers have lot of cultivated and wild species of cotton [28]. Since the 1970s India is the first country to grow hybrids of cotton. Hybrid-4, first intra-hirsutum hybrid was released in 1971. TCHB213, an inter-specific hybrid with high yield potential and superior qualities was released. Stable male sterile plants have been identified from hybrids involving *G.*

hirsutum, *G. raimondii*, *G. barbadense*, and *G. thurberi*. Cytoplasm conferring male sterility from *G. harknesii* was transferred in *G. hirsutum* and restorer genes for *G. harknesii* conferring fertility has been transferred successfully to Indian genotypes and they were utilized for development of cotton hybrids such as PKV Hyb-3.

2.6 Wild cotton species of America and Australia

Due to their superior yield and lint quality, the New World tetraploid cotton cultivars are presently known to dominate cotton production. Over 90% of global cotton production today is supplied by the modern cultivars of *G. hirsutum*, and the other 10% is supplied by *G. barbadense*: which has fine, strong, and long fiber making it ideal for specialized use, yet produces lower yields when compared to *G. hirsutum* [30]. Some researchers suggest that *G. barbadense* had likely been domesticated in northwest South America. The oldest archaeobotanical remains (including seed, fruit, fiber, fishing nets, fabrics, and yarn) of *G. barbadense* date from 5500 to 5000 BP and were recovered from central coastal Peru. The primitive characteristics that were seen through the analysis of morphology on archaeobotanical material indicates that they were derived from early domesticates and suggests that the original domestication of *G. barbadense* is likely to have occurred within the region. It is suggested that following the initial domestication of *G. barbadense,* it is likely that major route of dispersal would have been across the Andes to North-eastern South America [30]. A proceeding dispersal would have occurred after this secondary stage in agronomic evolution, where we begin to find *G. barbadense* in Central America and the Caribbean, as well as the Pacific where it had come into contact with *G. hirsutum* [31]. Archaeological evidence of *G. barbadense* in Real Alto of Coastal Ecuador dates from 3500 to 1500 BC and is associated with Valdvia culture. Early Valdiva cotton was recovered from Real Alto's domestic area within hearths, house floors, and other features within the household. The macro botanical remains uncovered through excavation consisted of 133 carbonized cottonseed fragments or whole seeds; and when compared to modern samples, showed the morphological characteristics of *G. barbadense* [31]. The evidence from Real Alto indicates that *G. barbadense* had likely been domesticated even earlier in southwest Ecuador *G. hirsutum* is unique among the *Gossypium* sp., as it has an immense indigenous range, occupying most of the Caribbean and Mesoamerica, where it shows a variation in its morphological forms [32]. The primary center of morphological diversity in *G. hirutum* is likely to have been in Mesoamerica, which represents the original geographic point of domestication [33]. A secondary center of diversity seems in the Caribbean; where primitive forms of *G. hirutum* dispersed into the Caribbean and were hybridized with *G. barbadense* cultivars [34]. This interpretation is supported by evidence of a high frequency of alleles from *G. barbadense* in Caribbean *G. hirutum* cultivars. Although archaeological evidence is limited, Wendel suggests that site of original domestication for *G. hirutum* in Yucatan peninsula of Mesoamerica, where wild variety; race *yucatanense* was found, and where *G. hirutum* represents a dominant component of indigenous vegetation [33].

2.7 Conclusion and future prospects

In this chapter, the wild species originated in various regions are presented and list of various genepool sources are also mentioned. The presence of wild relatives of cotton provide an opportunity to research workers for their utilization in breeding programs, where inter-specific hybridization is the first step and then duplication to make them fertile. It is reported through GWAS analysis that wild relatives are the reservoir of genes useable for the development and improvement of fiber related traits. In addition, these relatives also provide opportunity for the introduction of resistance/tolerance against biotic and abiotic stresses. The potential of those species of *Gossypium* is presented in cited paragraphs. Beside this, weather calamities are also affecting the potential due to change in climate pattern; but these deficiencies could be addressed through the utilization of treasure of wild relatives.

Acknowledgments

This research was supported by Centers for Advanced Studies-Punjab Agricultural Research Board (CAS-PARB) and Higher Education Commission (HEC), Pakistan.

References

[1] Myles S. Improving fruit and wine: what does genomics have to offer? Trends Genet 2013;29(4):190–6.
[2] Meyer RS, DuVal AE, Jensen HR. Patterns and processes in crop domestication: an historical review and quantitative analysis of 203 global food crops. New Phytol 2012;196(1):29–48.
[3] Preece C, Peñuelas J. A return to the wild: root exudates and food security. Trends Plant Sci 2020;25(1):14–21.
[4] Brubaker C, Paterson A, Wendel J. Comparative genetic mapping of allotetraploid cotton and its diploid progenitors. Genome 1999;42(2):184–203.
[5] Majumdar R, Lebar M, Mack B, Minocha R, Minocha S, Carter-Wientjes C, et al. The *Aspergillus flavus* spermidine synthase (spds) gene, is required for normal development, aflatoxin production, and pathogenesis during infection of maize Kernels. Front Plant Sci 2018;9:317.
[6] Wendel JF, Olson PD, Stewart JM. Genetic diversity, introgression, and independent domestication of old world cultivated cottons. Am J Bot 1989;76(12):1795–806.
[7] Metcalf AA. The world in so many words: a country-by-country tour of words that have shaped our language. University of Hawaii Press; 1999.
[8] Sanamyan M, Petlyakova J, Rakhmatullina E, Sharipova E. Cytogenetic collection of Uzbekistan. World cotton germplasm resources. Intech; 2014. 247-87.
[9] Percival A, Wendel J, Stewart J. Taxonomy and germplasm resources. In: Smith WC, Cothren JT, editors. Cotton: Origin, history, technology, and production. New York, NY: John Wiley and Sons, Inc; 1999.

[10] Wendel JF, Cronn RC. Polyploidy and the evolutionary history of cotton. Adv Agron 2003;78:139.
[11] Kole C. Wild Crop relatives: genomic and breeding resources. Springer; 2011.
[12] Fryxell P. A revised taxonomic interpretation of *Gossypium* L (Malvaceae). Rheedea 1992;2:108–16.
[13] Phillips LL. The cytogenetics of Gossypium and the origin of New World cottons. Evolution 1963;460–9.
[14] Keerio AA, Shen C, Nie Y, Ahmed MM, Zhang X, Lin Z. QTL mapping for fiber quality and yield traits based on introgression lines derived from *Gossypium hirsutum* × *G. tomentosum*. Int J Mol Sci 2018;19(1):243.
[15] Shim J, Mangat P, Angeles-Shim R. Natural variation in wild *Gossypium* species as a tool to broaden the genetic base of cultivated cotton. J Plant Sci Curr Res 2018;2(005):1–9.
[16] Kulkarni VN, Khadi BM, Maralappanavar MS, Deshapande LA, Narayanan S. The world-wide gene pools of *Gossypium arboreum* L. and *G. herbaceum* L., and their improvement. Genetics and genomics of cotton. Springer; 2009. p. 69–97.
[17] Wendel JF, Brubaker CL, Seelanan T. The origin and evolution of *Gossypium*. Physiology of cotton. Springer; 2010. 1–18.
[18] Cherry J, Katterman F, Endrizzi J. Comparative studies of seed proteins of species of *Gossypium* by gel electrophoresis. Evolution 1970;431–47.
[19] Narayanan S, Singh J, Varma P. A brief review of the genetic wealth in *Gossypium* with special reference to wild species. ISCIJ 1985;10:51–64.
[20] Khan SA, Hussain D, Askari E, Stewart JM, Malik K, Zafar Y. Molecular phylogeny of *Gossypium* species by DNA fingerprinting. Theor Appl Genet 2000;101(5–6):931–8.
[21] Rana M, Bhat K. A comparison of AFLP and RAPD markers for genetic diversity and cultivar identification in cotton. J Plant Biochem Biotechnol 2004;13(1):19–24.
[22] Yasmin T, Tabbasam N, Ullah I, Asif M, Zafar Y. Studying the extent of genetic diversity among *Gossypium arboreum* L. genotypes/cultivars using DNA fingerprinting. Genet Resour Crop Ev 2008;55(3):331–9.
[23] Rahman M, Shaheen T, Tabbasam N, Iqbal M, Ashraf M, Zafar Y, et al. Genetic resources in cotton and their improvement. Agron Sustain Dev 2012;32:419–32.
[24] Jia Y, Sun J, Du X. Cotton germplasm resources in China. World cotton germplasm resources. IntechOpen; 2014. 35-53.
[25] Feng HJ, Sun JL, Wang J, Jia YH, Zhang XY, Pang BY, et al. Genetic effects and heterosis of the fibre colour and quality of brown cotton (*Gossypium hirsutum*). Plant Breed 2011;130(4):450–6.
[26] Sheng W, ZHAO. G-h, JIA Y-h, DU X-m. Molecular cloning, and characterization of an *Adenylyl cyclase*-associated protein from *Gossypium arboreum* L. Agr Sci China 2009;8(7):777–83.
[27] Cai Y, Xiaohong H, Mo J, Sun Q, Yang J, Liu J. Molecular research and genetic engineering of resistance to Verticillium wilt in cotton: a review. Afr J Biotechnol 2009;8(25):7363–72.
[28] Santhanam V, Sundaram V. Agri-history of cotton in India. Asian Agri-History (India) 1997;1(4):235–51.
[29] Endrizzi J, Turcotte E, Kohel R. Genetics, cytology, and evolution of *Gossypium*. Adv Genet 1985;23:271–375.
[30] Damp JE, Pearsall DM. Early cotton from coastal Ecuador. Econ Bot 1994;48(2):163–5.
[31] Hutchinson J. New evidence on the origin of the Old World cottons. Heredity 1954;8(2):225–41.
[32] Jenkins J. Traditional crop breeding practices: an historical review to serve as a baseline for assessing the role of modern biotechnology. OECD; 2003.

[33] Moulherat C, Tengberg M, Haquet J-F, Mille Bt. First evidence of cotton at Neolithic Mehrgarh, Pakistan: analysis of mineralized fibres from a copper bead. J Archaeol Sci 2002;29(12):1393–401.

[34] Stephens S, Moseley ME. Early domesticated cottons from archaeological sites in central coastal Peru. Am Antiq 1974;39(1):109–22.

[35] Hutchinson JB, Ghose R. The classification of the cottons of Asia and Africa 1937. Indian J Agric Sci. 1937;7:233–57.

Chapter 3

Wheat Wild Germplasm: A Hidden Treasure

Muhammad Kashif[a], Shabir Hussain Wani[b] and Shadab Shaukat[a]
[a]Department of Plant Breeding & Genetics, University of Agriculture, Faisalabad, Pakistan; [b]Mountain Research Center for Field Crops, Sher-e-Kashmir University of Agricultural Sciences and Technology of Kashmir, Jammu and Kashmir, India

3.1 Introduction

Wheat belongs to the grass family and is being cultivated for its edible grain utilization. It is a common cereal grown across the globe [1]. Wheat was previously cultivated in the Fertile Crescent region which was considered to be the most expected center of origin for this crop. Three different inter-related diploid genomes were combined and resultantly an allohexaploid species (2n = 6x = 42, AABBDD genomes) was developed, which is commonly called as bread wheat *Triticum aestivum* [2–5]. Wheat has been grown since ancient times as mankind needed food. It has been domesticated since 8,000 years. It is a major staple crop of many civilized nations like North Africa, West Asia, and Europe. Globally wheat production is controlled by 12 major regions (Countries), which is about 90% of world wheat production. Nowadays, wheat grain is very important and grown on major commercial scale than any other crop. Wheat production is larger over all crops, including rice, maize, and potatoes and its world trade is more than any other crop [6]. In wheat, initial cultivated forms were landraces and farmers those who cultivated selected them in wild forms and wild germplasm have resistance. Selection was practiced more on yield and other plant characters before the start of plant breeding [1]. Several billions of people globally depend upon wheat for their major portion of diet. It is grown worldwide and is consumed by 33% of the population of this planet owing to its nutritional significance; range of end-use products and especially for its storage qualities. Bread making primarily utilize wheat flour. Its grains contain most of the essential components such as carbohydrates (60%–80%), moisture (12%), fats (1.5%–2%), and proteins (8%–15%) [7]. In developing countries, 4.5 billion people mostly rely on wheat for their daily caloric (21%) requirement and protein (20%) intake [8].

Wild Germplasm for Genetic Improvement in Crop Plants. http://dx.doi.org/10.1016/B978-0-12-822137-2.00003-5

Globally, wheat wild relatives are present in various world germplasm resources, it is poorly characterized and there may be numerous causes for this. First, it is difficult to utilize wild relatives in conventional breeding programs rather than cultivars of the same species. These wild relatives need more time for the germplasm improvement and few institutes are involved with their interspecific hybridization. Second, as compared to cultivars, wild relatives shatter more seeds and it is difficult to maintain and increase seeds of wild relatives. Finally, crop wild species in wheat produce many generations rapidly and face less danger of extinction as compared to cultivated landraces. In wheat, many annual wild relatives are under extinction threat due to changing pattern land use, monotype wheat cultivation, climate change, and stresses. Emmer wild wheat species have involvement from different wild populations to develop wild relatives [9,10]. Evidence from literature showed that multiple wild population contribution toward modern wheat is not so fast but this domestication was slow with the passage of time [11] while the domestication cumulative ratio was not on constant [12].

Emerging technologies improved the role of annual wild relatives as a germplasm source for their utilization in future breeding programs. Owing to great genetic variability, wild species of wheat attain a great priority among other species of wheat. Some of the wild relatives of that crop were used for the improvement against abiotic stresses [13]. Global warming is a serious threat of this era and wild plants try to acclimatize in hot environment and how such plants genetically respond to global warming is yet to be investigated properly. Due to great genetic diversity [14] and best genetic makeup, wild species of wheat are among the best cultivars to fulfill the food production of human beings [15,16]. Since 10,000 years, emmer wheat was domesticated and *Triticum dicoccoides*, was its wild progenitor. It was the result of hybridization of two diploid progenitors, (*T. uratu,* and *Ae. Speltoides).* Thus, wheat was concluded to be derived from three diploid, each having seven chromosomes with similar organization.

In order to improve and develop resistant and high yielding wheat varieties, it is necessary to use different wild wheat relatives in crop breeding. By adapting this program, several Asian countries have increased their yield and disease resistant wheat varieties have been developed and helped large numbers of farmers. Wild relatives have many useful genes to improve traits of economic importance if those could be transferred to cultivated species to enhance yield. Wild relatives have potential to boost up the climatic resilience and yield of crop plants. Polyploid wheat species have more diversity in domestication and have large portion of variability of its tetraploid wild progenitor [17].

3.1.1 History of Green Revolution

The developing world has witnessed that food crop productivity from last 50 years was extraordinary although scarcity of land was feared to beocme major problem. Increasing population is more than doubled this time and cereals production increased 3 times in this period with only one-third increase in land area cultivated.

Major breakthrough for world food was Green Revolution that was started in 1960s to overcome the malnourishment problem of the developing world.

At the start of Green Revolution (1935), scientist from Japan Gonjiro Inazuka working on wheat, crossed semi-dwarf local landraces with American and, two cultivated varieties that gave upgraded semi-dwarf variety, that was, Norin 10. In this semi-dwarf wheat plant, height was reduced to 60–110 cm with the Rht1 and Rht2 genes of Norin 10 as compared to other varieties, which had 150 cm average height. Wheat breeders from Washington State University (WSU), Orville Vogel used Norin 10 in late 1940s which help for the improvement of winter varieties. Dr. Norman E. Borlaug worked in Mexico for the development of rust resistant wheat used Vogel's varieties. In 1953, Borlaug started making crosses between the Vogel's semi-dwarf varieties along with Mexican varieties. Result of these new crosses improved spring wheat with improved special characteristics of short and firm straw, less lodging due to semi-dwarfness and more tillers per plant that give high yield. This semi-dwarf Mexican wheat progeny was disseminated in Mexico, which doubled the national wheat yield. Dr. Norman E. Borlaug named these effective genotypes as Lerma Rojo 64 and Sonora 64. These varieties showed tremendous potential and have revolutionized the production of wheat in Indo-Pak sub-continent. Millions of people were saved from starvation through this revolutionary breeding program, and he was awarded with Nobel Peace award.

FAO-Rockefeller Foundation scientists Dr Norman E. Borlaug visited Indo-Pak sub-continent in 1960 [18], and he trained first Pakistani wheat scientist Manzoor A. Bajwa at Mexico during 1961. Later on, a team of wheat scientists including Dr. Borlaug and Manzoor A. Bajwa recognized wheat variety that contained higher gluten contents and was having white grains (medium to hard). Such varieties were best for *chapatti* making and also showed tolerance against powdery mildew and rust. The new line named Mexi-Pak to show the collaboration between Mexico and Pakistan and this line showed a breakthrough in Green Revolution.

3.1.2 Collections of wheat and their wild relatives

There are about 800,000 accessions of wheat reported. International Maize and Wheat Improvement Center (CIMMYT), Australian Grains Gene bank, USDA-ARS Gene banks, and ICARDA played an important role in collection, maintenance, and conservation of germplasm of wheat. Other research institutes from Russia, Italy, and India also maintained 30,000 accessions. It is the demand of this era to develop and maintain small collections of genetic stocks for breeding and research.

Global ex situ collections of wild relatives are poorly represented due to:

1. Wild relatives shed their seeds by shattering more than crop cultivars;
2. In field, weeds are problem for regeneration of wild relatives;
3. Wild relatives seem less challenges due to their improved reproductive ability.

3.2 Breeding for disease resistance in wheat

Complete and clear-cut understanding for the plant defense mechanisms is required for the disease resistance manipulation of plant genome. Therefore, new emerging technology which will help for the production of new disease-free plant, is the prerequisite. Plant genetic resistance is mainly dependent on the plant/pathogen interaction at molecular level that give assistance to the crop for disease resistance and by improving plants at molecular level will achieve this goal against different diseases.

The success of any breeding program depends on the following elements:

1. The nature of the pathogen and diversity of virulence in the population;
2. Availability, diversity, and type of genetic resistance; and
3. The method used for screening and selection of environment for tracking resistance.

For the successful breeding program, disease resistance is most vibrant component. Different immune receptors (natural or genetic) can identify the range of plant pathogens that are present in plants on individual basis or may be in the form of combinations for vigorous, broad-spectrum disease resistance [19]. In plants, host pathogen susceptible target can be altered to hamper the effects of pathogens [20]. Host decoy proteins are serving to pathogens as a trap, which can be modified by using the genetic engineering technique and changing their specificity for pathogen recognition [21].

In wheat, there are different transgenically expressing alleles showing resistance at locus Pm3 that have specific type of disease resistance for powdery mildew in wheat field [22]. RenSeq of wild wheat varieties and *Aegilops tauschii* ssp. *strangulata* for trait-linked were explored [23] and resultantly, four stem rust (Sr) resistance genes were cloned [23]. Another study represents the application of RenSeq to produce stem rust free varieties [24].

3.3 Breeding for insect resistance

Globally wheat is facing number of insect-pest and these are causing significant damage to yield. Wild progenitors of wheat (West and Central Asia) are facing major insect pest attack in rainfed grassland [25]. With transportation of wheat through national boundaries and ocean, new insect pest hitch hikers to that area are introduced. The mapping of areas where wheat is grown can help to localize the center from where the insects of wheat are originated. Different sources of plant genetic resistance to insect pest have been recognized in cultivated cultivars [26,27] and many wild relatives of wheat [28,29], mostly resistance sources are used to develop resistant varieties.

Comparative analysis showed that insect threatened the yield more in post Green Revolution in 1990, where losses increased from 5.1% to 9.3% [30]. The number of Aphid generations was increased in wheat growing cycles and extended the distribution further in warmer winters. Aphids can modify their behavior if temperature becomes low or high [4] [31,32], and adapt the novel environment. Thus, it has become the best selection for genetic variation. The entry of any novel gene can positively impact the yield the wheat crop and protect it against insects and other pathogens.

3.4 Breeding for drought tolerance

Water shortage exerts negative impact on molecular, biochemical, physiological, and morphological processes and traits, respectively in wheat. Such shortage of water is called drought and it reduces the plant growth and reduce the crop yield. Global warming is one of the reason to cause drought and about 65 million hectares area was affected by drought stress in 2013 [33]. Physiologically, drought needs a context-dependent view to understand the ability of plants to make important changes that alleviate the effect of drought stress [34]. Drought-tolerant plants have less reduction in water content, membrane stability, and photosynthetic activity. The tolerant group accumulate soluble sugars, proline content, amino acids, chlorophyll content, and enzymatic and non-enzymatic antioxidant activities [35].

To understand the relationship between stem characters and grain weight under drought stress, it is important to know the sources of carbohydrates that support grain growth and development in wheat. Three sources that the carbohydrates availability can be obtained from (1) postanthesis synthesis directly transferred to the grains, (2) postanthesis synthesis, but stored temporarily in the stem before remobilization to the grains, and (3) preanthesis synthesis stored primarily in the stem and remobilized to the grains during the grain-filling stage [36].

3.5 Breeding for heat tolerance

"Heat stress is a function of the magnitude and rate of temperature increase, as well as duration of exposure to the raised temperature" [37]. Heat stress can also be described as "The increase in temperature above a threshold level for the duration of time which is sufficient to cause irreversible damage to plant growth and development" [38]. It has been reported that on an average 0.3°C temperature per decade is rising globally [39]. It is expected that temperature will increase approximately 1°C (till 2025) and 3°C (till 2100) above the present temperature [37]. Wheat is more delicate to cold (frost) and hot (heat stressed) conditions at reproductive stage [40,41]. Heat stress at grain filling stage disturbs grain maturity as well as grain weight which eventually causes deterioration of quality and grain yield [37,42]. According to an estimate increase of 1°C temperature causes 3%–10% loss in grain yield of wheat [43] and approximately 40% of wheat (36 million hectares) grown in temperate environment is facing terminal heat stress [44]. Series of experiments were conducted to evaluate the hypothesis that heat stress has impacted on different growth as well as developmental stages of wheat with respect to quality traits.

There is the need of varietal development that has not only the capability to avoid, escape, and tolerance against different stress but also have the potential to give more grain yields in heat stressed conditions [10]. High temperature cause reduced crop yield, short life cycle, compact biomass, less grains/spike, low grain weight, and test weight. All of these factors will ultimately reduce the yield of wheat and the target to attain food security will be compromised.

3.6 Breeding for quality traits

Diverse food products can be formulated from the flour obtained from milling of wheat. The typical characteristics of wheat flour are due to its proteins such as gliadins and glutenin. In products development and formulation, glutenin which is 40% of the wheat protein plays significant role in flour [45,46].

Grain hardness as well as protein content in wheat, both play major role in quality related characteristics of wheat flour. Although, grain harness is hereditary, severe environmental conditions have strong impact on it. It can be affected by heaving rains at the time of harvesting stage. Likewise, literature is available to prove that although nitrogen is not much influenced by hereditary characteristics, but other climatic factors during growing season such as moisture and availability of soil nitrogen etc. [47]. The efforts of breeders to develop wheat varieties having good nutritional quality, high yield, quality protein with high content, and other processing friendly characterizes is a challenging task, to obtain the long-term goal of alleviating micronutrient deficiencies from the masses. Wheat is almost consumed in all continents (except Antarctica) of the world due to its carbohydrates and protein contents since long time [8]. *Chapatti*, bread, cake, biscuit, pasta, macaroni are all formulated from wheat. Studies showed that breeding have impact on the income of farmers. Like in America, farmers cultivated varieties having 1.5% more proteins and about 10% income was increased [48]. Various genetic resources, that is, wild relatives and landraces contributed significantly in improving grain quality [49]. The synthetic hexaploid developed by hybridization among *T. dicoccum* or *T. durum* and *Ae. Tauschii* are supposed to have good nutritional characteristics and exhibit large variability for agronomic characteristics. Scientists also evaluated Zn and Fe concentrations in synthetic hexaploids [50]. Such investigations further paved the toward release of certain varieties which contained zinc level 20%–40% higher incomparison to existing [51].

References

[1] Shewry PR. Wheat. J Exp Bot 2009;60(6):1537–53. https://doi.org/10.1093/jxb/ erp058 PMID: 19386614.

[2] El Baidouri M, Murat F, Veyssiere M, et al. Reconciling the evolutionary origin of bread wheat (*Triticum aestivum*). New Phytol 2017;213:1477–86. https://doi.org/10.1111/nph.14113.

[3] Marcussen T, Sandve SR, Heier L, et al. Ancient hybridizations among the ancestral genomes of bread wheat. Science 2014;345:1250092. https://doi.org/10.1126/science.1250092.

[4] Petersen G, Seberg O, Yde M, Berthelsen K. Phylogenetic relationships of *Triticum* and *Aegilops* and evidence for the origin of the a, B, and D genomes of common wheat (*Triticum aestivum*). Mol Phylogenet Evol 2006;39:70–82. https://doi.org/10.1016/j.ympev.2006.01.023.
[5] Pont C, Salse J. Wheat paleohistory created asymmetrical genomic evolution. Curr Opin Plant Biol 2017;36:29–37. https://doi.org/10.1016/j.pbi.2017.01.001.
[6] Curtis BC, Rajaram S, Macpherson HG, editors. Bread wheat improvement and production No. 30. Rome Italy: Food and Agriculture Organization of the United Nations; 2002. p. 554.
[7] Anjum FM, Ahmad I, Butt MS, Sheikh MA, Pasha I. Amino acid composition of spring wheat and losses of lysine during chapatti baking. J Food Comp Anal 2005;18:523–32.
[8] Braun HJ, Atlin G, Payne T. Multi-location testing as a tool to identify plant response to global climate change. In: Reynolds MP, editor. Climate change and crop production. Wallingford (UK): CABI Publishers; 2010.
[9] Poets AM, Fang Z, Clegg MT, Morrell PL. Barley landraces are characterized by geographically heterogeneous genomic origins. Genome Biol 2015;16:173.
[10] Pankin A, Altmüller J, Becker C, Korff M. Targeted resequencing reveals genomic signatures of barley domestication. New Phytol 2018;218:1247–59.
[11] Allaby RG, Fuller DQ, Brown TA. The genetic expectations of a protracted model for the origins of domesticated crops. Proc Natl Acad Sci USA 2008;105:13982–6.
[12] Allaby RG, Stevens C, Lucas L, Maeda O, Fuller DQ. Geographic mosaics and changing rates of cereal domestication. Philos Trans R Soc B Biol Sci 2017;372:20160429.
[13] Hajjar R, Hodgkin T. The use of wild relatives in crop improvement: A survey of developments over the last 20 years. Euphytica 2007;156:1–13.
[14] Krugman T, Nevo E, Beharav A, Sela H, Fahima T. The institute of evolution wild cereal gene bank at the University of Haifa. Isr J Plant Sci 2018;65:129–46.
[15] Nevo E. Origin, evolution, population genetics and resources for breeding of wild barley, *Hordeum spontaneum*, in the Fertile Crescent. In: Shewry PR, editor. Barley: Genetics, biochemistry, molecular biology and biotechnology: C.A.B. International; 1992. p. 19–43.
[16] Lobell DB, Schlenker W, Costa-Roberts J. Climate trends and global crop production since 1980. Science 2011;333:616–20.
[17] Dubcovsky J, Dvorak J. Genome plasticity a key factor in the success of polyploid wheat under domestication. Science 2007;316:1862–6.
[18] CIMMYT. Wheat research and development in Pakistan; 1989.
[19] Fuchs M. Pyramiding resistance-conferring gene sequences in crops. Curr Opin Virol 2017;26:36–42.
[20] Li T, Liu B, Spalding MH, Weeks DP, Yang B. High-efficiency TALEN-based gene editing produces disease-resistant rice. Nat Biotechnol 2012;30:390–2.
[21] Kim SH, Qi D, Ashfield T, Helm M, Innes RW. Using decoys to expand the recognition specificity of a plant disease resistance protein. Science 2016;351:684–7.
[22] Brunner S, Stirnweis D, Diaz Quijano C, Buesing G, Herren G, Parlange F, et al. Transgenic Pm3 multilines of wheat show increased powdery mildew resistance in the field. Plant Biotechnol J 2012;10:398–409.
[23] Arora S, Steuernagel B, Gaurav K, Chandramohan S, Long Y, Matny O, et al. Resistance gene cloning from a wild crop relative by sequence capture and association genetics. Nat Biotechnol 2019;37:139–43.
[24] Steuernagel B, Periyannan SK, Hernández-Pinz6n I, Witek K, Rouse MN, Yu G, et al. Rapid cloning of disease-resistance genes in plants using mutagenesis and sequence capture. Nat Biotechnol 2016;34:652–5.
[25] Harlan JR. Crops and man. Madison, WI, USA: American Society of Agronomy and Crop Science Society of America; 1992.

[26] Hatchett JH, Gallun RL. Genetics of the ability of the Hessian fly, Mayetiola destructor, to survive on wheats having different genes for resistance. Ann Entomol Soc Am 1970;63:1400–7.
[27] Cox TS, Hatchett JH. Genetic model for wheat/Hessian fly (*Diptera: Cecidomyiidae*) interaction: strategies for deployment of resistance genes in wheat cultivars. Environ Entomol 1986;15:24–31.
[28] Hatchett JH, Gill BS. Expression and genetics of resistance to Hessian fly in *Triticum tauschii* (Coss) Schmal. In: Proceeding of the sixth international wheat genetics symposium. Kyoto, Japan: Kyoto University; 1983. p. 807–811.
[29] Cox TS, Hatchett JH. Hessian fly-resistance gene H26 transferred from *Triticum tauschii* to common wheat. Crop Sci 1994;34:958–60.
[30] Macfadyen S, Kriticos DJ. Modelling the geographical range of a species with variable life history. PLoS ONE 2012;7:e40313. doi: 10.1371/journal.pone.0040313.
[31] Ma G, Ma CS. Climate warming may increase aphids' dropping probabilities in response to high temperatures. J Insect Physiol 2012;58:1456–62. doi: 10.1016/j.jinsphys. 2012. 08. 012.
[32] Alford L, Andrade TO, Georges R, Burel F, van Baaren J. Could behavior and not physiological thermal tolerance determine winter survival of aphid since real fields? PLoS ONE 2014;9:e114982. doi: 10.1371/journal.pone.0114982.
[33] FAO. World food and agriculture. Statistical Yearbook. Available from: http://www.fao.org/3/i3107e/i3107e.pdf (accessed on 20 May 2019).
[34] Vinocur B, Altman A. Recent advances in engineering plant tolerance to abiotic stress: Achievements and limitations. Curr Opin Biotechnol 2005;16f:123–32.
[35] Abid M, Tian Z, Ata-Ul-Karim ST, Cui Y, Liu Y, Zahoor R, et al. Nitrogen nutrition improves the potential of wheat (*Triticum aestivum* L.) to alleviate the effects of drought stress during vegetative growth periods. Front Plant Sci 2016;7:981. doi: 10.3389/fpls.2016.00981.
[36] Kobata T, Palta JA, Turner NC. Rate of development of post anthesis water deficits and grain filling of spring wheat. Crop Sci 1992;32:1238.
[37] Wahid A, Gelani S, Ashraf M, Foolad MR. Heat tolerance in plants: an overview. Environ Exp Bot 2007;61:199–223.
[38] Hall AE. Crop responses to environment. Boca Raton, Florida: CRC Press LLC; 2001.
[39] Jones PD, New M, Parker DE, Mortin S, Rigor IG. Surface area temperature and its change over the past 150 years. Rev Geophys 1999;37:173–99.
[40] Alghabari FM, Lukac HE, Jones MJ, Gooding. Effect of Rht alleles on the tolerance of wheat grain set to high temperature and drought stress during booting and anthesis. J Agro Crop Sci 2014;200:36–45.
[41] Vara PV, Djanaguiraman M. Response of floret fertility and individual grain weight of wheat to high temperature stress: sensitive stages and thresholds for temperature and duration. Funct Plant Bio 2014;41:1261–9.
[42] Khan MA, Ahmad N, Akbar M, Rehman A, Iqbal M. Combining ability analysis in wheat. Pak J Agric Sci 2007;44:1–5.
[43] You L, Rosegrant MW, Wood S, Sun D. Impact of growing season temperature on wheat productivity in China. Agr Forest Meterol 2009;149:1009–14.
[44] Reynolds MP, Bonnett D, Chapman SC, Furbank RT, Manès Y, Mather DE, et al. Raising yield potential of wheat. Overview of a consortium approach and breeding strategies.. J Exp Bot 2011;62:439–52.
[45] Qi PF, Wei YM, Yue YW, Yan ZH, Zheng YL. Biochemical and molecular characterization of gliadins. Mol Biol 2006;40(5):713–23. DOI 101134/S0026893306050050.

[46] Metakovsky E, Melnik V, Quijano MR, Upelniek V, Carrillo JM. A catalog of gliadin alleles: polymorphism of 20th-century common wheat germplasms. Crop J 2018;6(6):628–41. Doi 101016/j.cj.2018.02.003.
[47] Belderok B, Mesdag H, Donner DA. Bread-making quality of wheat. Springer; 2000. 3.
[48] US Wheat Associates. Weekly price report. Available from: https://www.uswheat.org/market-and-crop-information/price-reports, 2019 [accessed 01.05.].
[49] Ogbonnaya FC, Abdalla O, Mujeeb-Kazi A, Kazi AG, Gosnian N, Lagudah ES. Synthetic hexaploids: harnessing species of the primary gene pool for wheat improvement. Plant Breed Rev 2013;37:35–122.
[50] Velu G, Guzman C, Mondal S, Autrique JE, Huerta J, Singh RP. Effect of drought and elevated temperature on grain zinc and iron concentrations in CIMMYT spring wheat. J Cereal Sci 2016;69:182–6. doi: 10.1016/j.jcs.2016.03.006.
[51] Singh RP, Velu G. Zinc-biofortified wheat: harnessing genetic diversity for improved nutritional quality. Science Brief: Biofortification. Available from: http://www.harvestplus.org/sites/default/files/publications/ScienceBrief-Biofortification1_ZincWheat_May2017.pdf.

Further reading

[1] Hullé M, Coeurd AA, Bankhead-Dronnet S, Harrington R. Aphids in the face of global changes. C R Biol 2010;333:497–503. doi: 10.1016/j.crvi.2010.03.005.
[2] Sareen S, Tyagi BS, Tiwari V, Sharma I. Response estimation of wheat synthetic lines to terminal heat stress using stress indices. J Agric Sci 2012;4:97–104.

Chapter 4

Emerging Avenues for the Exploitation of Wild Relatives of Rice in Plant Breeding

Muhammad Abu Bakar Saddique[a], Muhammad Saeed Ahmad[b], Muhammad Ali Sher[a], Asif Ali Khan[a] and Zulfiqar Ali[a]

[a]*Institute of Plant Breeding and Biotechnology, MNS University of Agriculture, Multan, Pakistan;* [b]*Department of Plant Breeding and Genetics, University of Agriculture, Faisalabad, Pakistan*

4.1 Setting the scene; what is wild germplasm

The wild germplasm of a crop is defined as the plants that have not gone through the process of domestication to attain the domestication syndrome. These plants are not fit for the general cultivation due to the presence of noxious or uneconomical characters. Domesticated plants differ from their wild progenitors in several morpho-physiological traits, particularly seed retention, seed germination, seed size, plant and seed growth habit, the coloration of seed, flowers and other vegetative parts, and/or edibility of economically important organs [1]. Another point needs to be understood before ending the given concept of the wild germplasm, that is, only traits showing a perfect dichotomy (e.g., hard seeded vs. free germinating legumes, brittle vs. non-brittle rachis cereals, bitterness vs. sweetness in almond [2,3], melon [4], cucumber [5], watermelon [6], pumpkin, and squash [7]) between the wild and cultigens are describing the actual episode of the domestication [8]. These domestication syndromes are mono- or di-genic in inheritance [9]. On the other hand, the traits showing continuous variation (e.g., seed size) between wild and cultigen (mostly polygenic) are called improvement traits that evolved under the process of domestication [8,10].

4.2 The value of wild germplasm

Crop yields are facing stagnancy and the use of wild progenitors in breeding programs is a workable option to create exploitable variability for genetic gains. These wild genotypes have the potential to make modern cultivars fit for the future's unpredicted climate scenarios. The exploitation of these ancient

Wild Germplasm for Genetic Improvement in Crop Plants. http://dx.doi.org/10.1016/B978-0-12-822137-2.00004-7

progenitors of the modern crops can provide new cultivars having more yield, tolerance to biotic and abiotic stresses, biofortified edible organs, and the capacity to harvest increasing CO_2 in the environment.

Modern plant breeding started in the 19th century, it brought huge success to the genetic improvement of the crop plants. During the first decades of the 20th century, new cultivars were developed for the various agro-ecological zones by using landraces as a genetic base. By the midcentury, wide hybridization became part of crop variety development programs. At the same time, the world realized that the food security of developing countries has prime importance. That put forth an era of the green revolution, having semi-dwarf genotypes for the global farming systems. At this time of the third agricultural revolution (1960), the plant breeder was the main driver of food security that saved humanity from hunger, conflicts, diseases, and demise [11]. But, adoption of modern high yielding cultivars led to the erosion of crop diversity. Landraces, heirloom, ecotypes, and wild progenitors were replaced by modern cultivars which is the ultimate product of a plant breeder [12,13]. The development of modern cultivars and the replacement of landraces is not accomplished by choice rather than the main force after it, is global food security. The exponential increase in the population while only arithmetic increase in the food supply has put a lot of pressure on the plant breeders [14,15]. They are providing high yielder modern cultivars that are bringing genetic uniformity and eliminating the diversity that was a buffer against many unseens. Modern cultivars have made our crops genetically vulnerable that is evident from a glaring example of the 1970 Southern corn leaf blight outbreak in the United States [1,16]. That outbreak forced plant breeders to rethink for bringing genetic diversity at farmer field through breeding for specific adaptation, systematic temporal and spatial gene placement, integration of vertical and horizontal resistance, use of interspecific hybridization, and varietal mixtures [17]. It would not be a balance statement if we say that the sole responsible for the genetic erosion is plant breeder. There are different other factors involved in this menace, namely, crop domestication, urbanization, farm mechanization, legislation and policy, wars, changing dietary habits, and many more [18,19]. Regardless of the reason for the genetic erosion, it is now the responsibility of plant breeder to restore the genetic diversity by using different resources. While, the use of crop wild progenitors and landraces in modern breeding programs can best restore the genetic diversity and broaden the genetic base of the crops [20]. We have to go back into the wild to prevent crop vulnerability. Wild relatives do have the valuable variation which is an important source of adaptation loci for changing farming systems, market demand, and climatic conditions [18].

4.3 Genus *Oryza* and its importance

The use of morphometry, cytogenetics, isozymes methods, and the molecular techniques helped to identify 139 species of the genus *Oryza*. From taxonomical point of view, it is one of the most difficult genera to categorize its species.

Scientists keep on adding and subtracting the species in this genus. Many species which were the part of this genus are now regarded as better suited to other genera, subspecies, or the synonym of the specie. Only 12 species are unanimously accepted as the member of the genus *Oryza* including *O. sativa*, *O. rufipogon*, *O. glaberrima*, *O. longiglumis*, *O. barthii*, *O. longistaminata*, *O. minuta*, *O. meyeriana*, *O. latifolia*, *O. officinalis*, *O. schechteri*, and *O. ridleyi* [21]. The most efficient sub-generic classification of the genus Oryza includes three sections and eight series having 26 species. It was proposed by Sharma and Shastry in 1965 (Table 4.1) [22]. These species are the source of biotic and abiotic stress tolerance, lodging and shattering resistance, biofortified genotypes, and weed competitiveness (Table 4.1) [23–26]. In short, Oryza species are the rich source of almost all traits needed by the modern rice to better survive in the present and future farming systems of the globe.

4.4 The barrier in using rice wild germplasm

The use of rice wild species in breeding programs is not an easy task. It is limited by several factors, some of these are easy to resolve while others are complicated. Easy to resolve factors are (1) collection of wild species from their natural habitat, (2) regeneration and the seed multiplication, (3) morpho-physiological and molecular characterization, (4) conservation and storage, and (5) seed testing and distribution. These problems can be resolved by following simple measures like the deployment of skillful scientists who know well about the members of *Oryza* species, seed regeneration under containments, timely harvesting, seed cleaning, and packaging for storage [60]. While some other problems need special measures to be resolved. These hurdles include (1) maladaptation, (2) tracing the beneficial loci present in the wild genome, and (3) the integration of beneficial alleles in modern genotypes [61]. The solutions to these complicated problems are given later.

4.5 Evolving avenues for the use of wild rice germplasm

As discussed earlier, the use of rice wild genotypes is hurdled by some complicated problems. These issues can be resolved by adopting modern technologies including breeding informatics, bioinformatics, biotechnology, etc.

4.5.1 Breeding informatics

The success of plant breeding is associated with the generation, storage, retrieval, and interpretation of the big quantitative data developed from the cultivated and wild species of crops. The improvement in the selection efficiency of a required genotype advances the plant breeding program and it is possible through breeding informatics. Rice breeding is changing thoroughly with the addition of new planting techniques, experimental designs, the genotypic selection based on morpho-physiological and molecular data, development of a large number of

TABLE 4.1 Species of the genus *Oryza* and their potential traits.

Section	Series	*Oryza* species	Genome	Chromosomes	Distribution	Traits
Padia	Meyerianae	*O. abromeitiana*	GG	2n = 2x = 24	Philippines	Resistant to bacterial blight [27]
		O. meyeriana	GG	2n = 2x = 24	Southeast Asia	Resistant to bacterial blight, shade tolerance, and adaptation to aerobic soil [28]
		O. granulate	GG	2n = 2x = 24	Southeast Asia	Resistant to bacterial blight, brown planthopper. Tolerant to drought and shade. Adapted to aerobic soils, plasticity against temperature and moisture [28,29].
	Ridleyanae	*O. ridleyi*	HHJJ	2n = 4x = 48	Thailand South Asia	Resistant to yellow stem borer, bacterial blight, bacterial blast, and whorl maggot. Tolerant to salinity and aerobic soil [30–33].
		O. longiglumis	HHJJ	2n = 4x = 48	Indonesia (Irian Jaya, Papua New Guinea)	Resistant to bacterial blight and bacterial blast. Adaptive to tropical and extreme temperatures (5°C–37°C) [28].
	Schlechteriane	*O. schlechteri*	HHKK	2n = 4x = 48	Indonesia (Irian Jaya, Papua New Guinea)	Flooding tolerant [29].
Angustifolia	Perrierianae	*O. tisseranti*	Unknown	2n = 2x = 24	Tropical West Africa	
		O. perrieri	Unknown	2n = 2x = 24	Madagascar Malagasy	Submergence tolerance and flooding escape. Tolerant to shade [34,35].
	Brachyanthae	*O. angustifolia*	Unknown	2n = 2x = 24	Tropical South Africa	
		O. brachyantha	FF	2n = 2x = 24	The West and Central Africa (Republic of Congo, Zambia)	Resistant to broad-spectrum bacterial blight. Also tolerant to many abiotic stresses [36].

Oryza	Sativae	*O. barthii*	AA	2n = 2x = 24	Africa	High amylose content, high nitrogen uptake efficiency, tolerant to silicone toxicity, and drought, resistance to green leafhopper, bacterial blight, bacterial leaf streak, blast, and sheath blight. Resistant to the nematode, tungro virus and borer attack [37–41].
		O. cubensis	AA	2n = 2x = 24	American	Resistant to rice leaf blast [42].
		O. balunga	AA	2n = 2x = 24	Southeast Asia	
		O. nivara	AA	2n = 2x = 24	Southeast Asia (Bangladesh, Cambodia, China, India, Laos, Malaysia, Myanmar, Nepal, Sri Lanka, Thailand, and Vietnam)	High photosynthetic and water use efficiency. Resistant to grassy stunt virus, bacterial blight, blast and brown plant hopper. Have yield-enhancing loci and high iron and zinc level [43–46].
		O. sativa	AA	2n = 2x = 24	Southeast Asia	Cultigen, highly productive [47].
		O. breviligulata	AA	2n = 2x = 24	Tropical Africa	Resistant to green leafhopper and bacterial blight. Drought avoidance [48].
		O. glaberrima	AA	2n = 2x = 24	Tropical West Africa	Tolerant to drought, acidity, and iron toxicity. Resistant to blast, rice yellow mottle virus, African gall midge, and nematodes. Have weed competitiveness [43,49–51].

(Continued)

TABLE 4.1 Species of the genus *Oryza* and their potential traits. (*Cont.*)

Section	Series	*Oryza* species	Genome	Chromosomes	Distribution	Traits
	Latifoliae	*O. latifolia*	CCDD	2n = 4x = 48	Central and South America	Resistant to brown plant hopper, white-backed plant hopper. Tolerant to bacterial blight. Lodging resistant. High iron and zinc contents [52,53].
		O. alta	CCDD	2n = 4x = 48	Central and South America	Resistant to striped and stem borer [54–56].
		O. grandiglumis	CCDD	2n = 4x = 48	South America	Tolerant to physical wounding and flooding [57,58].
		O. officinalis	CC	2n = 2x = 24	Southeast Asia	Resistant to thrips, brown and white-backed plant hopper, green leaf hopper. Tolerant to bacterial blight and stem rot. Have high iron and zinc content [46, 59].
		O. punctata	BB	2n = 2x = 24	Sudan	Resistant to zigzag leaf hopper and brown plant hopper [47].
		O. eichingeri	EE	2n = 2x = 24	East Africa	Resistant to brown plant bopper, white-backed plant hopper, and green leaf hopper [47].
		O. malampuzhaensis	BBCC	2n = 4x = 48	South India	Resistant to trips, green leafhopper, white-backed plant hopper, bacterial blight, and stem rot [47].
		O. minuta	BBCC	2n = 4x = 48	The Philippines and Papua New Guinea	Resistant to bacterial blight, blast, brown plant hopper and green leafhopper [47].
	Australiensis	*O. australiensis*	EE	2n = 2x = 24	Australia	Resistant to brown planthopper, bacterial blight, and bacterial blast. Drought avoidance [47].

molecular markers, modern software associated with molecular and other data analysis. Information is pyramiding with every passing day and making a mountain of wisdom. The precise information helps to develop efficient methods for the genotypic selection and its utility in the breeding scheme [62,63]. Breeding informatics is a science that deals with the collection, storage, retrieval, and interpretation of genotype, phenotype, and pedigree data of cultivated and wild species of a crop. Breeding informatics makes it easy to use crop elite genotypes and wild species in breeding programs. It includes software and web services to help a plant breeder. Like, Breeding-for-Result (B4R) has revolutionized the rice breeding system, it helps the exploitation of the *Oryza* species based on the data storage, retrieval, and interpretation. It is a cloud-based breeding management system and a source of applications and knowledge work activities that help decision making during rice breeding programs [64]. Similarly, the Genomic Open Source Breeding Informatics Initiative (GOBii) is a leading platform where the data of both cultivated and wild species of different food crops can be stored and analyzed. It is the first large-scale public-sector effort to systematically apply high-density genotypic information to the breeding of food crops in the developing countries. The Bill and Melinda Gates foundation is the main supporting body of GOBii platform. The ultimate target of the GOBii is to develop and implement genomic data management systems to support public sector plant breeding programs so that they can have success in the genetic manipulation of crops. It involves a multi-disciplinary team of software developers, curators, geneticists, breeders, molecular biologists, and bioinformatics expert from Cornell University, IRRI, CIMMYT, and ICRISAT. The target crops of the GOBii program are rice, wheat, maize, sorghum, and chickpea [65,66]. Such software programs, applications, and web platform are an excellent source to submit, retrieve and analyze the genotypic data of both cultivated and wild species of crop plants [67]. These internet-based services have opened new avenues where a plant breeder can select the genotypes of choice and bring them in the breeding program without much effort. These smart techniques are the real future of plant breeding. Breeding informatics tools are available to the plant breeders for shifting plant breeding to the next level (Table 4.2). These tools help to bring elite genotypes and useful wild progenitors in modern plant breeding programs.

4.5.2 Maladaptation syndrome and its solutions

Maladaptation is an obstacle to the use of wild germplasm in plant breeding. When some *Oryza* species are moved from their natural habitat to a different climatic zone, many of the beneficial alleles are suppressed by maladapted genes. The flowering in *O. sativa*, *O rufipogon* is photoperiod sensitive and plants only produce flowers under short-days [68,69]. The regions of long days or the season of long-day do not favor the flowering in many Oryza species. These species are maladaptive under long-days environments and are unable to shift toward

TABLE 4.2 Breeding informatics tools.

Software	Function	URLs
GOBii Genomics Data Manager (GDM)	The GOBii-GDM is a genomic data management system. It has a BrAPI (an application programming interface for plant breeding applications) enabled web-service layer, integrated web-based file management, combined genotyping data QC, and a Marker Tools Portal to access all GOBii-GDM tools and other breeding tools developed in Flapjack and Galaxy.	https://gobiiproject.atlassian.net/wiki/spaces/GD/overview
KDCompute data analysis	KDCompute is a web application and platform that enables users to conduct extensible data analysis of plant breeding within a safe, customizable and cooperative plugin framework.	http://cbsugobii05.biohpc.cornell.edu/wordpress/index.php/kdcompute/
GOBii JHI Tools	GOBii and James Hutton Institute (JHI) jointly developed an open-source platform (Marker-Assisted Back Crossing (MABC) module) for completing marker-assisted backcross visualization and selection analysis.	http://cbsugobii05.biohpc.cornell.edu/wordpress/index.php/flapjack/
GOBii Genomic Selection (GS)	The GS-Galaxy platform is used for uploading and access to source input files, imputation, encoding, model selection, model training, principal component analysis (PCA), clustering analysis, cross-validation, genomic estimated breeding value (GEBV) calculator and best linear unbiased prediction (BLUP)/ best linear unbiased estimation (BLUE) calculator.	http://cbsugobii05.biohpc.cornell.edu/wordpress/index.php/galaxy-gs/
Breeding4Results (B4R)	It deals with data recording of the rice genotypes from germination till maturity. Currently, the IRRI rice breeding and initiative research support platform is using it.	https://breeding4rice.irri.org/

the reproductive phase so many genes could not express in such situations. Different strategies are in practice for bridging this gap of maladaptation. One of the best ways is to grow these wild genotypes in controlled growth-chambers, greenhouse, and field environments. Whereas an artificial short-day environment may be produced for the induction of flowers. Another way to overcome the maladaptation syndrome is the application of plant growth regulator, for example, application of gibberellic acid induces flowering in rice even under long days [70,71].

4.5.3 Beneficial allele identification and cloning

Crop wild germplasm is a source of many beneficial genes but the unlocking of these loci is not an easy job. With the advent of new techniques and protocols, the identification and cloning of the valuable genes have become easy and practical. The quantitative trait locus (QTL) mapping is being used for tracing the desired loci for a long time but this technique is now being replaced by many modern methods. Now in modern days, the genome-wide association studies (GWAS) are used for the identification of beneficial alleles present in the wild species of the rice. Unlike the classical QTL mapping, which is dependent on the bi-parental crosses, the GWAS applies to the unrelated accessions. The GWAS helps to identify the quantitative trait nucleotide (QTN), which is otherwise not possible by QTL mapping [72]. One drawback of the GWAS is that it cannot help identify the rare alleles. This gap of the GWAS can be filled by the focused identification of germplasm strategy (FIGS). It is an efficient tool that helps to find accurately the useful genes in plants. It is much better than traditional approaches, which are largely based on hit and miss strategy. It identifies those accessions that have candidate genes. The FIGS is based on a powerful algorithm that matches plant characters with geographic and agroclimatic information of the places from where the samples were collected. This allows the rapid searching of thousands of plant samples to locate a few high potential genotypes that can meet the breeder's demand [73,74]. Other useful techniques used for the gene identification from the cultivated and wild species are resequencing or RNA-seq methods [75], progeny testing of intercrossed derivatives in conjunction with advanced backcross QTL (for the identification of genes having epistemic interactions) [76], and allele mining via ecotype targeting induced local lesions in genomes (EcoTILLING) [77,78]. The genetic basis of a QTL can be traced out by evolve and resequencing studies (E&R) [78,79]. Moreover, subjecting a population to truncating selection for several generations and then determining the differences of allele frequency between non-selected and selected populations using next-generation sequencing (NGS) is one of the excellent techniques for the identification of QTLs [72]. These techniques and many other evolving procedures can help identify beneficial genes from the rice wild progenitors, which can be further pyramid in modern cultivars to meet the challenges of global food security.

4.5.4 Introgression of beneficial alleles

Domestication of rice has narrowed its genetic base, many vital traits are not the part of modern days' rice, which otherwise were present in its wild progenitors, heirloom, and landraces of the *O. sativa*. This domestication bottleneck has made modern-day rice vulnerable to environmental stresses. During the early time, the Oryza species complex present in farmer field and natural landscapes produced new species and robust genotypes those laid the foundation of civilizations and transformed the hunter-gatherer to stay at home farmers. Now, again the wild relatives of the rice are in need to save the civilizations when modern rice is not much diverse to bare the shocks of climate change. Different Oryza species have various important traits, which if somehow transferred to *O. sativa*, can make an ideal rice plant that will provide high yield under all types of climatic adversities. But this adaptive introgression is not simple because of many barriers that make it difficult to transfer the required genes from wild species to the cultivated one [80,81]. Hybridization is one of the best techniques for the transfer of genes from wild to domesticated species. If the species are cross-compatible, then the sexual hybridization is used to get an adaptive introgression. In case, if species are not cross-compatible then bridge crossing or the somatic hybridization through protoplast is performed to transfer required loci in recipient plants. These techniques are very much helpful for transferring the candidate genes but also have many limitations as well [82–84]. Over time, many modern molecular techniques have been evolved making introgression an easy task. Like, marker-assisted backcross technique efficiently transfers adaptive loci from wild to cultivated genotypes. Moreover, the most used present-day technique is clustered regularly interspaced short palindromic repeats/CRISPR associated protein 9 (CRISPR Cas-9). This has been extensively used for the introgression of favorable genes in rice and many other crops. This technique is also used for the multiplex gene editing that helps breeder to address linkage drag and introgression of multiple genes at the same time [85–87]. All these techniques are very much useful for the introgression of beneficial alleles from the wild relatives of rice into the modern cultivars.

4.6 Conclusion

Rice wild germplasm is a term used for seed, plant, or plant part that is used in plant breeding, research, and conservation. The *Oryza* wild species are the rich source of beneficial loci. Their niches are prevailed by harsh climatic conditions so these plants have abundant resilient genes. But the introgression of these adaptive genes from wild to modern genotypes in not an easy job, it is a lengthy process with many hurdles. Plant breeders have been struggling for long time to overcome the barriers of introgression. Sometimes they got success but mostly the genotypic incompatibility hindered their path. Now, with the advancement in breeding informatics and biotechnology new protocols have been developed for the gene mining and their introgression from wild to modern rice. Moreover, these techniques are evolving and one day they will thoroughly change our rice breeding system and provide ideal genotypes for various geographical zones.

References

[1] Lenne JM, Wood D. Plant diseases and the use of wild germplasm. Ann Rev Phytopathol 1991;29(1):35–63.
[2] Heppner MJ. Further evidence on the factor for bitterness in the sweet almond. Genetics 1926;11(6):605.
[3] Ladizinsky G. On the origin of almond. Genet Resour Crop Evol 1999;46(2):143–7.
[4] Jayakodi M, Schreiber M, Mascher M. Sweet genes in melon and watermelon. Nat Genet 2019;51(11):1572–3.
[5] Chomicki G, Schaefer H, Renner SS. Origin and domestication of *Cucurbitaceae* crops: insights from phylogenies, genomics and archaeology. New Phytol 2020;226(5):1240–55.
[6] Navot N, Sarfatti M, Zamir D. Linkage relationships of genes affecting bitterness and flesh color in watermelon. J Heredity 1990;81(2):162–5.
[7] Paris HS, Brown RN. The genes of pumpkin and squash. HortScience 2005;40(6):1620–30.
[8] Abbo S, van-Oss RP, Gopher A, Saranga Y, Ofner I, Peleg Z. Plant domestication versus crop evolution: a conceptual framework for cereals and grain legumes. Trends Plant Sci 2014;19(6):351–60.
[9] Ladizinsky G. Plant evolution under domestication. Springer Science & Business Media; 2012.
[10] Evans LT. Crop evolution, adaptation and yield. Cambridge University Press; 1996.
[11] Evenson R, Gollin D. The Green Revolution: An end of century perspective. Department of economics working papers 2002-2007, Department of Economics, Williams College; 2002.
[12] Tripp R, Heide WM. The erosion of crop genetic diversity: challenges, strategies and uncertainties. Overseas Development Institute; 1996.
[13] Khan AA, Iqbal A, Awan FS, Khan IA. Genetic diversity in wheat germplasm collections from Baluchistan province of Pakistan. Pak J Bot 2010;42(1):89–96.
[14] Malthus T. An essay on the principle of population. London: J. Johnson; 1798. 1-126.
[15] Osei M, Asante M, Agyeman A, Adebayo M, Adu-Dapaah H. Plant breeding: a tool for achieving food sufficiency. Sustainable horticultural systems. Springer; 2014. 253-274.
[16] Keneni G, Bekele E, Imtiaz M, Dagne K. Genetic vulnerability of modern crop cultivars: causes, mechanism and remedies. Int J Plant Res 2012;2(3):69–79.
[17] Brozynska M, Furtado A, Henry RJ. Genomics of crop wild relatives: expanding the gene pool for crop improvement. Plant Biotechnol J 2016;14(4):1070–85.
[18] Dempewolf H, Baute G, Anderson J, Kilian B, Smith C, Guarino L. Past and future use of wild relatives in crop breeding. Crop Sci 2017;57(3):1070–82.
[19] Thormann I, Engels JM. Genetic diversity and erosion–A global perspective. Genetic diversity and erosion in plants. Springer; 2015. 263-294.
[20] Simmonds N. Variability in crop plants, its use and conservation. Biol Rev 1962;37(3):422–65.
[21] Nayar N. *Oryza* species and their interrelationships. Origin and phylogeny of rices. Elsevier; 2014. 59-115.
[22] Sharma S, Shastry S. Taxonomic studies in genus Oryza L. 6. A modified classification. Indian J Genet Plant Breed 1965;25(2):173–8.

[23] Futakuchi K, Fofana M, Sié M. Varietal differences in lodging resistance of African rice (*Oryza glaberrima* Steud). Asian J Plant Sci 2008;7(6):569–73.
[24] Watarai M, Inouye J. Internode elongation under different rising water conditions in African floating rice (*Oryza glaberrima* Steud.). J Faculty Agric Kyushu Univ Japan 1998;42(3–4):301–7.
[25] Moukoumbi Y, Sie M, Vodouhe R, Bonou W, Toulou B, Ahanchede A, et al. Screening of rice varieties for their weed competitiveness. Afr J Agric Res 2011;6(24):5. 5446-5456.
[26] Montcho D, Futakuchi K, Agbangla C, Fofana M, Dieng I. Yield loss of *Oryza glaberrima* caused by grain shattering under rainfed upland conditions. Int J Biol Chem Sci 2013;7(2):535–43.
[27] Vaughan DA, Ge S, Kaga A, Tomooka N. Phylogeny and biogeography of the genus *Oryza*. Rice biology in the genomics era. Springer; 2008. 219-234.
[28] Sanchez PL, Wing RA, Brar DS. The wild relative of rice: genomes and genomics. Genetics and genomics of rice. Springer; 2013. 9-25.
[29] Atwell BJ, Wang H, Scafaro AP. Could abiotic stress tolerance in wild relatives of rice be used to improve *Oryza sativa*? Plant Sci 2014;215:48–58.
[30] Padhi G, Sen P. Evaluation of wild rice species against yellow stem borer, *Scirpophaga incertulas* Walk. J Appl Zool Res 2002;13:147–8.
[31] Aggarwal R, Brar D, Khush G. Two new genomes in the *Oryza* complex identified on the basis of molecular divergence analysis using total genomic DNA hybridization. Mol Gen Genet MGG 1997;254(1):1–12.
[32] Brar D, Khush G. Alien introgression in rice. *Oryza*: From molecule to plant. Springer; 1997. 35-47.
[33] Ge S, Sang T, Lu B-R, Hong D-Y. Phylogeny of rice genomes with emphasis on origins of allotetraploid species. Proc Nat Acad Sci 1999;96(25):14400–5.
[34] Hattori Y, Nagai K, Furukawa S, Song X-J, Kawano R, Sakakibara H, et al. The ethylene response factors SNORKEL1 and SNORKEL2 allow rice to adapt to deep water. Nature 2009;460(7258):1026–30.
[35] Khanna A, Ellur RK, Krishnan SG, Mondal TK, Singh AK. An account of unclassified species (*Oryza schlechteri*), subspecies (*Oryza indandamanica* Ellis and *Oryza sativa* f. *spontanea* Baker), and ortho-group species (*Leersia perrieri*) of *Oryza*. The Wild Oryza Genomes: Springer; 2018. 295-300.
[36] Ram T, Laha G, Gautam S, Deen R, Madhav MS, Brar D, et al. Identification of new gene introgressed from Oryza brachyantha with broad-spectrum resistance to bacterial blight of rice in India. Rice Genet Newslett 2010;25:57.
[37] Hamaoka N, Uchida Y, Tomita M, Kumagai E, Araki T, Ueno O. Genetic variations in dry matter production, nitrogen uptake, and nitrogen use efficiency in the AA genome *Oryza* species grown under different nitrogen conditions. Plant Prod Sci 2013;16(2):107–16.
[38] Mitani-Ueno N, Ogai H, Yamaji N, Ma JF. Physiological and molecular characterization of Si uptake in wild rice species. Physiol Plant 2014;151(3):200–7.
[39] Brar D, Khush G. Transferring genes from wild species into rice. In: Quantitative genetics, genomics, and plant breeding. United Kingdom: CABI; 2002. p. 197–217.
[40] Prasad B, Eizenga G. Rice sheath blight disease resistance identified in *Oryza* spp. accessions. Plant Dis 2008;92(11):1503–9.
[41] Reversat G, Destombes D. Screening for resistance to *Heterodera sacchari* in the two cultivated rice species, *Oryza sativa* and *O. glaberrima*. Fundamental Appl Nematol 1998;21(4):307–18.
[42] Asuyama H. The rice blast disease. In: Proceedings of a symposium, IRRI. Baltimore, Maryland: The Johns Hopkins University Press; 1963.

[43] Lu B, Naredo ME, Macatangay M, Alvarez MT. Determination of chromosome numbers of wild *Oryza* species conserved in the International Rice Gene bank at IRRI. Int Rice Res Inst 1997;22(2):5–6.
[44] Hamaoka N, Yasui H, Yamagata Y, Inoue Y, Furuya N, Araki T, et al. A hairy-leaf gene, Blanket Leaf, of wild *Oryza nivara* increases photosynthetic water use efficiency in rice. Rice 2017;10(1):20.
[45] Samal R, Roy PS, Sahoo A, Kar MK, Patra BC, Marndi BC, et al. Morphological and molecular dissection of wild rices from eastern India suggests distinct speciation between *O. rufipogon* and *O. nivara* populations. Sci Rep 2018;8(1):1–13.
[46] Majumder S, Datta K, Datta SK. Rice biofortification: High iron, zinc, and vitamin-a to fight against "Hidden Hunger". Agronomy 2019;9(12):803.
[47] Brar DS, Khush GS. Wild relatives of rice: a valuable genetic resource for genomics and breeding research. The wild oryza genomes. Springer; 2018. 1-25.
[48] Wambugu PW, Henry RJ. Oryza barthii A. Chev. The wild oryza genomes. Springer; 2018. 67-74.
[49] Kreft H, Jetz W. Global patterns and determinants of vascular plant diversity. Proc Nat Acad Sci 2007;104(14):5925–30.
[50] Sarla N, Swamy BM. *Oryza glaberrima*: a source for the improvement of *Oryza sativa*. Curr Sci 2005;89:955–63.
[51] Ndjiondjop MN, Wambugu P, Sangare JR, Dro T, Kpeki B, Gnikoua K. *Oryza glaberrima* Steud. The wild oryza genomes. Springer; 2018. 105.
[52] Gireesh C. *Oryza latifolia* Desv. The wild oryza genomes. Springer; 2018. 151-57.
[53] Multani D, Khush G, Delos Reyes B, Brar D. Alien genes introgression and development of monosomic alien addition lines from *Oryza latifolia* Desv. to rice, *Oryza sativa* L.. Theoret Appl Genet 2003;107(3):395–405.
[54] Gireesh C. *Oryza alta* swollen. The wild oryza genomes. Springer; 2018. 55-59.
[55] Mizobuchi R, Fukuoka S, Tsushima S, Yano M, Sato H. QTLs for resistance to major rice diseases exacerbated by global warming: brown spot, bacterial seedling rot, and bacterial grain rot. Rice 2016;9(1):23.
[56] Eizenga G, Agrama H, Lee F, Jia Y. Exploring genetic diversity and potential novel disease resistance genes in a collection of rice (*Oryza* spp.) wild relatives. Genet Resour Crop Evol 2009;56(1):65–76.
[57] Kim KM, Cho SK, Shin SH, Kim G-T, Lee JH, Oh B-J, et al. Analysis of differentially expressed transcripts of fungal elicitor-and wound-treated wild rice (*Oryza grandiglumis*). J Plant Res 2005;118(5):347–54.
[58] Niroula RK, Pucciariello C, Ho VT, Novi G, Fukao T, Perata P. SUB1A-dependent and-independent mechanisms are involved in the flooding tolerance of wild rice species. Plant J 2012;72(2):282–93.
[59] Ray S, Vijayan J, Chakraborti M, Sarkar S, Bose LK, Singh ON. *Oryza officinalis* complex. The wild oryza genomes. Springer; 2018. 239-258.
[60] Borromeo T, Sanchez P, Quintana E, Engle L. Manual on rice genetic resources conservation and genebank management. Philippine Rice Research Institute; 1994.
[61] Wang C, Hu S, Gardner C, Lübberstedt T. Emerging avenues for utilization of exotic germplasm. Trends Plant Sci 2017;22(7):624–37.
[62] Ye G, Collard B, Zhao X, Nissila E. Enhancing rice breeding efficiency: The role of breeding. Sabrao J Breed Genet 2013;45(1).
[63] Nepolean T, Kaul J, Mukri G, Mittal S. Genomics-enabled next-generation breeding approaches for developing system-specific drought tolerant hybrids in maize. Front Plant Sci 2018;9:361.

[64] Collard BC, Gregorio GB, Thomson MJ, Islam MR, Vergara GV, Laborte AG, et al. Transforming rice breeding: Re-designing the irrigated breeding pipeline at the International Rice Research Institute (IRRI). Crop Breed Genet Genom 2019;1(1):e190008.

[65] Santantonio N, Atanda SA, Beyene Y, Varshney RK, Olsen M, Jones E, et al. Strategies for effective use of genomic information in crop breeding programs serving Africa and South Asia. Front Plant Sci 2020;11:353.

[66] Varshney RK, Sinha P, Singh VK, Kumar A, Zhang Q, Bennetzen JL. 5Gs for crop genetic improvement. Curr Opin Plant Biol 2020;56:190–6.

[67] Olsen M, Dreisigacker S, Dreher K. GOBII challenges and opportunities: CIMMYT perspectives; 2017.

[68] Greenup A, Peacock WJ, Dennis ES, Trevaskis B. The molecular biology of seasonal flowering-responses in Arabidopsis and the cereals. Ann Bot 2009;103(8):1165–72.

[69] Kim H, Jung J, Singh N, Greenberg A, Doyle JJ, Tyagi W, et al. Population dynamics among six major groups of the *Oryza rufipogon* species complex, wild relative of cultivated Asian rice. Rice 2016;9(1):1–15.

[70] Kwon C-T, Paek N-C. Gibberellic acid: A key phytohormone for spikelet fertility in rice grain production. Int J Mol Sci 2016;17(5):794.

[71] Mutasa-Göttgens E, Hedden P. Gibberellin as a factor in floral regulatory networks. J Exp Bot 2009;60(7):1979–89.

[72] Vlachos C, Kofler R. Optimizing the power to identify the genetic basis of complex traits with evolve and resequence studies. Mol Biol Evol 2019;36(12):2890–905.

[73] Platform CG. CGIAR Genebank Platform 2018 Annual Report. 2019.

[74] Bari A, Street K, Mackay M, Endresen DTF, De Pauw E, Amri A. Focused identification of germplasm strategy (FIGS) detects wheat stem rust resistance linked to environmental variables. Genet Resour Crop Evol 2012;59(7):1465–81.

[75] Wang Z, Gerstein M, Snyder M. RNA-Seq: a revolutionary tool for transcriptomics. Nat Rev Genet 2009;10(1):57–63.

[76] Tanksley S, Nelson J. Advanced backcross QTL analysis: a method for the simultaneous discovery and transfer of valuable QTLs from unadapted germplasm into elite breeding lines. Theoret Appl Genet 1996;92(2):191–203.

[77] Bajaj D, Srivastava R, Nath M, Tripathi S, Bharadwaj C, Upadhyaya HD, et al. Ecotilling-based association mapping efficiently delineates functionally relevant natural allelic variants of candidate genes governing agronomic traits in chickpea. Front Plant Sci 2016;7:450.

[78] Cooper JL, Henikoff S, Comai L, Till BJ. Tilling and ecotilling for rice. Rice protocols. Springer; 2013. 39-56.

[79] Vlachos C, Kofler R. MimicrEE2: Genome-wide forward simulations of evolve and resequencing studies. PLoS Comput Biol 2018;14(8):e1006413.

[80] Burgarella C, Barnaud A, Kane NA, Jankowski F, Scarcelli N, Billot C, et al. Adaptive introgression: An untapped evolutionary mechanism for crop adaptation. Front Plant Sci 2019;10:4.

[81] Velázquez-López R, Wegier A, Alavez V, Pérez-López J, Vázquez-Barrios V, Arroyo-Lambaer D, et al. The mating system of the wild-to-domesticated complex of *Gossypium hirsutum* L. is mixed. Front Plant Sci 2018;9:574.

[82] Waara S, Glimelius K. The potential of somatic hybridization in crop breeding. Euphytica 1995;85(1–3):217–33.

[83] Tan GX, Xiong ZY, Jin HJ, Li G, Zhu LL, Shu LH, et al. Characterization of interspecific hybrids between *Oryza sativa* L. and three wild rice species of China by genomic in situ hybridization. J Integr Plant Biol 2006;48(9):1077–83.

[84] Mwangangi IM, Muli JK, Neondo JO. Plant hybridization as an alternative technique in plant breeding improvement. Asian J Res Crop Sci 2019;4(1):1–11.
[85] Fiaz S, Ahmad S, Noor MA, Wang X, Younas A, Riaz A, et al. Applications of the CRISPR/Cas9 system for rice grain quality improvement: Perspectives and opportunities. Int J Mol Sci 2019;20(4):888.
[86] Ma X, Zhang Q, Zhu Q, Liu W, Chen Y, Qiu R, et al. A robust CRISPR/Cas9 system for convenient, high-efficiency multiplex genome editing in monocot and dicot plants. Mol Plant 2015;8(8):1274–84.
[87] Wolter F, Schindele P, Puchta H. Plant breeding at the speed of light: the power of CRISPR/Cas to generate directed genetic diversity at multiple sites. BMC Plant Biol 2019;19(1):176.

Chapter 5

Maize Genetic Resources and Pre-Breeding

Flavia França Teixeira and Claudia Teixeira Guimarães
Maize Germplasm Bank at Embrapa Maize and Sorghum, Sete Lagoas, Minas Gerais, Brazil

5.1 Maize genetic resources

Domesticated maize (*Zea mays* L. ssp. *mays*) belongs to the genus *Zea*, which includes four other species of perennial and annual grasses that are known collectively as Teosinte (*Zea* ssp.) that have Mesoamerican origin [1]. Possibly, maize originates from an evolution of teosinte that took place about 8,700 years ago, which are now called Lowlands of Mexico [2]. Mexico still preserves a great genetic diversity of maize. A deep study of the geographic maize racial dispersion in Mexico was carried out with more than 18,000 geo-referenced observations in 59 races of maize [3]. This study revealed that 54% of the observations comprised of six races, whereas some races have very few observations, few races are widely dispersed and many races are highly localized. The enriched areas of maize races and diversity in Mexico are in the mountains from northwest, south-central and southern Mexico, Oaxaca, Jalisco, Michoacán, and Chiapas States.

The maize domestication started in Central America and was spread by pre-Columbian peoples to other regions. Archaeological data indicates that maize was present in the Southwestern of the United States 4,000 years ago [4], crossed Panama 7,500 years ago [5], reached the Peruvian Coast 4,000 years ago [6], and reached the Andes [7] and the lowlands of the Bolivian Amazon 6,500 years ago [8]. Landraces and archaeological records suggest that the maize ancestral population in South America is derived from the Mexican domestication center and became isolated from the teosinte gene pool during the domestication process. Thus, this observation together with data related to genomics, linguistics, archeology, and paleoecology suggested that the Southwest of the Amazon is a secondary domestication center of maize [9].

Maize dispersion across the Americas is associated with a large number of adaptive changes that, together with its numerous uses, contributed to its great genetic variability among the cultivated plants [10]. In the late 1950s and early 1960s, maize from Central and South America were classified into races by

Wild Germplasm for Genetic Improvement in Crop Plants. http://dx.doi.org/10.1016/B978-0-12-822137-2.00005-9

regions: Mexico presented 32 races [11], Central America had 16 races [12], Colombia included 23 races [13], Cuba had 7 races [14], Brazil and adjacent regions had 52 races [10,15], Bolivia presented 32 races [16], Peru had 48 races [17], Chile had 19 races [18], Ecuador had 29 races [19], and Venezuela had 19 races [20]. Currently, there are about 300 races of maize identified and each race has thousands of varieties [21]. Classifying maize into races provides valuable information on the origin of cultivated materials and illustrates the variability of this crop. In addition to its genetic variability, maize reached the highest level of domestication among the cultivated species. Maize is a crop of great economic and food importance, exceeding 100 million tons of grain world production in 2018/2019 (http://www.worldofcorn.com), rated in US$ 51.9 billion. The major world producers of maize are United States, China, and Brazil, with a yield average in the United States in 2018 of about 11 tons/hectare. This large production is obtained due to high productivity in extensive areas, which implicates in narrow genetic diversity of the crop.

There are some ways to maintain the maize agrobiodiversity, one of which is the on-farm conservation, which is found in several regions of Mexico [22,23] and in other countries [24,25]. The maize genetic variability preservation through *in situ* conservation is widespread in Mexico, where many varieties, classified into races, are maintained on more than 400,000 ha spread across various regions in the country. The maize genetic variability conserved by Mexican farmers is associated with the gastronomic quality and local cuisine, which is very diverse [26]. The Chihuahua State in Mexico harbors a large number of producers of traditional maize varieties that are characterized by subsistence agriculture, in which the use of agricultural inputs is minimal. The genetic variability of more than 200 Chihuahua accessions of maize has been studied to form a collection of 48 accessions, representing these local races [27]. This collection has a potential use in breeding to obtain varieties that combine the genetic variability of maize local races and better agronomic performances, allowing the fortification of the farmers who maintain them. The genetic diversity of several races in the Oaxaca State in Mexico was studied, which allowed the formation of intra-racial groups with a potential to identify gaps in ex situ preservation and improvement actions [26]. The authors highlight that plant improvement can strengthen the socio-economy of local communities, achieving better conditions of *in situ* preservation. The preservation of maize races by farmers in Oaxaca State has also been studied [28]. The authors identified that farmers employ ancient practices in the selection, maintaining the phenotypic variation, the local adaptation and differentiation between varieties and races. Many other studies have been conducted in order to understand maize genetic diversity in terms of its adaptability to local growing conditions [29,30] and, in particular, regarding grain attributes [31–33]. The Mexican tradition of using maize for human consumption leads to the cultivation of traditional varieties due to the association between grain quality and gastronomic quality [34].

In addition to *in situ* conservation, the genetic diversity of maize is maintained in several collections conserved ex situ. The search for "Zea mays" on GRIN Global website (https://npgsweb.ars-grin.gov/gringlobal/search.aspx) in January 2020 found 14,921 accessions. The same search on GENESYS (https://www.genesys-pgr.org/a/v2eEqR0Kzjz) also performed in January 2020, returned 129,510 accessions, and this search in the GB database of CIMMYT (http://mgb.cimmyt.org/gringlobal/search.aspx) returns the maximum limit of search on the site, which is 20,000 accessions. The search on GENESYS identified the countries with the largest number of maize accessions, which are Mexico (31,983 accessions), United States (30,851 accessions), Ukraine (9,050 accessions), Brazil (7,755 accessions), and Romania (7,190 accessions). Out of more than 120,000 accessions of maize registered in GENESYS, approximately 60% (72,000 accessions) are classified as traditional cultivars/landraces, and the others are classified as inbred lines or improved materials. Duplications are possibly found among these collections, for example, a traditional Mexican landrace "Zapalote Chico" is found in the collection of at least seven countries around the world. The Brazilian maize GB has a large part of its accessions duplicated in more than one institution in Brazil, such as in CIMMYT and in the United States.

Among the GBs that preserve the biodiversity of maize, the collection of the International Center for the Improvement of Maize and Wheat (CIMMYT), located in El Batán, Mexico stands out to include accessions germplasm from Central America, Caribbean, as well as introductions from 64 countries, since 1943 [35]. Currently, CIMMYT conserves around 28,000 maize accessions (https://www.cimmyt.org/work/genetic-resources/). The maintenance of GBs includes a series of activities that are carried out systematically, including: accessions introduction, seed exchange, seed conservation, accessions regeneration and multiplication, germplasm characterization, and evaluation and data storage [36]. The introduction of an accession into a GB bank can be done by collecting directly from farmers and/or exchanging with other institutions. In an active GB, constant introductions should be made in order to form the most representative collection of the variability of the crop, without repetition. The germplasm exchange is carried out through the exchange of seed samples between GBs to expand the genetic variability available for research programs. The germplasm exchange in most countries follows "The International Treaty on Plant Genetic Resources for Food and Agriculture," which was adopted by the Thirty-First Session of the Conference of the Food and Agriculture Organization of the United Nations on November 3rd, 2001. According to Food and Agriculture Organization (FAO-http://www.fao.org/plant-treaty/overview/en/), the treaty aims at: recognizing the enormous contribution of farmers to the diversity of crops that feed the world; establishing a global system to provide farmers, plant breeders, and scientists with accessions to plant genetic materials and ensuring that recipients share the benefits of these genetic materials with the countries where they have been originated. Maize is one of the crops included at the Annex I of the Treaty [37], in which 146 countries are contracting parties (http://www.fao.org/fileadmin/user_upload/legal/docs/033s-e.pdf). Unfortunately, compliance with The International Treaty is still failing even among the signatory countries, with only 44 (36%) of the 121 signatory countries answering the requests of seed samples [38].

The seed conservation at maize GBs can be done in cold and dry chambers (5°C–8°C and 25%–30% RH) for medium term preservation or in chambers with negative temperatures, for a long-term preservation. Germplasm collections should not be kept as a single copy. An option to keep a backup copy of the germplasm collection is the world seed reserve Svalbard Global Seed Vault (https://www.croptrust.org/our-work/svalbard-global-seed-vault/) where there are already almost 1 million of samples preserved at −18°C. During the medium term preservation of maize accessions, the seeds viability should be monitored every 4 or 5 years. The samples with germination below 80% must be regenerated in the field, and seeds with low stored amount should be multiplied. The replenishment of the germplasm stock can also be carried out by additional collection at the original collection site, but this can often be impossible, as the farmer may no longer maintain the accession and its characteristics in the field. Maize accessions must be regenerated/multiplied under soil and climate conditions similar to their original collection sites. In the regeneration/multiplication of maize accessions, a minimum of 16 rows of 5 m each should be used, and the size of the regeneration/multiplication plot must be approximately 100 m^2 per accessions, with a minimum of 300 plants per plot. Regeneration/multiplication of accessions of maize GB not locally adapted can be carried out in greenhouses with controlled pollination [39].

The characterization and evaluation of accessions in a GB aim to expand the information about the collection to enable its use in research programs. The characterization describes each accession based on the maize descriptors [40]. These descriptors tend to be traits easy-to-measure such as kernel color and type, plant height, number of days to flowering, number of leaves above the ear and tassel type. Some descriptors are nominal, such as, kernel color and type; others are ordinals, such as abiotic and biotic stress susceptibility, whereas others are quantitative, such as plant and ear heights and number of days to flowering. The evaluations focus on more complex traits related to agronomic performance that often need to be observed in various environments. Tolerance to biotic and abiotic stresses, stability and adaptability, and combining ability are some of the characters that are frequently considered in the evaluation of maize accessions [41].

The maintenance and availability of information from each accession is essential for the efficient use of the available genetic variability preserved in GB. It is important that GB data is stored properly in digital media. There are at least three different categories of GB-related data: passport data, conservation data, and characterization data. Passport data includes information obtained during the collection, such as: date, vegetation type, soil conditions, geographic coordinates of the collection site, name of the farm and farmer, variety denomination, and its form of use. Conservation data should include information for each accession on: date and site of the last regeneration/multiplication, quantity of seeds in stock, storage place in the cold chamber, germination percentage, date of the last germination test, regeneration conditions (soil, climate) and number of regenerations/multiplications already carried out. Characterization/evaluation data contains information related to accessions performance. The morpho-agronomic used in characterization [40] may vary according to the objectives of the study, some commonly evaluated are: grain yield, response to biotic and abiotic stresses, oil and amino acid contents, and combinatory capacity [41].

5.2 Demands and structure of maize breeding programs

Pre-breeding activities must be planned according to the seed industry and the structure of the breeding program, because their products only will reach the market after being accepted for plant breeding and seed industry. The use of seeds of traditional maize varieties produced on the property has been gradually being replaced in the last 90 years and this substitution was mainly due to the development of maize hybrids. This history began on 20th century, when North American farmers were looking for increasing the maize productivity due to the demands for food and feed, associated with the low availability of new arable land [42]. In 1908, two young scientists, George Harrison Shull and Edward Murray East, independently rediscovered the phenomenon of inbreeding depression and hybrid vigor. Maize has important features such as separate male and female organs and easy pollination by the wind, allowing a large-scale hybridization. However, the inbred lines available at that time had low yield, reducing the seed production and increasing the seed costs, which delayed the commercial adoption of maize hybrids. A few years later, Donald F. Jones proposed the development of double-cross hybrids to reduce the costs of seeds. In the early 1930s, the demand for maize hybrids increased among the US farmers, and the single-cross hybrid only became commercially feasible in the 1960s [42]. The average maize yield in the United States increased from 1 ton/ha in 1930 to 8 ton/ha in 1997 [43] and it is still growing, with some hybrids exceeding 11 ton/ha in 2018 (http://www.worldofcorn.com). The increases on maize yield are also due to the use of more efficient inputs and agricultural practices [42].

The use of hybrids and synthetic varieties turned the classification of maize germplasm into heterotic patterns, which was exploited in maize breeding programs in the mid-1960s in the US Corn Belt. The heterotic pattern called Iowa Stiff Stalk Synthetic (BSSS, or simply SS) generated good inbred lines used as females in the production of commercial single-cross hybrids, whereas the inbred lines that combined well with this group were called non-stiff Stalk (NSS) [44]. Other heterotic patterns were employed in Europe, Asia [45], and Brazil

[46]. Maize can also be used for specific purposes, such as popcorn, sweet corn, baby corn, forage, green corn, and high nutritional quality corn. In these cases, the pre-breeding must consider, in addition to agronomic aspects, characters related to sensorial quality, required by consumers of each type of corn.

5.3 Integration between genetic resources and breeding

Maize genetic resources and plant breeding often tend to have few points in common, but it is necessary to bring both research themes closer together. The majority of Brazilian maize breeders do not use GB regularly [47]. The main causes of the low use of GB accessions were raised as the lack of documentation and adequate description of the germplasm collections; poor seed availability; poor agronomic adaptation; lack of information requested by the breeders; breeders' satisfaction with the genetic variability found in elite germplasm; difficulty in identifying potentially useful genes; and absence of pre-breeding programs [48]. In 1970, the National Academy of Sciences of United States designated a committee to quantify the genetic vulnerability of crops. About 70% of the growing area with maize was occupied by cultivars derived from only six inbred lines, which indicated that maize production in United States was genetically vulnerable [49]. However, the genetic basis of maize cultivated in the United States was in the 1980s broadened compared to the survey carried out in the 1970s, but a small number of cultivars was still responsible for most of the maize production [50]. The low variability between cultivars leads to the genetic vulnerability of crops, which caused several problems in the past. The restricted source of male-sterility used in maize hybrids cultivated in 1970 in United States was associated with susceptibility to the fungus *Helminthosporium maydis*. Thus, this genetic vulnerability reduced greatly maize production of the United States caused by this devastating disease in 1970 [48].

The use of maize accessions from GB should be increased in breeding to extend the genetic basis of maize cultivars. One way to expand the use of the germplasm bank through breeding is to develop participatory breeding programs. Maize breeding offers cultivars that are widespread among farmers, but some farmers need cultivars with specific attributes that are not covered by improved genotypes. Participatory breeding is a practical option to meet special demands with joint actions between breeders, agricultural technical assistance, and farmers. Participatory breeding should be employed in cases that the crop has local importance, for crops adapted to marginal environments, for low-input agriculture, for specific condition with high genotype-by-environment interactions, and for special traits with high value. Among the advantages of participatory breeding are the increase in the genetic diversity of cultivated plants, the rapid adoption of varieties from the GB, and the strengthening of rural communities. However, there is a relative high cost of farmer's participation on time, area, labor, and capital in the development of new varieties that have a restricted use for a few farmers. The ways of farmer and breeder participation in participatory breeding is quite variable, there are models in which both participate, from the germplasm selection to new cultivars testing, and there are also models in which the farmer's participation occurs only in final stages of cultivars selection. The most efficient participatory breeding programs are those that the breeder participates in germplasm selection, pre-breeding, cultivar development, and evaluation, and that the farmer participates only in the germplasm selection and final evaluation [51].

The maize participatory breeding is widely used in Mexico for traits related to flour quality and special climate conditions, with a favorable contribution of farmers in the conduction and selection of segregating populations [26,27,52–58]. Participatory breeding has also been used in Brazil where it was used to develop maize cultivars combining yield and nitrogen efficiency use [59] and for the selection of maize straw traits for handicrafts purposes [60]. Participatory breeding was also used to select open pollinated varieties in South Africa [61] and China [62] and for selection of local maize races in Portugal [63].

Another option for integration between breeding and the GB activities that also promotes the use of accessions is the pre-breeding. Pre-breeding can be divided into two types of activities, the first is the identification of genes and/or traits that have potential for breeding use in exotic germplasm, and the second is the incorporation of these into adapted elite germplasm [47]. In this case, exotic genotypes are not only those from other countries, but also the materials from regions with different environmental conditions, or not improved. Several methods can be used in the development of these two activities, among them the establishment of core collections, the identification of potentially useful genes, the classification of accessions into heterotic patterns, the synthesis of base populations, the improvement of GB bank accessions *per se*, and in crossings based on their heterotic patterns [48]. The choice of the most suitable methodology for pre-breeding programs depends on the number of genes and the heritability of the trait and also the attributes required to evaluate the trait expression. The backcross method is indicated when mono or oligogenic inheritance traits are considered. For quantitative traits, it is indicated to obtain populations through crossing with adapted materials and massive selection or recurrent intra or interpopulation selection [48].

For example, the gene conferring aluminum (Al) tolerance in maize, *ZmMATE1*, was mapped in a segregating population derived from Cateto Al237 x L53 and was cloned using molecular and physiological strategies [64]. Cateto Al237, a Brazilian maize line with poor agronomic attributes, was the donor of the superior allele of *ZmMATE1*, which was introgressed in maize elite lines using marker-assisted backcross [65]. Thus, non-elite maize germplasm is a powerful source of important alleles that have to be mined and characterized in order to be used in elite germplasm.

Some considerations must be done when planning a maize pre-breeding program. If the objective is to develop varieties, the intra-population recurrent selection may be more successful, whereas for hybrids development, the inter-population recurrent selection is a better option. The comparison among *in situ* conservation, participatory breeding, and pre-breeding allows the observation of some particularities. *In situ* conservation may or may not be associated with changes in the allelic frequency; the selection should be avoided to maintain the original genetic constitution of the population. However, in participatory breeding the selection should be done since the objective is not only to preserve the maize genetic variability, but also to ensure the livelihood of the farmer. Participatory breeding differs from these approaches due the participation of the farmers in the selection process. Regarding to the potential utilization of GB, *in situ* conservation or participatory breeding are practices that, compared to pre-breeding, lead to higher expectations to be used by farmers who use ancient or from traditional varieties or even accessions conserved in the GB. However, these practices serve a small number of farmers. On the other hand, materials from the GB that go through pre-breeding, breeding, and post-breeding will have superior agronomic performance and, consequently, are more likely to be chosen by farmers. In this way, all of these methodologies offer advantages in relation to the expansion of the genetic variability of cultivated plants, but with different purposes.

5.4 Evaluation of the variability maintained in maize GB

The first step toward germplasm evaluation is its characterization, which adds information to the accessions, following the descriptors for maize published by the International Board for Plant Genetic Resources [40]. A large proportion of maize accessions available in GRIN, GENESYS, and CIMMYT are characterized. Characterization of GB accessions is very important, as this information allows the planning of specific evaluations, such as for biotic and abiotic stresses resistance, and nutritional value. However, this characterization is not an easy task, especially regarding the step of determining which accessions, among thousands of options will be evaluated. The GB accessions included in the evaluation tests must have a certain similarity to each other. This is even more important in maize, given the great genetic variability present in the collection.

The core collections are representative samples of GB constituted in order to maintain genetic variability with a minimum of repeatability. The main advantage of core collections development is the faster assessment of their accessions and the GB evaluation. The core collections also provide greater availability of seeds for evaluation tests and for sample exchange. Core collections are developed according to sampling techniques that enable the maintenance of variability; therefore, it is necessary to consider the stratification of groups in the GB, the number of representatives and the methodology for selecting accessions in each group [66]. Several maize GBs have core or set collections, including CIMMYT [67], Uruguay [68], Brazil [66], China [69], and Europe [70].

TABLE 5.1 Homologous areas (HA) and number of sites included in LAMP [71].

	Altitude range	Latitude range	Number of sites
HA1	Below 1,200 m	from 26° N to 26° S	14
HA2	From 1,200 to 1,900 m	from 26° N to 26° S	4
HA3	From 1,900 to 2,600 m	from 26° N to 26° S	5
HA4	Above 2,600 m	from 26° N to 26° S	3
HA5	—	above 26° N or 26° S	8

Despite the difficulty to evaluate thousands of accessions, structured evaluation programs have already been successful in adding value to maize germplasm. The evaluation of about 12,000 maize accessions from different countries was carried out by the Latin American Maize Project (LAMP). This project brought together 12 countries: Argentina, Bolivia, Brazil, Colombia, Chile, Guatemala, Mexico, Paraguay, Peru, Uruguay, United States of America, and Venezuela for the evaluation of maize accessions in five stages. The regions where the evaluations were conducted were classified into homologous areas (HA) according to their latitude and altitude. Table 5.1 shows the definitions of these HA, as well as the number of regions used for the evaluations in each area. In the first stage of LAMP, accessions were evaluated in just one site within their original regions. In this phase, 20% of best performing accessions were selected for the second stage, in which the evaluations were conducted in two sites for the selection of 5% of the accessions with superior performance. In the third stage, selected accessions were exchanged between sites in the same HA. After selecting the best performing accessions, they were crossed with testers in each site. In the LAMP fourth stage, evaluation trials were conducted with hybridized accessions in order to estimate their combining ability. In the fifth stage, best performances accessions were integrated into the breeding programs of the participating countries [71].

5.5 Development of elite populations

The main objective of the plant genetic breeding is to develop varieties and/or hybrids with high yield, and also resistant to stress and adapted to target regions. Improved varieties are obtained by changing the allele frequency from an original population using the selection. However, a hybrid represents the first generation of a cross between inbred lines, whose superiority, called heterosis or hybrid vigor, manifests due heterozygous combinations of the alleles. The parental lines are obtained from populations, and their performances are conditioned to the genetic constitution and allele frequencies of the original populations [72]. Thus, in pre-breeding, new elite synthetic populations development methods vary according to the objective of breeding and genetic control of the focused traits. The backcross method is the fastest and least expensive way to improve populations originating from the GB. However, the percentage of alleles of unimproved populations to be incorporated into the elite germplasm will depend on performance of both the donor and the recurrent parent. Although the backcross first generation is the base population for starting the selection [73], it depends on the degree of dominance of the trait, the diversity between the parents and the performance of the donor parent [74].

Selection methods reduce the time to obtain cultivars in breeding programs should be used to save time and resource. In addition, agility in the breeding process is especially important when the objective is to obtain new pathogen-resistant cultivars, since they are expected to have their sources of resistance broken in a few years. The use of molecular markers to assist the recovery of the recurrent parent genotype in backcross programs is one of the molecular techniques with the greatest potential for routine use in marker-assisted selection [75]. The number of backcross generations required to recover the genotype of the recurrent parent using molecular markers assisted selection varies according to the number of chromosomes of the species, genome size in cM, number of markers employed, size of the population, and the inheritance of the trait to be introgressed. In a simulation study, species with a genome consisting of 10 chromosomes, with 200 cM each, genotyped by 80 markers and employing a population with 100 progenies, molecular markers assisted selection can reduce the number of backcross generations from 7 to 3, with 99% recovery of the recurrent parental genome [75]. Simulation was also used to assess variations in population size and the efficiency of molecular markers assisted selection [76]. The authors concluded that increases in population size lead to a reduction in the number of markers to be genotyped.

The use of elite populations for hybrids development must consider for the exploitation of heterosis. For this purpose, reciprocal recurrent selection is a widely used methodology, and aims, in addition to improving populations *per se*, to increase the heterotic response between two populations, through the selection of families from these populations tested-cross, concomitantly. In each population and each cycle, genotypes are selected based on their performance in crossings with genotypes from the other population, thus forming populations that will serve as a basis for the extraction of lines with greater heterotic potential [72].

5.6 Maize pre-breeding programs

In addition to evaluate germplasm for traits relevant for breeding, pre-breeding develops new populations that can be used in advanced stages of the breeding program. Some programs have already been carried out to develop new maize populations for improvement, from unimproved germplasm, among which stands out: the fourth stage of LAMP, the Germoplasm Enhancement of Maize (GEM) program, the program developed by the Nucleus Maize Support Project in Brazil and the CIMMYT pre-breeding program. As previously mentioned, the fourth stage of LAMP evaluated maize germplasm accessions crossed with testers from the breeding programs of the participating institutions [71]. In this way, LAMP developed segregating populations, evaluated their combining abilities in relation to the usual breeding programs testers and classified the accessions in heterotic groups.

The GEM project started in 1993 and included the participation of public and private sectors with the support of the American Seed Trade Association (ASTA). The main objective of GEM was to improve and to expand the genetic basis of hybrids cultivated by American farmers, with a focus on yield, disease and insect resistance, and nutritional value. The activities developed by the GEM project considered the information obtained by LAMP and had as starting point in the crossed of 51 LAMP best yield accessions with the public North American inbred lines B73 and Mo17, from opposite heterotic patterns. The populations obtained from these crosses were backcrossed with the respective inbred lines and S_3 progenies were obtained with 25% and 50% of unimproved germplasm. GEM Project also counted with the participation of private improvement companies that crossed the germplasm accessions with inbred lines from their programs and returned the seeds of these crossings to the GEM coordinator. The S_3 progenies were crossed with testers from opposite heterotic patterns for the evaluation of a total of 564 testcrosses in at least six sites. These evaluations allowed the identification of hybrids with exceptional performance and inbred lines derived from GB accessions. Additionally, the best materials were even superior to commercial checks in terms of yield in several locations [77]. In the GEM project, the grains were also evaluated for starch, protein, and oil content, which allowed the identification of accessions that, in some hybrid combinations, can equal or exceed the commercial values of the cultivars (Table 5.2). Inbreed lines that exceed the maximum percentages of the protein, oil, and starch ranges of the North American Corn Belt hybrids were obtained [77].

The Nucleus Maize Support Project was carried out in Brazil and evaluated more than 1,200 accessions of Maize Brazilian GB, which allowed the formation of composites with sources of resistance to several pathogens. This maize pre-breeding project was quite complete, once included both the germplasm evaluation and the development of new populations, generating five broad-based composites with resistance to *Puccinia polysora*, *Exserohilum turcicum*, *Physopella zeae*, *Phaeosphaeria maydis*, and maize stunts and viruses [78]. The broad genetic base of the composites obtained by Nucleus Maize Support Project is extremely favorable, since possibly the alleles responsible for resistance do not have the same origin, which increases the genetic diversity. The formation of composites also allowed the recombination among different sources of resistance, resulting in new gene combinations. Another advantage is that

TABLE 5.2 Maize grain composition in Corn Belt hybrids, GEM populations and GEM inbreed lines groups. [77]

	Amplitude of grain composition[a]		
Groups	Protein(%)	Oil (%)	Starch (%)
Corn Belt hybrids	9.0–11.0	3.0–4.0	69.0–70.0
GEM population	9.9–14.7	3.3–5.4	65.5–70.5
Inbreed line A group	10.4–15.1	3.3–5.0	64.7–70.0
Inbreed line B group	9.4–14.6	2.7–5.8	65.3–72.9
Inbreed line C group Puerto Rico	10.2–14.7	2.9–5.6	66.1–70.6
Inbreed line C group Hawaii	9.9–14.9	3.9–5.7	65.3–70.2

[a]*Percentage of dry matter.*

the great variability of the origin of the accessions increases the environmental adaptation of these composites. In contrast to the advantages of broad genetic base composites, these populations are uneven, which prevents the direct use in breeding programs, as they require populations with a certain genetic uniformity. However, the potential of composite as sources of resistance to pathogens is immense to develop new populations through crosses with elite genotypes and extraction of resistant inbred lines.

In the CIMMYT pre-breeding program, gene pools are divided according to heterotic patterns (flint and dent grains), cropping region (highlands, subtropical, and tropical), maturation cycle (early, intermediate, and late), and grain color (white and yellow), thus 34 different gene pools are considered. Reciprocal recurrent selection in half sib families and S_1 and S_2 lines, with selection between and within families, is used to improve the combining ability of these gene pools using inbred lines or other representatives of the opposite heterotic pattern as testers. Intrapopulation breeding is also carried out based on the selection of one ear per row with the objective to improve ear and plant traits controlled by genes with additive effects [79]. The pre-breeding methodology used by CIMMYT is outlined in Fig. 5.1, in which information about race, adaptation, plant, and ear traits are attributed to each accession. These accessions are genotyped by molecular markers in order to establish clusters through multivariate analyzes and to select 20% of the accessions within each cluster, to create representative gene pool for each cluster.

The accessions of each gene pool are evaluated in crosses with testers to classify them in heterotic patterns. In each heterotic pattern, accessions are crossed with elite inbred lines within the same heterotic pattern to form populations with, on average, 25% of alleles derived from GB accessions and 75% of alleles from elite germplasm. The selection is performed for agronomic traits in

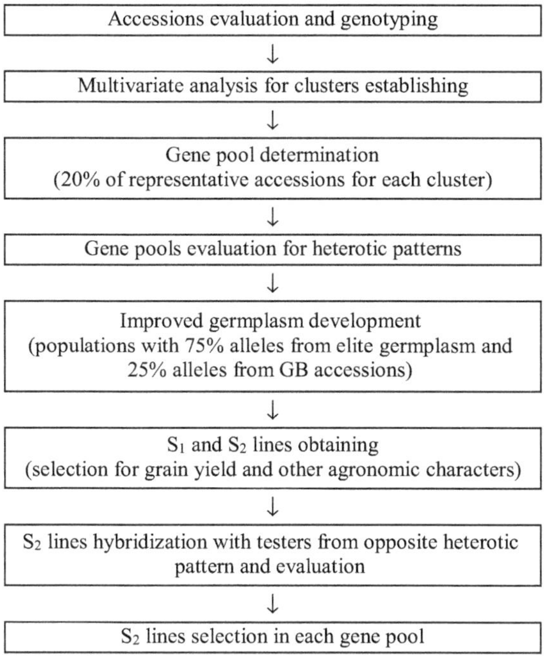

FIGURE 5.1 Methodology used in maize pre-breeding program at CIMMYT [80].

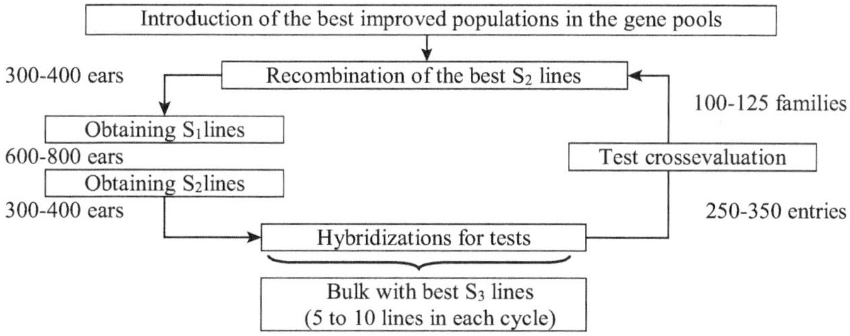

FIGURE 5.2 Population improvement methodology within each gene pool employed in the CIMMYT pre-breeding program [80].

S_2 lines derived from the improved populations, which are crossed with testers from opposite heterotic pattern, and these hybrids are evaluated for several traits in at least two locations. The S_2 lines selected are submitted to the procedure shown in Fig. 5.2, for gene pools improvement. These lines are recombined to obtain about 300–400 S_0 families that are self-fertilized to obtain 600–800 S_1 families and then, 300–400 S_2, using selection in these steps. These S_2 lines are crossed with testers from the opposite heterotic pattern and evaluated for grain yield and agronomic traits. In general, about 250–350 S_2 lines are evaluated in order to select from 100 and 125 S_2 lines for the next recombination cycle [80].

The unimproved germplasm preserved in GB and the elite germplasm used by breeding differ from each other in some aspects; however, the crucial point that makes a given germplasm classified as elite is the standardization of a series of traits that are required to develop commercial genotypes. In this way, genotypes without a standard performance cannot be classified as elite and therefore have no potential for immediate commercial use. The more improved is a crop, the more phenotypically distant is between elite genotypes and GB accessions and, consequently, the path to bring useful alleles from one collection to another becomes longer. Due to the high improvement degree of maize cultivars, pre-breeding can be divided into stages according to the proximity between these two poles: GB and breeding programs. GB maintenance activities, preliminary characterization of GBs accessions, and agronomic evaluations *per se* of the accessions can be classified as the first step of pre-breeding. In an intermediate stage, there are activities to form composites and accessions classified in heterotic patterns. The stage closest to the breeding involves obtaining inbreeding families to the extraction of inbreed lines and those that use populations with a higher percentage of elite germplasm alleles.

5.7 Concluding considerations

The expectations of maize breeding in relation to the GB, and vice versa, must be known to develop a pre-breeding program. The products obtained by the pre-breeding must serve two users: the breeding and the consumer market. Thus, it is important to consider relevant aspects to both in pre-breeding programs, such as grain yield, adaptability, stresses tolerance, nutritional quality, and classification in heterotic patterns. When relevant information is available, it is feasible to incorporate useful attributes derived from GB accessions in elite germplasm, increasing genetic diversity and allowing the development of new cultivars. Thus, from the point of view of genetic resources, it is important to determine the traits valued by the breeding, taking into account the demands of seed producers, farmers, and consumer market. Planning is a fundamental task for the success of breeding programs, since the activities that evaluate the GB must be conducted using appropriate methodologies and in the ideal environmental conditions so that the results can bring useful information. In addition, the information availability through databases and catalogs is essential so that knowledge about GB accessions is available to the scientific community. The time required between the identification of an accession with an adaptive advantage in the GB and its effective commercial use surely exceeds a decade. Thus, it is necessary that the activities developed in the scope of pre-improvement envision future demands, and for that they must be in tune with the demands of agribusiness.

References

[1] Buckler ES, Stevens NM. Maize origins, domestication, and selection. In: Motley TJ, Zerega N, Cross H, editors. New York: Columbia University Press; 2005. p. 67–90. Available from: http://www.maizegenetics.net/publications/Buckler2005DarwinHaverst.pdf.

[2] Piperno DR, Ranere AJ, Holst I, Iriarte J, Dickau R. Starch grain and phytolith evidence for early ninth millennium B.P. maize from the Central Balsas River Valley, Mexico. Proc Natl Acad Sci USA 2009;106(13):5019–24.

[3] Orozco-Ramírez Q, Perales H, Hijmans RJ. Geographical distribution and diversity of maize (*Zea mays* L. subsp. *mays*) races in Mexico. Genet Resour Crop Evol 2017;64:855–65.

[4] Merrill WL, Hard RJ, Mabry JB, Fritz GJ, Adams JR, Roney JR, et al. The diffusion of maize to the southwestern United States and its impact. Proc Natl Acad Sci 2009;106(50):21019–26.

[5] Piperno KH, Clary RG, Cooke AJ, Ranere D. Weiland, preceramic maize in Central Panama: phytolith and pollen evidence. Am Anthropol 1985;87:871–8.

[6] Grobman A, Bonavia D, Dillehay TD, Piperno DR, Iriarte J, Holst I. Preceramic maize from Paredones and *Huaca Prieta*. Peru. Proc Natl Acad Sci 2012;109(5):1755–9.

[7] Bush MB, Correa-Metrio A, McMichel CH, Sully S, Shadik CR, Valencia BG, et al. A 6900-year history of landscape modification by humans in lowland Amazonia. Quat Sci Rev 2016;141:52–64.

[8] Brugger SO, Gobet E, van Leeuwen JFN, Ledru MP, Colombaroli D, van der Knaap WO, et al. Long-term man-environment interactions in the Bolivian Amazon: 8000 years of vegetation dynamics. Quat Sci Rev 2016;132:114–28.

[9] Kistler L, Maezumi SY, Souza JG, Przelomska NAS, Costa FM, Smith O, et al. Multi-proxy evidence highlights a complex evolutionary legacy of maize in South America. Science 2018;362:1309–13.

[10] Brieger FG, Gurgel JTA, Paterninai E, Blumenschein A, Alleoni MR. Races of maize in Brazil and other eastern South American countries. Washington, DC: National Academy of Sciences; 1958. p. 283. (Publication 593).

[11] Wellhausen EJ, Roberts LM, Hernandez EX, Mangelsdorf PC. Races of maize in Mexico their origin, characteristics and distribution. Cambridge: Bussey Institution of Harvard University; 1952. p. 223.

[12] Wellhausen EJ, Fuentes AO, Corzo AH, Mangelsdorf PC. Races of maize in Central America. Washington, DC: National Academy of Sciences, National Research Council; 1957. p. 28. (Publication 511).

[13] Roberts LM, Grant UJ, Ramirez RE, Hatheway WH, Smith DL, Mangelsdorf PC. Races of maize in Colombia. Washington, DC: National Academy of Sciences, National Research Council; 1957. p. 153. (Publication 51).

[14] Hatheway WH. Races of maize in Cuba. Washington, DC: National Academy of Sciences; 1957. p. 75. (Publication 453).

[15] Paterniani E, Goodman MM. Races of maize in Brazil and adjacent areas. Mexico: Centro Internacional de Mejoramineto de Maiz e Trigo; 1977. p. 95.

[16] Ramirez RE, Timothy DH, Diaz EB, Grant UJ, Calle EN, Anderson E, et al. Races of maize in Bolivia. Washington, DC: National Academy of Sciences, National Research Council; 1960. p. 159. (Publication 747).

[17] Grobman A, Salhuana W, Sevilla R, Mangelsdorf PC. Races of maize in Peru their origin, evolution and classification. Washington, DC: National Academy of Sciences; 1961. p. 374. (Publication 915).

[18] Timothy DH, Peña BV, Ramirez RE, Brown WL, Anderson E. Races of Maize in Chile. Washington, DC: National Academy of Sciences, National Research Council; 1961. p. 84. (Publication 847).
[19] Timothy DH, Hatheway WH, Grant UJ, Torregroza MC, Sarria DV, Varela DA. Races of maize in Ecuador. Washington, DC: National Academy of Sciences, National Research Council; 1963. p. 147. (Publication 975).
[20] Grant UJ, Hatheway WH, Timothy DH, Cassalett CD, Roberts LM. Races of maize in Venezuela. Washington, DC: National Academy of Sciences; 1963. p. 92. (Publication 1136).
[21] Paterniani E., Nass LL, Santos M.X. O valor dos recursos genéticos de milho para o Brasil. In, Udry CV, Duarte W. Uma história brasileira do milho, o valor dos recursos genéticos. Brasília: Paralelo; 15, 2000. p.11–41.
[22] Leyva-Madrigal KY, Báez-Astorga PA, Negrete-Yankelevich S, Núñez-de la Mora A, Amescua-Villela G, Maldonado-Mendoza IE. Maize genetic diversity in traditionally cultivated polycultures in an isolated rural community in Mexico: implications for management and sustainability. Plant Ecol Divers 2020;13(1):15–28.
[23] Alvarado-Beltrán G, López-Sánchez H, Santacruz-Varela A, Muñoz-Orozco A, Valadez-Moctezuma E, Gutiérrez-Espinosa MA, et al. Morphological variability of native maize (*Zea mays* L.) of the west highland of Puebla and east highland of Tlaxcala, Mexico. Rev FCA UN Cuyo 2019;51(2):217–34.
[24] Modi AT. Short-term preservation of maize landrace seed and taro propagules using indigenous storage methods. S Afr J Bot 2004;70(1):16–23.
[25] Canci A, Vogt GA, Canci IJ. A diversidade das espécies crioulas em Anchieta-SC; diagnóstico, resultados de pesquisa e outros apontamentos para a conservação da agrodiversidade. McLee: São Miguel do Oeste; 2004. p. 112.
[26] Aragón-Cuevas F, Taba S, Castro-Garcia FH, Hernández-Casillas JM, Cabrera-Toledo JM, Alcalá LO, et al. In situ conservation, use of local maize races in Oaxaca, Mexico, a participatory, decentralized, approach., In: Taba, S., editor. Proceedings of Latin American maize germplasm conservation: regeneration, in situ conservation, core subsets, prebreeding, 2003 Apr 7-10. Mexico: D.F., CIMMYT; 2005. p. 83–92.
[27] Vega SR, Taba S, Solís ED, Díaz JC. Conservation and use of highland maize races in Chihuahua, Mexico. In: Taba, S., editor. Proceedings of Latin American maize germplasm conservation: regeneration, in situ conservation, core subsets, prebreeding, 2003 Apr 7-10. Mexico: D.F., CIMMYT; 2005. p.39–50.
[28] Rendón-Aguilar B, Aguilar-Rojas V, Aragón-Martínez MC, Ávila-Castañeda JF, Bernal-Ramírez LA, Bravo-Avilez D, et al. Diversidad de maíz en la sierra sur de Oaxaca, México: conocimiento y manejo tradicional. Polibot 2015;3(9):151–74.
[29] Ruiz-Corral JA, Ramírez-Díaz JL, Hernández-Casillas JM, Aragón-Cuevas F, Sánchez-González JJ, Ortega- Corona A, et al. Razas mexicanas de maíz como fuente de germoplasma para la adaptación al cambio climático. Rev Mexicana Cienc Agric 2011;2:365–79.
[30] Cabrera-Toledo JM, Carballo-Carballo A, Aragon-Cuevas F. Agronomic assessment of tiny Zapalote race maize in the Isthmian region of Oaxaca. Rev Mex Cienc Agríc 2015;6:2075–82.
[31] Salinas-Moreno Y, Pérez-Alonso JJ, Vázquez-Carrillo G, Aragón-Cuevas F, Velázquez-Cardelas GA. Antocianinas y actividad antioxidante en maíces (*Zea mays* L.) de las razas chalqueño, elotes cónicos y bolita. Agrocienc 2012;46(7):693–706.
[32] Salinas-Moreno Y, Aragón-Cuevas F, Ybarra-Moncada C, Aguilar-Villarreal J, Altunar-López B, Sosa-Montes E. Caracterización física y composición química de razas de maíz de grano azul/morado de las regiones tropicales y subtropicales de Oaxaca. Rev Fitotec Mex 2013;36(1):23–31.

[33] Figueroa-Cárdenas JD, Narváez-González DE, Mauricio-Sánchez A, Taba S, Gaytán-Martínez M, Véles-Medina J, et al. Propiedades físicas del grano y calidad de los grupos raciales de maíces nativos (criollos) de México. Rev Fitotec Mex 2013;36:305–14.
[34] Fernández-Suárez R, Morales-Chávez LA, Gálvez-Mariscal A. Importancia de los maíces nativos de México en la dieta nacional: una revisión indispensable. Rev Fitotec Mex 2013;36:275–83.
[35] Taba S., Shands HL, Eberhart S.A. The growth of CIMMYT's maize collection with the introduction of Latin American Maize Landraces Accessions through the cooperative regeneration project. In: Taba S., editor. Proceedings of Latin American maize germplasm conservation, regeneration, in situ conservation, core subsets, prebreeding, 2003 Apr 7-10. Mexico: DF, CIMMYT; 2005. p. 1–8.
[36] Teixeira FF, Souza BO, Andrade RV, Padilha L. Boas práticas na manutenção de germoplasma de variedades crioulas de milho.Sete Lagoas: Embrapa Milho e Sorgo; 2005. (Comunicado Técnico, 113).
[37] FAO. International treaty on plant genetic resources for food and agriculture. Rome: Food and Agriculture Organization of the United Nations; 2009. p. 56. Available from:: http://www.fao.org/3/a-i0510e.pdf.
[38] Bjørnstad Å, Tekle S, Göransson M. "Facilitated access" to plant genetic resources: does it work? Genet Resour Crop Evol 2013;60:1959–65.
[39] Pardey PG, Koo B, Van Dusen ME, Skovmand B, Taba S, Wright BD. CIMMYT Genebank. In: Koo B, Pardey PG, Wrigth BD, editors. Saving seeds: the economics of conserving crop genetic resources ex situ in future harvest centres of the CGIARCambridge: CABI Publishing; 2004. p. 21–47.
[40] IBPGR. Descriptors for maize. Mexico City: International Maize and Wheat Improvement Center; Rome: International Board for Plant Genetic Resources; 1991.
[41] Teixeira FF, Guimarães LJM, Guimarães PEO, Pacheco CAP, Parentoni SN, Silva AR. Pré-melhoramento de milho. In:Lopes MA, Fávero AP, Ferreira MAJF, Faleiro FG, Folle SM, Guimarães EP, editors. Pré-melhoramento de plantas: estado da arte e experiências de sucesso. Brasília, DF: Embrapa Informação Tecnológica; 2011. p. 571–614.
[42] Duvick DN. Biotechnology in the 1930s: the development of hybrid maize. Nat Rev Genet 2001;2:69–74.
[43] USDA-NASS. Crops production data, by states, 1866-1997.Washington: USDA-NASS; 2000.
[44] Reif JC, Hallauer AR, Melchinger AE. Heterosis and heterotic patterns in maize. Maydica 2005;50:215–23.
[45] Reif JC, Fischer S, Schrag TA, Lamkey KR, Klein D, Dhillon BS, et al. Broadening the genetic base of European maize heterotic pools with US Cornbelt germplasm using field and molecular marker data. Theor Appl Genet 2010;120:301–10.
[46] Parentoni SN, Magalhães JV, Pacheco CAP, Santos MX, Abadie T, Gama EEG, et al. Heterotic groups based on yield-specific combining ability data and phylogenetic relationship by RAPD markers for 28 tropical maize open pollinated varieties. Euphytica 2001;121:197–208.
[47] Nass LL, Paterniani E. Pre-breeding: a link between genetic resources and maize breeding. Sci Agric 2000;57(3):581–7.
[48] Nass LL, Nishikawa MAN, Fávero AP, Lopes MA Pré-melhoramento de germoplasma vegetal. In: Nass LL editor. Recursos genéticos vegetais. Brasília, DF: Embrapa Recursos Genéticos e Biotecnologia; 2007. p. 683–716.
[49] National Academy of Sciences. Genetic vulnerability of major crops: committee on genetic vulnerability of major crops. Washington, DC: National Research Council, National Academy of Sciences; 1972.

[50] Duvick DN. Genetic diversity in major farm crops on the farm and in reserve. Econ Bot 1984;38:161–78.
[51] Morris ML, Bellon MR. Participatory plant breeding research: opportunities and challenges for the international crop improvement system. Euphytica 2004;136:21–5.
[52] Smale M, Bellon MR, Aguirre JA, Manuel Rosas I, Mendoza J, Solano AM, et al. The economy costs and benefits of a participatory project to conserve maize landraces on farms in Oaxaca, Mexico. Agric. Econom 2003;29:265–75.
[53] Smith ME, Castillo FG, Gómez F. Participatory plant breeding with maize in Mexico and Honduras. Euphytica 2001;22:551–65.
[54] Soleri D, Smith SE, Cleveland DA. Evaluating the potential for farmer and plant breeder collaboration: a case study of farmer maize selection in Oaxaca, Mexico. Euphytica 2000;116:41–57.
[55] Louette D, Smale M. Farmer seed selection practices and traditional maize varieties in Cuzalapa, Mexico. Euphytica. 2000;113:25–41.
[56] Figueroa JD, Mauricio A., Taba S., Morales E., Mendonza A., Gaytán M., et al. Kernel characteristics, tortilla making quality of maize accessions from Mexico, the Caribbean, South, Central America. In: Taba S., editor. Proceedings of Latin American maize germplasm conservation, regeneration, in situ conservation, core subsets, pre-breeding, 2003 Apr 7-10. Mexico: DF, CIMMYT; 2005. p. 39–50.
[57] Rincón-Sanchez F., Ruiz-Torres N.A. Enhacement of farmers' maize varieties with na improved in Saltillo, Coahuila, Mexico. In: Taba S., editor. Proceedings of Latin American maize germplasm conservation, regeneration, in situ conservation, core subsets, pre-breeding, 2003 Apr 7-10. Mexico: DF, CIMMYT; 2005. p. 59–62.
[58] Rincón-Sanchez F, Ruiz-Torres NA. Assessing maize diversity and defining a representative subset by means of selected morphological ear traits. Plant Genet Resour 2018;16(6):488–97.
[59] Machado AT, Fernandes MS. Participatory maize breeding for low nitrogen tolerance. Euphytica 2001;122:567–73.
[60] Teixeira FF, Vasconcellos JH, Andrade RV, Santos MX, Netto DAM, Novotny EH, et al. Desempenho de variedades de milho quanto a qualidade da palha para artesanato. Rev Bras Milho Sorgo 2007;6(1):84–94.
[61] Chimonyo VGP, Mutengwa CS, Chiduza C, Tandzi L. Participatory variety selection of maize genotypes in the eastern cape province of South Africa. S Afr J Agric Ext 2019;47(1):103–17.
[62] Song Y, Zhang Y, Song X, Vernooy R. Access and benefit sharing in participatory plant breeding in Southwest China. Farm Matt 2016;18–24.
[63] Mendes-Moreira P, Satovic Z, Mendes-Moreira J, Santos JP, Santos JPN, Pego S, et al. Maize participatory breeding in Portugal: comparison of farmer's and breeder's on-farm selection. Plant Breed 2017;136:861–71.
[64] Maron LG, Piñeros MA, Guimarães CT, Magalhaes JV, Pleiman JK, Mao C, et al. Two functionally distinct members of the MATE (multi-drug and toxic compound extrusion) family of transporters potentially underlie two major aluminum tolerance QTLs in maize. Plant J 2010;61(5):728–40.
[65] Guimarães CT, Simões CC, Pastina MM, Maron LG, Magalhães JV, Vasconcellos RCC, et al. Genetic dissection of Al tolerance QTLs in the maize genome by high density SNP scan. BMC Genom 2014;15:153.
[66] Abadie T, Cordeiro CMT, Andrade RV, Parentoni SN, Magalhães JR. A coleção nuclear de milho para o Brasil. Brasília: Embrapa Recursos Genéticos e Biotecnologia; 2000.
[67] Bänziger M, Long J. The potential for increasing the iron and zinc density of maize through plant-breeding. Food Nutr Bull 2000;21(4):397–400.

[68] Malosetti M, Abadie T. Sampling strategy to develop a core collection of Uruguayan maize landraces based on morphological traits. Genet Resour Crop Evol 2001;48:381–90.
[69] Yao QL, Fan P, Zou SX. Constructing a core collection for maize (Zea mays L.) landrace from wuling mountain region in China. Agric Sci China 2008;7(12):1423–32.
[70] Malvar RA, Butrón A, Alvarez A, Ordás B, Soengas P, Revilla P, et al. Evaluation of the European Union maize landrace core collection for resistance to *Sesamia nonagrioides* (Lepidoptera: Noctuidae) and *Ostrinia nubilalis* (Lepidoptera: Crambidae). J Econ Entomol 2004;97(2):628–34.
[71] Salhuana W, Sevilla R, Eberhart SA. Latina American Maize Project: final report. Johnston: Pioneer Hi-Bred International; 1997. p. 147.
[72] Paterniani E, Miranda Filho JB. Melhoramento de populações. In: Paterniani E, Viégas GP, editors. Melhoramento e produção de milho. Campinas: Fundação Cargill; 1987. p. 217–274.
[73] Eberhart SA. Regional maize dialles with U. S. and semi-exotic varieties. Crop Sci 1971;11:911–4.
[74] Dudley JW. Theory for transfer of alleles. Crop Sci 1982;22:631–937.
[75] Openshaw SJ, Jarboe SG, Beavis WD. Marker assisted selection in backcross breeding. In: Lower R, editor. Join plant breeding symposium on analysis of molecular marker data. Corvallis: Oregon State University; 1994.
[76] Frisch M, Bohn M, Melchinger AE. Comparison of selection strategies for marker-assisted backcrossing of a gene. Crop Sci 1999;39(5):1295–301.
[77] Pollak LM. The history and success of public-private project on germplasm enhancement of maize (GEM). Adv Agron 2003;18:45–87.
[78] Miranda Filho JB, Nass LL, Santos MX, Regitano Neto A. Avaliação de acessos de milho para resistência a doençasfoliares. Brasília: Embrapa Recursos Genéticos e Biotecnologia; 2000. p. 147. (Circular Técnica 3).
[79] Taba S, Krakowsky M, Kirubi D, Rivas M, Rodríguez M, Ramírez M, et al. Maize genetic resources: annual report 2002. México: CIMMYT; 2002. p. 147.
[80] Taba S, Krakowsky M, Rivas M, Rodriguez M, Diaz J. CIMMYT's Maize prebreeding program. México: CIMMYT; 2007. Available at: http://www.cimmyt.org/Research/Maize/symposium/posters/taba_prebred.pdf.

Chapter 6

Utilization of Wild Ancestors for Biotic and Abiotic Stress Tolerance in Barley

Shoaib Liaqat, Muhammad Abu Bakar Saddique, Muhammad Ali Sher and Zulfiqar Ali
Institute of Plant Breeding and Biotechnology, MNS University of Agriculture, Multan, Pakistan

6.1 Significance of wild germplasm

Plant breeders have exploited the genetic diversity to increase the crop performance over time but the annual yield increase is not sufficient to meet the need of human population [1]. Plant breeders have continuously enhanced the yield potential by improving the germplasm performance and also by adoption to various new farm management practices and mechanization. There is a need to approach potential genetic diversity of crop wild relative to gratify the increasing population demand for more food in the scenario of food security and climate change. These crop wild progenitors stored in seed banks or present in the natural environment are the actual required material to equip our crops for the changing climate [2].

The global increase in temperature and changing pattern of drought and salinity has adversely affected the plant growth and yield [3]. Genetic erosion in cultivated varieties such as loss of 50% genetic variability in cultivated soybean [4] and rice [5] give a significant reason for evaluating its gene pool and wild relatives. For most crops, the wide range of gene pool exist in wild species, from which transfer of desirable traits is quite possible in cultigens and it is believed that alleles contributing from wild relatives are always dominant than that of cultivated species [6]. Crop wild relatives possess significant genetic diversity at both population and individual levels in comparison with their domesticated genotypes and distributed in all continents except Antarctica and mostly found in the Vavilov Center of Diversity and their contiguous areas [7,8]. Worldwide distribution of crop wild relatives provide a valuable source for potential genetic variability in fact, among 50,000–60,000 total crop species, maximum have been suggested for direct exploitation in breeding programs for food security.

Wild Germplasm for Genetic Improvement in Crop Plants. http://dx.doi.org/10.1016/B978-0-12-822137-2.00006-0

But many finding suggested that use of crop wild germplasm for crop improvement have gradually adopted over the years [9] and there is an urgent need for collecting and conserving over 70% of total crop wild relative species because almost 95% of the total species have described insufficient geographic and ecological variability according to their native distribution [8].

6.2 Barley origin and distribution

In ancient times, people were foragers and dependent on wild cereals and nuts for their food [10]. About 10,000 years before, they started to settle down and left some remains of domesticated barley in Neolithic archaeological sites in the Fertile Crescent [11]. For adaptation and speciation, domestication is considered to be a major evolutionary trial that was practiced by humans during the last 10,000 years, so, it is better to understand the origin and early agrarian practices of barley for its domestication [12].

Barley was first appeared in Southwestern Asia and has been found in two-rowed form that has morphological resemblance with wild barley nowadays [13]. There are conflicting opinions regarding origin and evolution of cultivated barley as suggested that the center of origin of *Hordeum* was "The Fertile Crescent," including Asia-Minor, Caucasia, Iraq, Turkey, Jordan, Syria, Palestine, and Lebanon [14,15]. To determine the allelic frequency, different amplified fragment length polymorphism markers were used to test the polymorphic loci in both wild and cultivated barley and it was concluded that wild barley population that have resemblance to the cultivated gene pool was from Israel-Jordan, it clearly indicates that this is the area of primary center of domestication for cultivated barley [16]. But later studies did not support this idea of only one center of origin [17,18]. Wild barley distribution has wider range extending into Kyrgyzstan, Afghanistan, western Pakistan, and Tibetan plateau [17]. Haplotype frequency difference revealed the place of domestications of two barley from different geographic regions, one was from Fertile Crescent and second from 1500–3000 km far away and also suggested that American and European cultivars have major contribution from Fertile Crescent [19]. Despite this, some other regions like Ethiopia and Morocco have been considered as candidate regions for barley domestication [20] and key trait of hulled/naked barley provide morphological marker to identify the origin and domestication of barley [21]. The hulless grain belongs to a single origin in domesticated barley and the gene for this trait is present on long arm of 7H chromosomes which is controlled by single locus [22]. Genome-wide Diversity Array Technology (DArT) markers were utilized to differentiate the geographic dissemination of wild barley germplasm worldwide [17]. *H. spontaneum* is distributed in the east-Mediterranean and Near East Fertile Crescent, extending from Israel/Jordan to southern Turkey [13]. A small population is also associated with Central Asia and Tibet and considered as secondary centers of diversity [23]. Several gene banks collect and maintain wild barley germplasm while Israel, Canada, Syria, the United States, and Germany are prominent [24].

6.3 Genetic diversity of wild barley

Genetic variability is the basis of evolutionary process and development of new species [25]. Wild barley has variability in allelic content and due to high mean proportion of polymorphic loci, it represents high genetic variability when compared with mean genetic estimates naturally [26]. Wild barley accessions from 16 countries were characterized by using simple sequence repeat (SSR) markers and 359 alleles were identified at 7 chromosomes. The SSR analysis revealed that stratified sampling is more effective when practiced to assess country-wise SSR dissimilarity and frequency of polymorphic alleles is more than 7% [27]. They also concluded that barley accession from Israel and Jordan were found to be more diversified genetically.

Environmental heterogeneity is considered to be a key regulator that provoked genetic diversity as both at small and large geographic scale [28]. By using different types of molecular markers, natural areas of distribution of wild barley have indicated the pattern of genetic diversity and also revealed that remaining of wild barley at Fertile Crescent exhibit a maximum portion of diversity [12,29]. Materials obtained from Fertile Crescent reflects different behaviors as compare to Central Asia due to differential mechanism of genetic variation in various regions [19], also allelic variation at several SSR loci have represented phylogenetic variation between Fertile Crescent and North Africans [30]. Genotyping analysis through short sequence repeats and inter-simple sequence repeat (ISSR) loci exposed the double clustered structure of wild barley genotypes that separates the Tibet and Fertile Crescent ones [29]. Wild barley with maximum genetic diversity has associated with the western part of Israel, Syria, and Jordan along with regions of Central Asia [31]. In 1980, the wild barley identification, collection, and evaluation was started from Tibetan and it was also established that material from Tibetan plateau have potential of genetic variability especially regarding agronomic traits such as total biomass and chilling tolerance [32]. Thirty-three wild barley accessions were evaluated for genetic variability from Tibetan Plateau by using 52 SSR markers and it was concluded that out of 206 alleles, about 111 were found to be common in all material and SSR markers significantly differentiated the landraces and Tibetan wild barley with the existence of 24 unique alleles with very high frequency observed in Tibetan wild barley accessions [33]. It was observed that due to high genetic variability and potential characters, Tibetan wild barley had achieved maximum emphasis. Twenty-seven barley accessions were evaluated for resistance to leaf rust, net blotch, powdery mildew, and also against abiotic stresses, that is, drought, salinity, heat, and chilling stress [34].

6.4 Environmental adaptation of wild barley

Significant phenotypic and genotypic diversity in the population of wild barley suggest a positive effect on adaption to the environment. The important drivers show a blend of climate (including temperature and precipitation), pathogen pressure, and soil [35]. The barley population originated from Israel has wider variability as compared to Iran and Turkey [36]. Barley genotypes collected from Jordanian Mountains, Eastern Galilee, and Golan Heights were more differing from those originated from sites of lower altitude, showing the more intensity and frequency of high temperature, and drought stress at those sites of high altitude [37]. Allelic difference of subtilisin/α-amylase inhibitor gene *Isa* is also associated with the moisture availability, and very high diversity observed in the germplasm adapted to drought tolerance [38]. The inclination of both temperature and aridity showed a high influence on the allele frequency in the materials from Near East Fertile Crescent [39]. These results depicted that presence of genetic variability is more important for the population to adapt the natural climatic variations [38]. Detailed investigation of hereditary changes in the population of wild barley at Israeli sites showed that the climatic, topographic, and edaphic variables are the main driving forces and these populations represented high phenotypic variability across Israel according to edaphic and climatic variations from robust mesic phenotypes to slender xeric genotypes [13]. Germination pattern and desiccation of xeric and mesic phenotypes due to adaptive variability include seed dormancy for longer period, increased root development, and tolerance to desiccation in xeric plants [40]. Furthermore, tolerance of seedling to desiccation and genetic basis of caryopsis dormancy in wild barley were described [41].

6.5 Biotic stress resistance

Diseases can seriously influence the production of barley, either affecting the grain yield or the grain quality that lower net profit to the farmers. Almost 30 types of disease that includes prominently leaf rust, yellow rust, leaf scald, and net blotch are extensively damaging diseases on barley crop. To induce disease resistance, wild barley is an important and rich source of genes [42], and to achieve this goal, extensive efforts have been carried out to distinguish localize significant QTL and genes from wild germplasm [43]. Inheritance of heterozygosity at different loci in 116 wild barley accessions represented a high level of micro and macro scale disease resistance when screened at seedling stage [44]. The recurrence of resistance in Israel and Jordan genotypes was high for *Septoria* speckled leaf blotch that was 98% and 77%, respectively, about 68%–72% net smudge, 90%–70% for leaf rust, low for stem rust (26% and 2%), and intermediate for spot smear (46% and 53%). Significant level of variation for disease reaction and also heterozygosity at different loci has revealed that wild barley accession is an important and undiscovered source of distinct alleles for resistance to diseases and improvement of domesticated barley [13].

6.5.1 Powdery mildew (*Erysiphe graminis hordei*) resistance

Powdery mildew, most common disease of barley, results 5%–10% yield reduction in every region and in sever conditions may cause upto 30% yield losses.

It also deteriorates the grain quality of barley [45]. The powdery mildew resistance in barley is controlled by two genetically distinct pathways. One pathway for resistance is controlled by recessive alleles of *mlo* locus, and the second pathway for resistance is controlled by many race-specific resistance genes (R genes) like, *Mlk*, *Mla*, and *Mlg* [46]. These genes activate rapidly and cause the death of host cells at the site of infection, called cell death [47]. *Mla* locus is complex that is present on the barley 1HS chromosome and 31 alleles in the *Mla* locus were identified by RFLP, SSR, SCAR, and SNP (gene-specific) based markers. The increasing numbers of new alleles for that locus are mostly originated from the accession of wild from Israel [48]. In numerous countries, many accessions of barley having resistance to powdery mildew were recognized and extensively utilized in resistant cultivars development [42]. About 75% of 277 wild barley accessions obtained from Israel were found to be resistant for powdery mildew [49] and numerous accessions from Israel are profoundly resistant to powdery mildex [50]. The wild barley resistant genes also demonstrated diverse levels of resistance to powdery mildew: *mlt* with three alleles indicated a recessive inheritance mode on 7HS chromosome. *Mli* and *Mlf* restricted on chromosomes 5HL and 7HL were semi-dominant resistant genes [51] however, wild barley mainly possess the QTLs for powdery mildew resistance on 1H chromosome [43].

6.5.2 Leaf scald

Leaf scald of barley was brought about by the contagious pathogen *Rhynchosporium secalis* and they can easily break the newly adapted genetic resistance in barley population [52]. In this way, it is hard to track down a host that can completely oppose this pathogen. But it can be managed to find the resistant sources from the wild germplasm of barley. In cultivated barley, *Rrs1* locus on chromosome 3H long and *Rrs2* on chromosome 7H long arm were responsible for resistance to leaf scald [53]. While in the wild barley, these genes were present on five chromosomes (1HS, 3H, 6HS, 7HL, and 7HS). In another study, it was described that the *Rrs12* locus was mapped on the chromosome 7H short arm and expected to be allelic to *Rh2* in cultivated barley [54]. Further, four QTLs (Rrsq1, Rrsq2, Rrsq3, and Rrsq4) on chromosomes 2H, 3H, 4H, and 6H, respectively were recognized for scald resistance. A resistant QTL, QRh.S42-4H, and two susceptible QTLs, QRh.S42-2H and QRh.S42-3H, were distinguished from a BC2DH S42 (a cross between spring cultivated barley "Scarlett" and a wild barley germplasm "ISR42-8") [55]. Another QTL identified for resistance against scald, assigned as *Rrs*-1H-1-4, was found from a hybrid of OUH602 and Harrington [43].

6.5.3 Leaf rust

Another significant disease of barley in various global regions is due to *Puccinia hordei Otth*. There are two mechanisms that control the resistance by a few or several genes, and partial resistance is frequently viewed as more solid than race-specific resistance [56]. Seven different genes for leaf rust resistance were detected on all chromosomes except 4H and cultivated barley possess disease resistance that was specifically confined and constrained to the genes *Rph7* and *Rph3* [57]. Wild barley (*Hordeum spontaneum*) was enclosing the genes for leaf-rust (*Rph10, Rph11, Rph15, and Rph16*), and a massive diversity was found in wild barley genotypes from Israel. Responses to disease of 292 wild barley genotypes gathered at 16 localities incorporating the whole ecological range of wild barley in Israel [58]. The outcomes indicated that 52% were resistant to race Tel-Aviv, 67% were resistant to race *Puccinia hordei*. It affirmed the hypothetical gene and further exhibited that the Near East represents a significant center for the evaluation of *P. hordei Otth* resistance [58]. Several QTLs associated with leaf rust resistance were identified on chromosome 2H, 3H, 4H, 5H, and 7H. In cultivated barley, only six QTLs were related to resistance of that disease such as Rphq1 (1H), Rphq2 (2H), and Rphq3 (6H), Rphq3, Rphq4 (7H), and Rphq5 (4H) and contributed approximately 63% of the phenotypic variance and were more effective at the adult plant stage. But in case of wild barley, resistant QTLs were present on same chromosome especially at the QTL location QLr.S42-2H [55].

6.5.4 Barley yellow dwarf virus (BYDV)

Resistance against other pathogens including barley yellow dwarf virus (BYDV) in wild barley germplasm have been studied. It has been described that some wild barley genotypes exhibit resistance against BYDV [59]. The *Isa* gene is responsible for bifunctional amylase subtilisin inhibitor (BASI) that provides defense against parasitic pathogens; and diversity pattern distribution in the *Isa defense* locus in the population of wild barley recommend a versatile determination at this locus [60]. Different responses were obtained from 12 wild barley genotypes and 36 cultivated varieties that originated from different regions of the world by using RAPD markers; and genetic closeness of BYMV complex (BaMMV, BaYMV, and BaYMV-2) among all the genotypes was identified and it was also discovered that wild barley germplasms and cultivated varieties exhibited varied level of response to BYMV complex [61].

6.6 Abiotic stress tolerance

6.6.1 Drought stress tolerance

Drought is the most important abiotic stress that harms the plant development and growth. Tolerance against drought is a complex trait and likewise monogenic resistance, it cannot be genetically assessed. Barley wild relatives have a significant genetic potential for abiotic and drought stress tolerance and serve as an important source for adaptability of plant to drought tolerance [13,35]. The wild barley germplasm has the huge genetic diversities for abiotic stress tolerance [35], they can be extensively utilized in plant breeding to develop abiotic stress tolerant genotypes especially drought. Carbon isotopes as an indicator for drought tolerance were used in term of yield stability and water use efficiency and recorded 10 QTLs on all of chromosomes for drought tolerance [62].

Dehydrins (DHNs) are the proteins that are quite effective in adaptive defense of a plant for salinity, drought, and low temperature stress [63]. They significantly scavenge the plants from temperature and drought stress, and these are the most common proteins produced in plants under drought stress [64]. The domain analysis revealed that the length of DHN polypeptide use from 82 to 575 amino acid residues. On the four chromosomes of barley 13 *Dhn* genes were identified [65]. The *Dhn1* role in drought tolerance has been confirmed by many researchers and reported the QTL colocalization with *Dhns* like QTLs for relative water content [62], overlapping with *Dhn* genes cluster with chromosome 5H. The *Dhn5* showed the high polymorphism in the wild barley from South Western Asia and Zagros Mountains, and the *Dhn9* locus showed the moderate polymorphism [66]. Differential expression of *Dhn1* in resistant genotypes was observes as early expression (after 3 h of dehydration) and higher expression (after 12–24 h of dehydration). Similarly, in case of *Dhn3* when expression was induced through drought stress, both sensitive and resistant genotypes showed expression 3–12 h but minute difference was observed after 12 and 24 h. But in case of *Dhn5* when drought stress applied, expression started after 3 h in all genotypes and increasing pattern was observed after 12 h that was slightly high in resistant genotypes than sensitive [63]. Expression of *Dhn1* under drought stress showed the inter-slope divergent expression between barley genotypes, European slope (ES) and African slope (AS), showing the adaptive nature of *Dhn1* gene for drought tolerance. Microclimatic natural selection gives off an impression of the evolutionary driving force resulting inter-slope divergent evolution of *Dhn1* at ECI. The *Dhn6* genetic pattern was also studied in 48 wild barley genotypes at ECI. *Dhn6* differential expression pattern was detected at different intervals after dehydration and the inter-slope amino acid genetic divergence showed a positive and significant selection of *Dhn6* [67]. Furthermore, expression analysis of most sensitive and tolerant genotypes was conducted under severe drought stress condition at early growth stage and it was concluded that 70 out of 1100 transcript-derive fragments (TDFs) were varying expressions and 11 of them clearly indicated down-regulation for susceptible and as well as tolerant genotypes [68]. RT-PCR is used to test isolate and sequence of these TDFs. Out of seven TDFs, *Hsdr4* (*Hordeum spontaneum* dehydration-responsive 4) was found to be good for drought tolerance.

6.6.2 Heat stress tolerance

Sustainability of agriculture has greatly disturbed due to global rise in temperature all over the world [69] and this increase will start to reduce the total cereal production after 2030 [3]. Temperature tolerance have significant impact on crop yield as reported on desert conditioned plants when studied on grain filing stage [70] and also disturb the vegetative and reproductive plant growth stages especially at anthesis that results in delayed seed maturity [71]. Despite having economic significance, very few studies have been conducted to explore thermal tolerance mechanism in barley. Wild barley (*Hordeum spontaneum* L.) have cross compatibility and distributed over a larger area mainly originated from Mediterranean regions [72]. Wild barley can survive in various environmental conditions and regarded as drought and heat tolerant crop due to phenotypic plasticity [73]. Fifty-one wild barley genotypes were analyzed through molecular makers and it was suggested that temperature and precipitation have a significant role for the adaptation of wild barley in the Fertile Crescent [39].

Leaf photosynthesis is the major metabolic process that is disturbed due to high-temperature stress [74]. So, in the scenario of climate change and food security, it is urgently required to develop heat-tolerant material by exploiting the wild genetic resources and it mainly depends on some major factors such as cross-compatibility in both intra- and inter-genomic pairing of chromosomes. Use of *Hordeum vulgare* ssp. *spontaneum* appears to be significant as it could be resulted in heat tolerant barley cultivars with improved quality and yield due to complete genome homology between cultivated barley *H. vulgare* ssp. *vulgare* ($2n = 2x = 14$, HH) and *H. vulgare* ssp. *spontaneum* ($2n = 2x = 14$, HH). Also adaptability to harsh environmental factors mainly high temperature could be described for the reason of improved hybridization and gene introgression for temperature tolerance [75].

6.6.3 Aluminum stress tolerance

Toxicity due to aluminum (Al^{3+}) is the major factor that limits the crop growth and yield under acidic soils. Among all the cereals, barley planting is inhibited in areas with high Al^{3+} concentration and pH below 5.5 pH. Cultivated barley varieties also showed a wider genetic variability for tolerance against Al^{3+} toxicity but it is becoming narrow and being exhausted due to specificity.

Wild barley has the maximum genetic diversity and may be the potential source for the novel genes for tolerance against Al^{+3} in cultivated barley. To exploit wild barley genetic potential, a multidisciplinary program was initiated [32]. The maximum accession of wild barley was evaluated and observed that some accessions have a high tolerance against Al^{+3}. The protein associated with tolerance against Al^{+3} toxicity was also characterized and evaluated in the wild barley [76]. Genome-wide association analysis (GWAS) was used to find the variation among genes controlling Al^{3+} tolerance in wild barley, using 110 and 56 barley genotypes from Tibetan wild cultivated cultivars, respectively. Two new loci bpb08524 and bpb-9458 were recognized in wild Tibetan barley, present on chromosome 7H and 2H and contributed 9.7% and 12.9% of phenotypic variability, respectively. The studies on the genetic diversity among Tibetan wild barley accession against Al^{+3} stress tolerance may be helpful to understand the tolerance mechanism against Al^{3+}-stress [77].

6.6.4 Salinity stress tolerance

Soil salinity is a major abiotic stress which limits agricultural production and is an eco-environmental and agricultural issue around the globe. Wild barley has the maximum genetic diversity and the huge allelic variation which would be beneficial in the development of salinity tolerant cultivars. To understand the mechanism for salinity tolerance in barley, 189 wild barley genotypes were evaluated under salinity stress, and they determined the different concentration of Na^+ and K^+ ions of 48 genotypes for salinity tolerance [78]. The wild genotypes showed immense variation for salinity tolerance, which was closely linked with Na^+/K^+ ratio. Furthermore, *HvHKT1* and *HvHKT2* functional and allelic diversity was determined by gene expression and association analysis and it was observed that K^+ and Na^+ transport under salinity stress are mainly controlled by *HvHKT1* and *HvHKT2* [79]. The *HvCBF1*, *HvCBF3*, and *HvCBF4* showed significant diverse sequence variation in amplicon as evident by the identification of single nucleotide polymorphisms (SNPs) and 3, 8, and 13 haplotypes, respectively. *HvCBF4* gene demonstrated high salt resilience, keeping up altogether lower Na^+/K^+ proportions and high biomass showed first-class alleles having salinity tolerance from Tibetan wild barley and could be a rich source for improving salinity tolerance in cultivated barley. Homeostasis of ionic components and their relation to primary and secondary metabolites was observed with the help of change in ionic profile due to slat stress [80]. The results of metabolite profiling proposed that osmotic change is an essential component and polyols play a significant role in creating salt tolerance. These findings provides the opportunity to understand the mechanism involved in barley salt-resistance and give conceivable usage of Tibetan wild grain in creating grain cultivars with salt resistance [80].

6.7 Conclusion

To feed the increasing population on this planet, production of food ought to be reasonably increased. However, the challenge of food increment has been adversely inclined by the changing climatic conditions. So, the plant breeder has to focus toward the genetic diversity and those characters that will enhance the potential for high yield. For that purpose, new innovations must be developed for better phenotyping and genotyping of plant germplasm to enhance breeding proficiency. Identification, evaluation, and introgression of new gene/alleles source of genetic variability will be helpful for adaptation. Wild barley germplasm is the rich source for the genetic variability and the most important traits which are transferred from wild barley to cultivated barley are biotic resistance (herbicide, bacteria, virus, and fungus), abiotic (heat, cold, drought, mineral, and salinity tolerance), dormancy, and improved malting quality. The wild barley is the potential source for these genes of biotic and abiotic stress tolerance. Many QTLs have been identified in wild barley and the drought tolerant QTLs were present on most of the chromosome. Genomic approaches with advances to target gene of interest for specific response to climatic factors that maintain and enhance the quality and yield may be deployed for better adaptability of existing varieties into new environment. So, there is an urgent need to evaluate and conserve the wild germplasm both in situ and ex situ for barley improvement.

References

[1] Alston JM, Andersen MA, James JS, Pardey PG. Persistence pays: US agricultural productivity growth and the benefits from public R&D spending. Springer Science & Business Media; 2009.

[2] Meyer RS, Knapp S, Karol KG, Little DP, Nee MH, Litt AJMp, et al. Reply to J. Samuels: Taxonomic notes on several wild relatives of *Solanum melongena* L. Mol Phylogenet EvolV 69 2013;(1):306–7.

[3] Tester M, Langridge PJS. Breeding technologies to increase crop production in a changing world. Science 2010;327(5967):818–22.

[4] Zhou X, Carter TE, Cui Z, Miyazaki S, Burton JWJCS. Genetic diversity patterns in Japanese soybean cultivars based on coefficient of parentage. Crop Sci 2002;42(4):1331–42.

[5] Xu X, Liu X, Ge S, Jensen JD, Hu F, Li X, et al. Resequencing 50 accessions of cultivated and wild rice yields markers for identifying agronomically important genes. Nat Biotechnol 2012;30(1):105–11.

[6] Hawkes JG. The importance of wild germplasm in plant breeding. Euphytica 1977;26(3):615–21.

[7] Larson G, Piperno DR, Allaby RG, Purugganan MD, Andersson L, Arroyo-Kalin M, et al. Current perspectives and the future of domestication studies. Proc Natl Acad Sci USA 2014;111(17):6139–46.

[8] Castañeda-Álvarez NP, Khoury CK, Achicanoy HA, Bernau V, Dempewolf H, Eastwood RJ, et al. Global conservation priorities for crop wild relatives. Nat Plants 2016;2(4):1–6.

[9] Maxted N, Kell SCommission on Genetic Resources for Food and Agriculture. Establishment of a global network for the in situ conservation of crop wild relatives: status and needs Food and Agriculture Organization of the United Nation; 2009. p. 266.

[10] Fuller DQ, Allaby RG, Stevens C. Domestication as innovation: the entanglement of techniques, technology and chance in the domestication of cereal crops. J Water Archaeol 2010;42(1):13–28.

[11] Diamond J. Evolution, consequences and future of plant and animal domestication. Nature 2002;418(6898):700–7.

[12] Zohary D, Hopf M, Weiss E. Domestication of Plants in the Old World: The origin and spread of domesticated plants in Southwest Asia, Europe, and the Mediterranean Basin. Oxford University Press; 2012.

[13] Nevo E, Fu Y-B, Pavlicek T, Khalifa S, Tavasi M, Beiles A. Evolution of wild cereals during 28 years of global warming in Israel. Proc Natl Acad Sci USA 2012;109(9):3412–5.

[14] Vavilov N. Studies on the origin of cultivated plants. Leningrad* Dobzhansky, Th.(1951): Genetics and the origin of species. Columbia Univ. Press, New York, より引用; 1926.

[15] Vavilov NI. The origin, variation, immunity and breeding of cultivated plants. Science 1952;115:433–4.

[16] Badr A, Rabey HE, Effgen S, Ibrahim H, Pozzi C, Rohde W, et al. On the origin and domestication history of barley (*Hordeum vulgare*). Mol Biol Evol 2000;17(4):499–510.

[17] Dai F, Nevo E, Wu D, Comadran J, Zhou M, Qiu L, et al. Tibet is one of the centers of domestication of cultivated barley. Proc Natl Acad Sci USA 2012;109(42):16969–73.

[18] Brown TA, Jones MK, Powell W, Allaby RG. The complex origins of domesticated crops in the Fertile Crescent. Trends Ecol Evol 2009;24(2):103–9.
[19] Morrell PL, Clegg MT. Genetic evidence for a second domestication of barley (*Hordeum vulgare*) east of the Fertile Crescent. Agron Plants Genet 2007;104(9):3289–94.
[20] Molina-Cano J-L, Russell J, Moralejo M, Escacena J, Arias G, Powell W, et al. Chloroplast DNA microsatellite analysis supports a polyphyletic origin for barley. Theor Appl Genet 2005;110(4):613–9.
[21] Zohary D. Monophyletic vs. polyphyletic origin of the crops on which agriculture was founded in the Near East. Genet Resour Crop Evol 1999;46(2):133–42.
[22] Taketa S, Kikuchi S, Awayama T, Yamamoto S, Ichii M, Kawasaki S, et al. Monophyletic origin of naked barley inferred from molecular analyses of a marker closely linked to the naked caryopsis gene (nud). Theor Appl Genet 2004;108(7):1236–42.
[23] Azhaguvel P, Komatsuda T. A phylogenetic analysis based on nucleotide sequence of a marker linked to the brittle rachis locus indicates a diphyletic origin of barley. Ann Bot 2007;100(5):1009–15.
[24] Newton AC, Flavell AJ, George TS, Leat P, Mullholland B, Ramsay L, et al. Crops that feed the world 4. Barley: a resilient crop? Strengths and weaknesses in the context of food security. Food Secur 2011;3(2):141.
[25] Peng JH, Sun D, Nevo E. Domestication evolution, genetics and genomics in wheat. Mol Breed 2011;28(3):281.
[26] Nevo E, Beiles A, Kaplan D, Golenberg EM, Olsvig–Whittaker L, Naveh Z. Natural selection of allozyme polymorphisms: a microsite test revealing ecological genetic differentiation in wild barley. Evolution 1986;40(1):13–20.
[27] Fu Y-B, Horbach C. Genetic diversity in a core subset of wild barley germplasm. Diversity 2012;4(2):239–57.
[28] Nevo E, Beiles A. Genetic variation in nature. Scolarpedia 2011;6(7):8821.
[29] Wang A, Yu Z, Ding Y. Genetic diversity analysis of wild close relatives of barley from Tibet and the Middle East by ISSR and SSR markers. Comptes Rendus Biol 2009;332(4):393–403.
[30] Orabi J, Jahoor A, Backes G. Genetic diversity and population structure of wild and cultivated barley from West Asia and North Africa. Plant Breed 2009;128(6):606–14.
[31] Yin Y, Ma D, Ding YJT, Genetics A. Analysis of genetic diversity of hordein in wild close relatives of barley from Tibet. Theor Appl Genet 2003;107(5):837–42.
[32] Dai H, Shan W, Zhao J, Zhang G, Li C, Wu F, et al. Difference in response to aluminum stress among Tibetan wild barley genotypes. J Plant Nutr Soil Sci 2011;174(6):952–60.
[33] Gong X, Westcott S, Li C, Yan G, Lance R, Sun D. Comparative analysis of genetic diversity between Qinghai-Tibetan wild and Chinese landrace barley. Genome 2009;52(10):849–61.
[34] Sun D, Gong X. Barley germplasm and utilization. Genetics and improvement of barley malt qualitySpringer; 2009. p. 18–62.
[35] Nevo E, Chen G. Drought and salt tolerances in wild relatives for wheat and barley improvement. Plant Cell Environ 2010;33(4):670–85.
[36] Turpeinen T, Vanhala T, Nevo E, Nissilä EJT, Genetics A. AFLP genetic polymorphism in wild barley (*Hordeum spontaneum*) populations in Israel. Theor Appl Genet 2003;106(7):1333–9.
[37] Baek H, Beharav A, Nevo EJT, Genetics A. Ecological-genomic diversity of microsatellites in wild barley, *Hordeum spontaneum*, populations in Jordan. Theor Appl Genet 2003;106(3):397–410.
[38] Cronin JK, Bundock PC, Henry RJ, Nevo E. Adaptive climatic molecular evolution in wild barley at the Isa defense locus. Proc Natl Acad Sci USA 2007;104(8):2773–8.

[39] Hübner S, Höffken M, Oren E, Haseneyer G, Stein N, Graner A, et al. Strong correlation of wild barley (*Hordeum spontaneum*) population structure with temperature and precipitation variation. Mol Ecol 2009;18(7):1523–36.

[40] Chen G, Li C, Shi Y, Nevo E. Wild barley, *Hordeum spontaneum*, a genetic resource for crop improvement in cold and arid regions. Science Cold Arid Region 2008;1:0115–24.

[41] Zhang F, Chen G, Huang Q, Orion O, Krugman T, Fahima T, et al. Genetic basis of barley caryopsis dormancy and seedling desiccation tolerance at the germination stage. Theor Appl Genet 2005;110(3):445–53.

[42] Moseman J, Nevo E, Zohary D. Resistance of *Hordeum spontaneum* collected in Israel to infection with *Erysiphe graminis hordei*. Crop Sci 1983;23(6):1115–9.

[43] Yun S, Gyenis L, Bossolini E, Hayes P, Matus I, Smith K, et al. Validation of quantitative trait loci for multiple disease resistance in barley using advanced backcross lines developed with a wild barley. Crop Breed Genet 2006;46(3):1179–86.

[44] Fetch T Jr, Steffenson BJ, Nevo E. Diversity and sources of multiple disease resistance in *Hordeum spontaneum*. Am Phytopathol Soc 2003;87(12):1439–48.

[45] Czembor HJ, Czembor JH. Resistance to powdery mildew in barley cultivars and breeding lines included in 1998-2000 Polish registration trials. Plant Breed Seed Sci 2001;45(1):21–41.

[46] Jørgensen JH, Wolfe M. Genetics of powdery mildew resistance in barley. Crit Rev Plant Sci 1994;13(1):97–119.

[47] Freialdenhoven A, Scherag B, Hollricher K, Collinge DB, Thordal-Christensen H, Schulze-Lefert P. Nar-1 and Nar-2, two loci required for Mla12-specified race-specific resistance to powdery mildew in barley. Plant Cell 1994;6(7):983–94.

[48] Kinizios S, Jahoor A, Fischbeck G. Powdery-mildew-resistance genes Mla29 and Mla32 in *H. spontaneum* derived winter-barley lines. Plant Breed 1995;114(3):265–6.

[49] Corke H, Atsmon D. Wild barley (*Hordeum spontaneum* Koch) and its potential utilization in barley protein improvement. Israel J Bot 1990;39(3):271–86.

[50] Rubiales D, Brown J, Martín A. Hordeum chilense resistance to powdery mildew and its potential use in cereal breeding. Euphytica 1993;67(3):215–20.

[51] Schönfeld M, Ragni A, Fischbeck G, Jahoor A. RFLP mapping of three new loci for resistance genes to powdery mildew (*Erysiphe graminis* f. sp. *hordei*) in barley. Theor Appl Genet 1996;93(1–2):48–56.

[52] Genger R, Williams K, Raman H, Read B, Wallwork H, Burdon J, et al. Leaf scald resistance genes in *Hordeum vulgare* and *Hordeum vulgare* ssp. spontaneum: parallels between cultivated and wild barley. Aust J Agric Res 2003;54(12):1335–42.

[53] Genger R, Nesbitt K, Brown A, Abbott D, Burdon J. A novel barley scald resistance gene: genetic mapping of the Rrs15 scald resistance gene derived from wild barley, *Hordeum vulgare* ssp. *spontaneum*. Plant Breed 2005;124(2):137–41.

[54] Abbott D, Brown A, Burdon J. Genes for scald resistance from wild barley (*Hordeum vulgare* ssp *spontaneum*) and their linkage to isozyme markers. Euphytica 1991;61(3):225–31.

[55] Von Korff M, Wang H, Leon J, Pillen K. AB-QTL analysis in spring barley. I. Detection of resistance genes against powdery mildew, leaf rust and scald introgressed from wild barley. Theor Appl Genet 2005;111(3):583–90.

[56] Kicherer S, Backes G, Walther U, Jahoor A. Localising QTLs for leaf rust resistance and agronomic traits in barley (*Hordeum vulgare* L). Theor Appl Genet 2000;100(6):881–8.

[57] Borovkova I, Steffenson B, Jin Y, Kilian A, Kleinhofs A, Blake T. Identification and mapping of a leaf rust resistance gene in barley line Q21861. Genome 1997;40(2):236–41.

[58] Moseman J, Nevo E, El-Morshidy M. Reactions of *Hordeum spontaneum* to infection with two cultures of *Puccinia hordei* from Israel and United States. Euphytica 1990;49(2):169–75.

[59] Schooler A, Franckowiak J. Registration of North Dakota 497 and 586 barley germplasm (Reg. No. GP53 and GP54). Crop Sci 1981;21(2):352.
[60] Fitzgerald TL, Shapter FM, McDonald S, Waters DL, Chivers IH, Drenth A, et al. Genome diversity in wild grasses under environmental stress. Proc Natl Acad Sci U S A 2011;108(52):21140–5.
[61] Ordon F, Schiemann A, Friedt W. Assessment of the genetic relatedness of barley accessions (*Hordeum vulgare* s.l.) resistant to soil-borne mosaic-inducing viruses (BaMMV, BaYMV, BaYMV-2) using RAPDs. Theor Appl Genet 1997;94(3–4):325–30.
[62] Teulat B, Merah O, Sirault X, Borries C, Waugh R, This D, et al. QTLs for grain carbon isotope discrimination in field-grown barley. Theor Appl Genet 2002;106(1):118–26.
[63] Choi D-W, Close TJT, Genetics A. A newly identified barley gene, Dhn12, encoding a YSK2 DHN, is located on chromosome 6H and has embryo-specific expression. Theor Appl Genet 2000;100(8):1274–8.
[64] Suprunova T, Krugman T, Fahima T, Chen G, Shams I, Korol A, et al. Differential expression of dehydrin genes in wild barley, *Hordeum spontaneum*, associated with resistance to water deficit. Plant Cell Environ 2004;27(10):1297–308.
[65] Choi D-W, Zhu B, Close T. The barley (*Hordeum vulgare* L.) dehydrin multigene family: sequences, allele types, chromosome assignments, and expression characteristics of 11 Dhn genes of cv Dicktoo. Theor Appl Genet 1999;98(8):1234–47.
[66] Morrell PL, Lundy KE, Clegg M. Distinct geographic patterns of genetic diversity are maintained in wild barley (*Hordeum vulgare* ssp. *spontaneum*) despite migration. Proc Natl Acad Sci USA 2003;100(19):10812–7.
[67] Yang Z, Zhang T, Li G, Nevo E. Adaptive microclimatic evolution of the dehydrin 6 gene in wild barley at "Evolution Canyon", Israel. Genetica 2011;139(11–12):1429–38.
[68] Suprunova T, Krugman T, Distelfeld A, Fahima T, Nevo E, Korol A. Identification of a novel gene (Hsdr4) involved in water-stress tolerance in wild barley. Plant Mol Biol 2007;64(1–2):17–34.
[69] Battisti DS, Naylor RL. Historical warnings of future food insecurity with unprecedented seasonal heat. Science 2009;323(5911):240–4.
[70] Porporato A, Laio F, Ridolfi L, Caylor KK, Rodriguez-Iturbe I. Soil moisture and plant stress dynamics along the Kalahari precipitation gradient. J Geophys Res Atmos 2003;108(D3):4127.
[71] Giorno F, Wolters-Arts M, Mariani C, Rieu I. Ensuring reproduction at high temperatures: the heat stress response during anther and pollen development. Plants (Basel) 2013;2(3):489–506.
[72] Thormann I, Reeves P, Thumm S, Reilley A, Engels JM, Biradar CM, et al. Genotypic and phenotypic changes in wild barley (*Hordeum vulgare* subsp. spontaneum) during a period of climate change in Jordan. Genet Resour Crop Evol 2017;64(6):1295–312.
[73] Arzani A, Ashraf M. Smart engineering of genetic resources for enhanced salinity tolerance in crop plants. Crit Rev Plant Sci 2016;35(3):146–89.
[74] Wang N, Zhao J, He X, Sun H, Zhang G, Wu F. Comparative proteomic analysis of drought tolerance in the two contrasting Tibetan wild genotypes and cultivated genotype. BMC Genom 2015;16(1):432.
[75] Bahrami F, Arzani A, Rahimmalek M. Photosynthetic and yield performance of wild barley (*Hordeum vulgare* ssp. *spontaneum*) under terminal heat stress. Photosynthetica 2019;57(1):9–17.
[76] Dai H, Zhao J, Ahmed IM, Cao F, Chen Z-H, Zhang G, et al. Differences in physiological features associated with aluminum tolerance in Tibetan wild and cultivated barleys. Plant Physiol Biochem 2014;75:36–44.
[77] Cai S, Wu D, Jabeen Z, Huang Y, Huang Y, Zhang G. Genome-wide association analysis of aluminum tolerance in cultivated and Tibetan wild barley. PLoS One 2013;8(7):e69776.

[78] Qiu L, Wu D, Ali S, Cai S, Dai F, Jin X, et al. Evaluation of salinity tolerance and analysis of allelic function of HvHKT1 and HvHKT2 in Tibetan wild barley. Theor Appl Genet 2011;122(4):695–703.
[79] Wu D, Qiu L, Xu L, Ye L, Chen M, Sun D, et al. Genetic variation of HvCBF genes and their association with salinity tolerance in Tibetan annual wild barley. PLoS One 2011;6(7):e22938.
[80] Wu D, Cai S, Chen M, Ye L, Chen Z, Zhang H, et al. Tissue metabolic responses to salt stress in wild and cultivated barley. PLoS One 2013;8(1):e55431.

Chapter 7

Effects of Natural Variations on Biofortification

Amir Maqbool, Allah Bakhsh and Emre Aksoy
Niğde Ömer Halisdemir University, Faculty of Agricultural Sciences and Technologies, Department of Agricultural Genetic Engineering, Niğde, Turkey

7.1 Introduction

Global food demand is increasing due to overpopulation while its sustainability is under a tremendous risk caused by erratic global climate conditions. Plant producers and breeders focused mainly on developing crop varieties with higher yields that can withstand environmental constraints. However, the selection of high yielding varieties led to the ignorance of nutritious food production. Majority of the crop varieties available in the market at present have comparatively less nutrition in their edible parts as compared to the varieties available 50 years ago [1,2], since there is a negative correlation between the grain size and yield with nutritional quality [3,4]. Concentrations of various minerals have declined over time, and/or, in high-yielding genotypes due to the genetic [1] and/or environmental factors [5]. Enhancing the mineral nutrient content of plant edible parts is known as biofortification and there are two types of it: agronomic and genetic biofortification.

Agronomic biofortification has been used for years in the field to enhance the macro and micronutrient levels of plants with the aid of fertilization. It worked well in terms of yield and mineral content enhancement of different crops; however, there are natural limiting factors, such as mineral uptake and translocation machinery, which restrain our efforts to increase the grain mineral content above a threshold level. Besides these limiting factors, the over-utilization of synthetic fertilizers contaminates the soil and groundwater resources by soil leaching. Therefore, uncontrolled fertilization usage does not help biofortification of crops but on the contrary leads to metal toxicity with unprecedented results in the environment, wildlife, and human population.

Recently, plant breeders started to develop new varieties with higher nutritional qualities due to an increase in the public awareness of malnutrition. To this end, the interest in breeding crops with high yields, better root systems, higher mineral accumulation, and lower levels of anti-nutrients has been developed in

Wild Germplasm for Genetic Improvement in Crop Plants. http://dx.doi.org/10.1016/B978-0-12-822137-2.00007-2

recent years. Among others, this strategy showed a high potential to become the most sustainable, cost-effective, and easy to implement in developing countries [6–11].

As the genotypic variability of today's crops is low (due to founder effect) [12], the existence of sizeable genetic diversity should be investigated in the natural gene pool of wild relatives and accessions. As far as the biofortification is concerned, genetic variation reflects the upper limit of mineral concentration that could be achieved. Mineral concentrations in edible parts of crops as well as their relatives demonstrated a high level of natural genetic variation [6], indicating their incorporation into the breeding programs to enhance the mineral contents of today's crop varieties. Therefore, several studies have been performed in various crop species to determine the genetic variability in mineral contents. Biofortification of staple food crops with micronutrients and vitamins is one of the meticulous challenges for agricultural scientists at present and it will be more and more valuable in the coming years. This chapter compiles some of the phenotypic and genetic studies performed in the model plant, *Arabidopsis thaliana*, and different crop species to determine the natural variation in mineral contents, genetic factors controlling the variation, their incorporation into breeding programs for biofortification, and their dissemination in the world.

7.2 Biofortification

Enhancing crop plants with vitamin and mineral contents and increasing their bioavailability via conventional (plant breeding and agronomic practices) and non-conventional (transgenic approaches) can be defined as biofortification [13,14].

Our crop breeding system has canalized all the efforts to enhance the grain yield and produce more crops to meet the demand of the ever-increasing human population. However, many staple crop species that are used in present agriculture produce grains with micronutrient deficiencies since there is a negative correlation between the grain size and yield with nutritional quality. This is one of the major reasons that lead to human malnutrition. According to the literature, human beings need 49 nutrients to carry out normal metabolic activities. The deficiency of even one nutrient can cause retarded growth, sickness, and impaired growth in children [15]. Recently, Shukla and Mishra [16] reported that more than half of the world population is suffering from micronutrient malnutrition. The deficiency of nutrients and vitamins can be made up of the intake of different food products comprising fruits, vegetables, cereals, and pulses that are an integral part of the human diet. Owing to the prevailing situation of malnutrition in the world, the need of the hour is to produce biofortified staple crops. Thus, the biofortification of staple food crops demands careful and diligent effort from the researchers worldwide. The scientific developments are already in transition from producing more quantity of crops to producing nutrient-rich crops in sufficient quantities. The approaches to biofortify food crops can be classified into agronomic and genetic biofortification. Genetic biofortification is further achieved through either conventional breeding or transgenic approaches.

7.3 Agronomic biofortification

Agronomic biofortification is the intended use of mineral fertilizers to increase the concentration of any mineral in an edible part of the crop to enhance its dietary intake [17]. In other words, agronomic biofortification involves the physical application of fertilizers and other agronomic practices to ameliorate the nutrient acquisition and storage in crop plants [18]. By using these strategies, various staple food crops including wheat, maize, rice, barley, potato, pulses, and tomato have been biofortified with different micronutrients [19]. The synthetic fertilizers are applied to crops to enhance the availability of macronutrients such as nitrogen, phosphorous, and potassium in plants, which in turn lead to increase the crop productivity. Besides, plants also adsorb micronutrients (iron, zinc, copper, manganese, etc.) from the soil and accumulate in edible plant parts [20]. Plants uptake specific form of these nutrients from soil, however different agro-ecological zones have different soil conditions (high or low pH, salt accumulation, different oxidative form of nutrients, and their complexes with other substances), which create difficulties in the proper nutrient uptake by plants. In this regard, micronutrient fertilization through soil or by foliar means can contribute to the diminution in micronutrient deficiency problems in human beings by increasing nutrient uptake and accumulation in the plant's edible parts. Besides synthetic fertilizers, application of different beneficial microorganisms and fungi (such as Rhizobium, Bacillus, Azotobacter, Pseudomonas, and arbuscular mycorrhizal fungi—AMF) have been tested to demonstrate their efficient usage in the elimination of mineral deficiency symptoms in plants by increasing the phytoavailability of micronutrients and enhancement of micronutrient accumulation in edible parts of the crops [21].

7.4 Genetic biofortification

To enhance the essential micronutrient levels in crop grains, crop breeding, or transgenic approaches can be used. Breeding is considered one of the effective ways to produce biofortified crops with enriched nutrients and vitamins. For biofortification via conventional has been achieved in crops when genetic diversity is available in primary, secondary, and tertiary gene pool of targeted crops [19]. The process involves the screening of a gene pool for high nutritional value or mineral accumulation and then crossing the determined parent lines with high mineral levels to the recipient high yielding, stress-tolerant cultivars over generations to get the new cultivars with desired nutritional, and other agronomical characteristics. Finally, the stability of mineral levels in the edible parts of the crop should be tested in target environmental conditions, and seeds should be distributed to the farmers.

Genetic diversity of the main tool for plant breeders is to create better cultivars with improved nutritional composition. However, in the case of the absence of genetic diversity, the transgenic approach is considered as a better option to accomplish the task [19]. In the case of transgenic, enrichment of essential micronutrients in crop grains can be done by incorporation of genes involved in mineral uptake, translocation, distribution, and sequestration. Originated from different organisms, such as bacteria, archaea, or other plants, or the host plant itself, the genes involved in nutrient bioavailability, inhibition of nutrient-limiting factors, deposition of micronutrients in edible plant organs, re-engineering of biological pathways can also be done easily via genetic engineering and genome editing. Various crop species have been genetically engineered to enrich micronutrient [22], vitamin, essential amino acid, and fatty acid levels in edible parts by overexpression or suppression of at least one gene involved in mineral metabolism. The transgenic-based approach has advantages that a useful gene once discovered, can be utilized for targeting multiple crops.

7.5 Importance of natural variation in the development of new traits

Natural variation refers to the genetic diversity of a single plant species in the wild. Natural variations are valuable sources and powerful tools for a plant breeder to introduce or introgress beneficial traits in plant cultivar via hybridization. Humans have a long history of utilizing natural genetic variation in plants; one salient example is during the domestication of crops from wild relatives [23]. The domestication of modern bread wheat via two successive interspecies hybridizations resulted in a complex hexaploid genome [24].

The plant breeders exploit the natural variation available in the existing germplasm of a plant species. The backgrounds of interest are typically crossed to the domesticated strain via either intra- or interspecific hybridization. However, the inheritance of traits of interest is another difficult task once complex traits are introduced. Until now, many traits line plant architecture, fruit, and seed structure and morphology, nutritional composition, and other yield and quality traits have been introgressed in crop plants [25,26]. Further investigations of natural genetic polymorphism are helping researchers to understand how plant species adapt to changing environments. The new sequencing technologies play a pivotal role to explore plant variations at greater depth and accuracy. This is important for species with large and repetitive genomes and will help to obtain near-complete descriptions of genetic variations in natural populations. Owing to current global climate changes, a thorough understanding of natural variations and their role in plant adaption will of vital importance [27].

Some of the phenotypic differences (monogenic) observed in wild or cultivated plants occur by variations in a single gene. However, most of the natural traits are quantitative and controlled by molecular polymorphisms at multiple loci and genes (multigenic), which are called quantitative trait loci (QTL) and quantitative trait genes (QTGs). Studies in different domesticated crop cultivars, their relatives, and recombinant inbred lines (RILs) derived from two parents led to the identification of the genes and genomic regions underlying QTLs involved in yield, stress tolerance, and quality development [28,29]. QTLs were generally identified by analyzing the segregation of traits with markers in bi-parental populations. The success of this method is limited in a bi-parental population with a very limited recombinational history.

With recent advancements in whole genome sequencing and precise mass-phenotyping platforms, correlating genotypic and phenotypic variability in genome-wide association studies (GWAS) has become the main method for systematic identification of candidate loci in the genome that is contributory to the phenotypic differences [30]. One of the major rationales for using GWAS is the identification of novel functional variation that can be implemented in cultivar improvement through marker-assisted selection (MAS). However, the success of GWAS depends on genetic diversity. The broader the gene pool, the more variation can be discovered.

7.6 The exploitation of natural variations in *Arabidopsis* gene pool

Thale cress (*A. thaliana*) has been used as the model plant in modern plant studies for the last 6 decades due to its various advantages [31]. After its genome was sequenced in 2000 [32], many bioinformatic resources were developed and readily available on the internet [33]. They include TAIR (The Arabidopsis Information Resource) (arabidopsis.org, [34]), Araport (Arabidopsis Information Portal) (araport.org [35]), Virtual Plant (virtualplant.bio.nyu.edu, [36]), Bio-Analytic Resource (BAR) for Plant Biology (bar.toronto.ca [37]), ePlant (bar.toronto.ca/eplant [37]), eFP Browser (bar.utoronto.ca/efp/cgi-bin/efpWeb.cgi), and Genevestigator (genevestigator.org, [38]).

Due to its relatively smaller genome (135 Mb), and the utility of simple methods of genetic transformation and molecular marker development, functional characterization of thousands of genes have been completed in *Arabidopsis*. Forward genetic screenings were performed in the mutagenized population of *Arabidopsis* via radiation or chemicals to determine the function of genes. However, these screenings were tedious, and identification of the gene linked with the phenotype was laborious. To this end, the *Arabidopsis* research has made the use of plants harboring disruptive mutations in these genes. Some collections of *Arabidopsis* mutants that harbor Agrobacterial plasmid T-DNAs of known sequences were generated in the model *Arabidopsis* accession (ecotype) Colombia-0 (Col-0) background [39–43]. These lines, namely the SALK, SAIL, GABI-KAT, and WISC, contain in total over 260,000 individual mutants representing at least one insertion mutation in most *Arabidopsis* genes [44]. In addition to T-DNA insertion lines in Col-0 ecotype, other available mutant lines include a FLAG-tag collection in Wassilewskija (WS) [45] and transposon-insertion collections in the Landsberg erecta and Nössen ecotypes known as CSHL and RIKEN collections, respectively [46,47]. As Col-0 is the accession that was sequenced in 2000, it has been used as the model accession of *Arabidopsis* for decades. All these mutant lines are readily available through the Arabidopsis Biological Resource Center (ABRC) at Ohio State University (https://abrc.osu.edu), the Nottingham Arabidopsis Stock Centre (http://arabidopsis.info/), and RIKEN BioResource Research Center (BRC) (https://epd.brc.riken.jp/en/seed).

In nature, *Arabidopsis* is an inbred plant since its self-pollination capacity is high. Therefore, it persist its nearly complete homozygous genome for generations [48]. This property enables the study of diverse phenotypes of genetically identical plants under different environmental conditions. However, *Arabidopsis* outcross on average only about once in 10–100 generations [49], suggesting it as the model plant to study the dynamics of population diversity and evolution under different environmental conditions.

A. thaliana is an annual herbaceous plant that exists in patchy populations in the wild distributed mainly in Eurasia and the north of Africa. It is self-compatible and wild populations are highly homozygous with average outcrossing rates in the range of 0.3–2.5% [50–52]. Over the years, tens of studies have generated a large pool of phenotypic, genotypic, and molecular data across hundreds of different ecotypes, which can be used in studying the reasons for natural variation at the system level [53]. *Arabidopsis* possesses a large collection of naturally inbred lines (accessions) that are products of natural selection under diverse ecological conditions [54]. GWAS in *Arabidopsis* started with 96 accessions to determine the linkage disequilibrium and population structure in the gene pool [55]. Then, single nucleotide polymorphisms (SNPs) were determined in 20 accessions [56], and a 250 k SNP array was developed and deployed to genotype the RegMap collection of 1,307 accessions collected from different parts of the world [57].

To understand the natural distribution, adaptation, and the genes regulating the natural variations between *Arabidopsis* accessions, full genome sequences for 1,135 naturally inbred *Arabidopsis* accessions were determined in 1001 Genomes Project [25]. These natural *Arabidopsis* accessions represented a well-studied population history distributed all over the world under different climatic and environmental conditions [58,59]. This platform provides the densest variant map for any organism, with 1 genetic variant every 10 base pairs. Using the publicly available genomic data by 1001 Genomes Project (1001genomes.org), environment × genome associations and correlations between the local environments and phenotypes were determined in different studies to understand the genes and gene networks involved in natural distribution, adaptation, and evolution of *Arabidopsis* accessions and species [60–63]. For instance, epigenome [64], methylome [65], and mobilome (transposable element movements—[66] changes were investigated via GWAS in *Arabidopsis* accessions.

In recent years, a comparison of *A. thaliana* with its close relatives was used to understand the molecular variations contributing to tolerance mechanisms since different species in the genus have adapted to different environments and therefore evolved diverse mechanisms to cope with these challenging conditions. In one of these studies, drought tolerance was evaluated in three species, namely *A. thaliana*, *A. halleri*, and *A. lyrate*, and comparison of transcriptome identified previously unknown genes potentially demonstrate the natural divergence of drought tolerance mechanism in the same genus [67]. Natural genetic variation among accessions and species was also used in the characterization of root system architectural modifications under nutritional deficiencies [68].

New bioinformatic tools were developed to analyze the genome-wide association data in *Arabidopsis* to discover new genes associated with the investigated phenotypes. One of them is CLIMtools (github.com/CLIMtools), which is an interactive web-based database of the environment × genome associations and correlations between the local environments [62]. This study utilized the SNP data generated by the 1001 project and incorporated it in CLIMtools to determine the genes involved in environmental responses in a panel of *Arabidopsis* accessions. AraQTL (http://www.bioinformatics.nl/AraQTL/) is a database for comparative analysis and meta-analysis of all published *Arabidopsis* expression quantitative trait loci (eQTL) datasets [69]. AraPheno (https://arapheno.1001genomes.org/) was developed as a central database to gather and arrange phenotypes of *Arabidopsis* accessions [70]. Also, AraGWAS Catalog (https://aragwas.1001genomes.org/) was developed as the main reserve of genetic associations for all existing phenotypes from AraPheno [71], and it is like the NHGRI-EBI GWAS Catalog [72], which brings 5,687 GWAS together comprising 71,673 variant-trait associations from 3,567 publications.

7.7 Genetic diversity in *Arabidopsis* regarding micronutrient quantity

Arabidopsis shows an abundant natural variation in its ionome. The variation of essential element concentrations in the shoot ionome of *Arabidopsis* accessions is estimated to be more than fourfolds while it is over fivefolds in the seed [73].

With the advances in next-generation sequencing, GWAS of *Arabidopsis* accessions has become a hot topic under various environmental stress conditions in recent years. As expected, these studies also have extended in the field of mineral deficiency and biofortification. In one of the earliest studies, natural variation in shoot molybdenum content across 92 *Arabidopsis* accessions was shown to be regulated by a variation in *MOLYBDENUM TRANSPORTER1* (*MOT1*) belonging to the sulfate transporter family [74]. In a study, a large natural variation in micronutrient seed content was identified in a pool of 108 accessions, and several of them maintained high seed zinc (Zn) levels even when grown under Zn deficiency [75]. In another study, physiological and molecular parameters were examined in 19 *Arabidopsis* accessions under Zn deficiency

[76]. Accessions with a higher tolerance to Zn deficiency showed an increased expression of *ZINC TRANSPORTER4* (*ZIP4*) and *IRON-REGULATED TRANSPORTER3* (*IRT3*). Interestingly, shoot copper (Cu) concentration was able to distinguish plants regarding their Zn nutritional status in addition to shoot Zn concentration. Calcium (Ca) concentration was also shown to be used in a grouping of accessions according to their Zn deficiency tolerance. In a separate study, the natural variation in leaf Zn content was investigated in a global collection of 349 *Arabidopsis* accessions [77]. The researchers determined two accessions, namely Van-0 and Fab-2, as accumulating lower levels of Zn compared to Col-0. QTL and bulked segregant analysis (BSA) identified *HEAVY METAL-ATPASE4* (*HMA4*) as a strong candidate accounting for the phenotypic variation in leaf Zn concentration.

As the roots are the primary organ required for mineral uptake from the rhizosphere, their morphological and physiological responses to mineral deficiencies are very essential to determine the natural variations in plant responses. The significance of the hidden half of the plants has recently been recognized attention, which is reflected by a huge increase in GWAS publications about root system architectural dynamics under environmental stress conditions, including mineral deficiencies and toxicities. In an early study, natural variation in root system dynamics has been evaluated under potassium (K) deficiency in a panel of 24 accessions, and QTLs linked with tolerance were determined in a RIL population generated by Col-0 x Ct-1 cross [78]. Another study used the natural variation of early root development under iron (Fe) deficiency among 134 natural *Arabidopsis* accessions from Sweden to show a significant deviation for this trait [79]. Applying genome-wide association (GWA) mapping the scientists discovered *FERRIC REDUCTION OXIDASE2* (*FRO2*) locus as the major factor of root response to Fe deficiency. A recent study investigated the root growth rate of 227 genetically diverse natural accessions from the RegMap population under single or combinatory iron (Fe), phosphorous (P), and Zn deficiencies, and identified novel genes that can regulate the mineral metabolism in GWAS [80]. In a similar study, modulation of root elongation under mild nitrogen (N) deficiency was determined to be controlled by the brassinosteroid (BR) signaling kinase BSK3 in a GWAS mapping of a diverse panel of 200 *Arabidopsis* accessions [81]. In a physiological study, hormone levels were correlated with the root system architectural dynamics in 13 accessions [82], bringing a new dimension to the natural variation of plant nutrients and regulation of mineral uptake and distribution in plants.

7.8 Genetic diversity in crop species regarding micronutrient quantity

Natural variations in the germplasm of different crop species are the key constituent to develop biofortified crops via breeding. For effective biofortification, there must be a wide range difference between the accessions, cultivars, or wild relatives of any crop species regarding (macro- and micro-) nutrient concentrations, especially in the edible parts such as grains or seeds. In the last 2 decades, extensive studies have been done on several crop species to determine the genetic diversity related to nutrient concentrations in their edible parts. As a result, a wide range of diversity is found in the germplasm of several crop plants, suggesting an urgent requirement for the detailed investigation of genetic diversity within natural populations and their association with the phenotypes in future breeding programs.

7.8.1 Wheat

Wheat (*Triticum aestivum* L.) is a major crop constituting around 30% of the total cereals consumed in many parts of the world [83]. Despite a significant increase in wheat production in the last decades, a significant proportion of the world population which depend on wheat as a staple crop suffer from micronutrient deficiencies due to the low grain mineral concentrations [15]. Studies with a wide range of wheat germplasm demonstrated a genetic variation of grain mineral content in 2–3 orders of magnitude [84]. Present wheat cultivars accumulate limited levels of minerals in the grain [85] and demonstrate a narrow genetic variation for grain minerals compared to the wild species, landraces, and lines from pre-breeding program [86,87]. Studies also demonstrated a negative correlation between grain Fe and Zn levels and yields, suggesting lower mineral accumulation in high yield wheat cultivars [88]. This correlation was under the environmental influence.

The International Maize and Wheat Improvement Center (CIMMYT) together with the Consultative Group on International Agricultural Research (CGIAR) and HarvestPlus initiated a research program to document the genetic variability among wild relatives of wheat and other species in the genus *Triticum* from the stand point of grain mineral concentrations. This effort will select genotypes with higher grain mineral levels that can be integrated in future wheat breeding programs. As an example, developed by the cross between *T. turgidum ssp. durum* L. and *Aegilops tauschii* (Coss), synthetic hexaploid wheat (SHW) was used to introduce novel genes/genetic variations into bread wheat [89,90] and it can be a potential source of high grain mineral concentrations. A GWAS of a panel of 123 SHW genotypes identified 13 SHWs with the highest levels of essential minerals accumulating in the grain [91]. Moreover, they also identified new 92 genes and markers that can be used in future breeding programs for the purpose of biofortification. A recent study of 353 wheat varieties from European countries demonstrated a wide range of genetic variation between cultivars in grain Ca levels (288.20–647.50 $\mu g\ g^{-1}$ DW) with a heritability of 0.73 [92]. In the same wheat panel, grain Zn levels showed a moderate heritability of 0.54 and the values ranged between 25.05 and 52.67 $\mu g\ g^{-1}$ DW [93].

The wheat biofortification breeding program at CIMMYT highlights the advancement of new spring bread wheat genotypes that bring together the broad adaptability, high yields, enhanced grain mineral concentration, improved processing quality, disease, and stress resilience. Targeted breeding for superior nutritional quality has effectively enhanced the grain Zn levels by 30–40% in high yielding wheat cultivars [94]. These genotypes were tested at target locations (Pakistan and Northern India) in field trials led by HarvestPlus [95]. The wheat lines demonstrated a high heritability and a high genetic correlation between locations, and more importantly, many lines accumulated 28 and 25% higher levels of Zn and Fe in the grain, proving the target level of Zn (10 mg kg^{-1} in Pakistan and Northern India) was exceeded while the target level of Fe (25 mg kg^{-1}) was almost cached in trials. A second-year trial identified a few lines with an increment of 75–150% in grain Zn levels compared to the check varieties, suggesting the success of CIMMYT breeding program [96]. In 2013 season, selected varieties seeds enriched with Zn and Fe were pre-distributed to farmers in India. In 2014, these varieties were commercialized by a local seed company and distributed to over 10,000 farmers [97]. This effort demonstrated a good example of the public-private partnership (PPP) and fast-track commercialization of wheat varieties developed by CIMMYT. Two varieties, namely Abhay (ZincShakthi) and Akshai (BHU3), were widely accepted and produced by farmer in India in 2015–16 season.

7.8.2 Rice

Rice (*Oryza sativa*) is a major staple crop that provides essential dietary nutrients to about half of the global population. Unfortunately, rice grains are not enriched with the required nutrient concentrations by humans [98]. In addition to agronomic biofortification, the development of new cultivars (by exploiting genetic diversity present in the rice germplasm) with high nutrient concentrations is proposed as a solution to cope with malnutrition in human population [99]. Pinson and coworkers [100] checked the mineral concentrations of 1763 rice accessions of diverse geographic and genetic origins. As a result, a wide range of genetic diversity was found related to the nutrient concentrations of 16 elements. Most of the accessions with a high concentration of an element came from the same region giving the clue of their genetic similarity related to that trait. The concentrations of main micronutrients (Fe and Zn), which are most commonly deficient in humans, were found to be between 1.55 and 16.58 μg g^{-1} for Fe and 15.72–65.01 μg g^{-1} for Zn. However, the target set by HarvestPlus for zinc biofortification of rice is 28 ppm, which is much less than the Zn concentration present in many accessions identified by Ref. [100]. These accessions can be crossed with the cultivars having desirable traits (such as disease resistance and high yield) to develop new biofortified rice varieties. In Bangladesh, several varieties with high Zn concentration have been released up to now and most of them accumulated 92% of the target zinc level.

7.8.3 Sorghum

Sorghum is an important cereal mainly used and produced in Africa, Asia, and the Central America. The International Crops Research Institute for the Semi-Arid Tropics (ICRISAT) has screened 2200 sorghum genotypes for zinc and iron concentrations. The range of iron and zinc concentrations was found to be between 20 to 70 ppm and 13 to 47 ppm, respectively [22]. However, the South African lines accumulated up to 55 ppm of zinc in the grain [101]. Sorghum genotypes having high Zn and Fe concentrations were identified with low levels of tannin and phytate, indicating the high bioavailability of these micronutrients for the humans [102]. ICRISAT has developed new biofortified sorghum lines with 50–60% more Zn and Fe contents than the previous popular sorghum varieties.

7.8.4 Maize

Maize (*Zea mays*) is a cereal crop consumed by humans and animals and every part of it is consumed for different purposes and nothing goes as waste [9]. However, most of the maize cultivars grown in different regions of the world have low provitamin A concentration (2 $\mu g\ g^{-1}$). In developing countries, provitamin A deficiency is affecting millions of children and pregnant women leading to severe diseases [103]. Initial screening of the maize germplasm for provitamin A found the concentrations range between 0 and 19 ppm in the tested 1500 maize accessions [104–106]. Some accessions had more provitamin A accumulation than the target, which is 15 ppm. Breeding programs for provitamin A biofortification were initiated in 2007 by the joint force of three institutes, namely CIMMYT, International Institute of Tropical Agriculture (IITA), and Zambia Agriculture Research Institute (ZARI). Until now, more than 40 varieties have been released in many African countries with high levels of grain Zn. Cultivars released in 2012–13 had 6–8 ppm of additional provitamin A and the cultivars released in the year of 2015–16 showed 10 ppm additional provitamin A than the already existing local cultivars [107]. CIMMYT and IITA are also working on white maize for Zn biofortification. In the white maize germplasm, grain Zn concentration ranged between 17 and 42 ppm. At present, many varieties have been released by CIMMYT with the support of HarvestPlus. The latest Zn-biofortified variety released in Colombia is known as Bio-Mzn 01. This Bio-Mzn 01 has 36% more Zn than the other maize varieties while its yield is 6–8 $t\ ha^{-1}$ (tons per hectare), which is double the average yield in Colombia [108].

7.8.5 Potato

Improving the nutritional value of different crop species has become one of the major goals of current crop breeders. Potato (*Solanum tuberosum* L.) is an important part of the global food system. Regardless of other crop species potatoes are produced vegetatively from the other potatoes and its production and quality depend upon the previously harvested potatoes. Therefore, potato quality regarding nutrients and other perspectives is an important issue of the current scenario [109]. As we know, micronutrient deficiency in the major staple crops is an important problem related to the poor nutrient ingestion by humans and animals. Consumption of 150 g of potato may contribute to 6% of the daily required iron by humans [110]. Mainly potato is the modest source of the micronutrients, but a wide range of genetic variability related to micronutrient concentration is present among different genotypes of potato. The genetic variability in potato germplasm is the major tool used by the plant breeders to develop biofortified potatoes [111]. In a recent study conducted by Singh and colleagues [111], micronutrient concentration analysis was done in 37 (*S. tuberosum* ssp. *andigena*) accessions. As a result, significant variations were observed in the concentrations of iron, manganese (Mn), copper, and zinc. The concentration of Fe was varied between 18.03 and 45.97 mg kg^{-1}, Mn from 7.27 to 29.80 mg kg^{-1}, Cu from 4.43 to 22.47 mg kg^{-1}, Zn from 12.33 to 33.87 mg kg^{-1} of dry weight (DW). Furthermore, Andre and coworkers [112] reported genetic variability in the micronutrient concentrations of 74 tested Andean genotypes. Concentrations of Fe ranged between 29.87 and 157.96 μg g^{-1}, Ca from 271.09 to 1092.93 μg g^{-1}, and Zn from 12.6 to 28.83 μg g^{-1} of DW. These identified *Andigena* and Andean potato accessions having naturally high nutrient accumulation can be used as a source for the potato biofortification in future breeding studies.

International Potato Center (CIP) in Peru started the breeding program for zinc and iron biofortification. The initial screening of potato germplasm accessions found a wide range of genetic diversity in the potato germplasm with the values ranged between 8–25 ppm for zinc and 11–30 ppm for the iron [113,114]. Heritability of the accessions with high micronutrients was also found high and there was no negative correlation between the micronutrient concentrations and other important traits [115]. There is a significant genetic variation in macro and micronutrient contents in the tubers of *Solanum* species and wild relatives [6]. A large genetic variation was reported for tuber mineral concentrations in a gene pool comprised of tetraploid and diploid *Solanum* clones [116]. The broad-sense heritability was calculated as 84%, 82%, 65%, and 49% for Mn, Zn, Fe, and Cu, respectively.

In the last few years, CIP has initiated another project in which diploid Andean potato genotypes found to have high zinc and iron concentrations were crossed with disease-resistant tetraploids potato clones (developed by CIP). As a result, the new biofortified potatoes developed by the CIP have tubers with almost 60–80% higher zinc and iron contents than the local ones [117]. Biofortified potatoes were taken into adaptation studies in two target countries Rwanda and Ethiopia. These advanced lines accumulated up to 60% higher zinc and iron content in their tubers than the local cultivars, and they were applied for official release in Rwanda in 2017.

7.8.6 Sweet potato

Sweet potato (*Ipomœa batatas* L.) is a major food commodity in Africa where zinc and iron deficiencies are the eminent health problems. Like other crop species, a significant genetic diversity related to micronutrient quantity is present among the various genotypes of the sweet potato. Chipungu and coworkers [118] have evaluated micronutrient concentrations present in 15 different cultivars of sweet potato. A significant difference was found among the tested genotypes. Among 15 cultivars, the concentrations of β-carotene (precursor of vitamin A) and ascorbic acid (vitamin C) ranged between 0.19–10.9 mg $(100\ g)^{-1}$ and 13.3523.84 mg $(100\ g)^{-1}$ fresh weight (FW). Genotypes named Zondeni and Babache had the highest levels of β-carotene and ascorbic acid, that is, 10.9 and 23.84 mg $(100\ g)^{-1}$ FW, respectively. The iron content ranged between 0.09 and 0.67 mg $(100\ g)^{-1}$ DW among the 15 tested genotypes while the zinc concentration was observed in the range of 0.33–0.63 mg $(100\ g)^{-1}$ DW. Genotype named Mzgunu showed the highest levels of zinc (0.63 mg $(100\ g)^{-1}$ DW) and iron (0.67 mg $(100\ g)^{-1}$ DW). These variations in the naturally accumulated nutrient and mineral concentrations in different genotypes of sweet potato can be used as a potential source for sweet potato biofortification.

Vitamin A deficiency is a major health problem in Uganda and other African countries. Fish oil and other sources of vitamin A are expensive and unaffordable for many people in poor countries like Uganda. However, sweet potato is one of the major consumed food commodities in many African countries. Therefore, sweet potatoes can be the cheapest source of vitamin A. Orange flesh sweet potato cultivars having high levels of Vitamin A and carotenoids developed by the team HarvestPlus and CIP are now released in several African countries [119]. Varieties named Vita, Naspot 12O, Naspot 13O, Ejumula, Kabode, and Kakamega have been registered and released in Uganda. Three other varieties named Chiwoko, Kokota, and Twatasha are released in Zambia. Furthermore, many biofortified sweet potato varieties have been released in Latin America and in many Asian countries. According to a survey in northern Uganda, orange flesh sweet potatoes are contributing about 68% of the daily required vitamin A at the household level [120].

7.8.7 Common bean

Common bean (*Phaseolus vulgaris* L.) is a leguminous crop mainly grown in tropical areas. It is mainly cultivated in developing countries where it serves as a cheap source of protein and minerals for majority of the population. However, its quality-related to micronutrient intake is not satisfactory [121,122], when checked the iron and zinc levels in seeds of 90 local genotypes belonging to different regions of Tanzania. Results showed a wide range of genetic diversity in zinc and iron concentrations. The iron concentration ranged between 23.63and 105.5 ppm, whereas the zinc concentration was found between

19 and 56.13 ppm. The highest iron was found in the genotype Bangaya Akatebe (105.5 ppm) and the highest zinc concentration was found in the genotype named Shona Eugenia (56.13 ppm). Several studies have been present until now on the genetic diversity related to the micronutrient concentrations in the common bean.

Moreover, wild relatives can also be a source for producing biofortified common beans. Some wild and weedy common beans have more iron, zinc, and protein contents than the cultivated ones. However, in wild beans the digestibility of the protein was low which may be related to the higher concentration of tannins in them. By careful selection, these wild relatives can be used in the development of biofortified common beans having high protein, iron, and zinc contents [123]. Up to now by using these genetic differences related to the mineral concentrations of common bean, several biofortified common bean varieties have been released in many countries. HarvestPlus has released iron-biofortified beans in many African countries to deal with iron deficiency in the African population. Between 2010 and 2012, 10 varieties were released in Rwanda. They came in different colors, sizes, and other characteristics to appeal to bean producers and consumers. These iron biofortified beans have twofold more iron as compared to the local cultivars. Consumption of these iron beans showed satisfying results to deal with the iron deficiency anemia, as Rwandan women who used iron beans (compared to the unfortified beans) showed 4.2-g L^{-1} increase in hemoglobin for every 1 g Fe consumed from beans over the 128 study days [124]. HarvestPlus and its partners have distributed Fe-enriched beans to small landholders in Rwanda through formal delivery approaches, and as a result, approximately 28% of rural households in the country grew a Fe-enriched bean variety between 2012 and 2015 [125].

7.8.8 Cowpea

Cowpea (*Vigna unguiculata*) is the commonly produced legume crop mainly in Brazil, India, and savanna regions of the sub-Saharan Africa. This crop survives better than other crops in the extremely dry conditions of the region. The two main producers of the cowpea are Nigeria and the Niger Republic [126]. In these regions, the development of the nutrient-rich cowpea cultivars can be an immense benefit for the consumers to fulfill their nutrient requirements. The germplasm of cowpea is maintained by the International Institute of Tropical Agriculture (IITA) with more than 15,000 accessions collected from many countries. HarvestPlus is in the field of making nutrient-rich cowpea genotypes in the targeted countries including Nigeria, India, Brazil, and Bangladesh [117].

Until now, several studies have been done to examine the genetic variability in the cowpea germplasm related to nutrient concentrations. Ref. [126] checked the protein contents and mineral concentrations of 1541 accessions. These accessions belong to various regions of the world and the germplasm was maintained by the gene bank at IITA. A significant difference was found in the protein percentage and nutrient concentrations of the tested genotypes. The protein concentrations were found between 17.5% and 32.5%, and the micronutrient concentrations were in the range of 33.6–79.5 $mg.kg^{-1}$ for Fe, 22.1–58.0 $mg.kg^{-1}$ for Zn, and 310–1395 $mg.kg^{-1}$ for Ca. In another study, a wide range difference in the crude protein concentration (16.6–24.8%) and micronutrients (Fe—34.1–64.4 $mg.kg^{-1}$ and Zn–32.2–49.1 $mg.kg^{-1}$) were found among 240 tested genotypes [127]. Among all the tested genotypes, TVu-12802 showed the highest crude protein content and high concentrations of iron and zinc with a low ratio of the phytic acid. These genotypes having high nutritional value can serve as the potential candidates to breed biofortified cowpea varieties in poor African and Asian countries, where the majority of the population rely on the cheapest sources of the food. Iron cowpea is one of the best examples of the utilization of iron-enriched accessions developed by the HarvestPlus. Brazil is the second-largest producer of the cowpea in the world and most of the produced crop is consumed by the country itself. In Brazil, the Brazilian Agricultural Research Corporation initiated a project under the umbrella of "BioFort." Until now, the Brazilian Agricultural Research Corporation has released three biofortified cowpea varieties in the country having 40% more iron than the previous cultivars [117].

7.8.9 Lentil

Lentil (*Lens culinaris* L.) has high protein and carbohydrate content with essential dietary components for the humans [128]. Extensive research work has done on the genetic diversity related to nutrient concentrations in some lentil accessions. Mainly the Canadian-grown lentils are rich in the micronutrients having high Fe, selenium (Se), and Zn concentrations with low phytic acid content. Furthermore, we know that Se is an essential micronutrient required by humans for the functioning of enzymes involved in cellular antioxidant protection and many others. Uptake of suboptimal levels of Se increases the risk of cancer, HIV, and many other viral diseases in humans. Lentils can be a great source of Se because a wide range of genetic diversity is present in lentil germplasm related to the Se concentration with high Se levels in many accessions [129]. Se concentration is determined in the genotypes of six major growing regions of lentils from Nepal, Morocco, the United States of America, Syria, Turkey, and Australia. The highest Se concentrations were found in the lentils from Nepal (80 $\mu g\ kg^{-1}$) while the lowest was present in Syrian lentil (22 $\mu g\ kg^{-1}$) genotypes. However, all genotypes coming from six major growing regions had less Se concentrations as compared to the Se concentrations of Canadian genotypes (more than 650 $\mu g\ kg^{-1}$) [129]. The Nepalese and Canadian genotypes with high Se concentrations can be used for the breeding of biofortified lentils. Moreover, a high genetic diversity related to the zinc content is also present in lentil germplasm. Research lead by The International Center for

Agricultural Research in the Dry Areas (ICARDA) in collaboration with the national institutes of India, Nepal, and Bangladesh has released many varieties in the targeted countries (Bangladesh, Nepal, Bangladesh, Ethiopia, and Syria) with high Zn (51–64 ppm) concentrations [117].

7.9 Conclusion and future prospects

Crop breeding ignored the significance of nutritional quality development for decades since the breeders prioritize the yield improvement and governments supported mass productions to feed the world. However, hidden hunger is still one of the most important constraints of humanity in the developing world. To eradicate this notorious terminology from our knowledge, the most feasible and sustainable way is to canalize our efforts in genetic biofortification, especially breeding. Of course, a good breeding program requires parents originating from diverse genetic and/or phenotypic backgrounds. This, in turn, requires the screening of a large gene pool with a high degree of genetic variability. Overall, breeding programs for the development of cultivars with high yield and nutritional quality benefits from natural variations in mechanisms involved in nutrient uptake, translocation, and sequestration in edible parts.

The historical developments in genetic studies related to QTL and SNP identification have shown the power of natural genetic variations in discoveries of new genes linked with different environmental phenotypes. Recent advancements in next-generation sequencing, high-throughput phenotyping, and development of new bioinformatic tools that incorporate massive data generated in large-scale experiments demonstrate the potential success of future crop breeding programs to develop new high yielding cultivars with better nutritional quality.

Although this chapter mainly focused on GWAS of natural *Arabidopsis* accessions regarding their ionomic variations and root phenotypic diversifications under nutritional deficiencies, there are several GWAS done in other crop species to exploit the natural variation in ionomics or root system architecture (reviewed in Ref. [130]). Some of the GWAS examples include ionomic variations in *Brassca napus* [131], maize [132], soybean [133], rice [134,135], wheat [91–93], and peanut [136], and K uptake rate in Tibetan wild barley genotypes under K deficiency [137]. We expect to see a sharp increase in the number of GWAS studies related to biofortification in the coming decade. Hence, there is a great promise lies in studying different accessions to characterize the gene function of genetic variants by using natural variations in future studies. We expect to see the incorporation of genome editing technologies to be integrated into mutation development in natural wild relatives of crop species to evaluate the function of genes or genomic regions involved in the diversification of mineral metabolism in plants.

References

[1] Davis DR, Epp MD, Riordan HD. Changes in USDA food composition data for 43 garden crops, 1950 to 1999. J Am Coll Nutr 2004;23(6):669–82.
[2] Farnham MW, Keinath AP, Grusak MA. Mineral concentration of broccoli florets in relation to year of cultivar release. Crop Sci 2011;51(6):2721–7.
[3] Garvin DF, Welch RM, Finley JW. Historical shifts in the seed mineral micronutrient concentration of US hard red winter wheat germplasm. J Sci Food Agric 2006;86(13):2213–20.
[4] Murphy KM, Reeves PG, Jones SS. Relationship between yield and mineral nutrient concentrations in historical and modern spring wheat cultivars. Euphytica 2008;163(3):381–90.
[5] Jarrell WM, Beverly RB. The dilution effect in plant nutrition studies. Adv Agron 1981;34(1):197–224.
[6] White PJ, Broadley MR. Biofortification of crops with seven mineral elements often lacking in human diets–iron, zinc, copper, calcium, magnesium, selenium and iodine. New Phytol 2009;182(1):49–84.
[7] Gomez-Coronado F, Almeida AS, Santamaria O, Cakmak I, Poblaciones MJ. Potential of advanced breeding lines of bread-making wheat to accumulate grain minerals (Ca, Fe, Mg and Zn) and low phytates under Mediterranean conditions. J Agron Crop Sci 2019;205(3): 341–52.
[8] Rehman HM, Cooper JW, Lam HM, Yang SH. Legume biofortification is an underexploited strategy for combating hidden hunger. Plant Cell Environ 2019;42(1):52–70.
[9] Sah S, Singh N, Singh R. Iron acquisition in maize (*Zea mays* L.) using *Pseudomonas siderophore*. 3 Biotech 2017;7(2):121.
[10] Umar M, Nawaz R, Sher A, Ali A, Hussain R, Khalid MW. Current status and future perspectives of biofortification in wheat. Asian J Res Crop Sci 2019;4(4):1–4.
[11] Jha AB, Warkentin TD. Biofortification of pulse crops: status and future perspectives. Plants 2020;9(1):73.
[12] Ladizinsky G. Founder effect in crop-plant evolution. Econ Bot 1985;39(2):191–9.
[13] Hirschi KD. Nutrient biofortification of food crops. Annu Rev Nutr 2009;29:401–21.
[14] Bouis HE, Hotz C, McClafferty B, Meenakshi JV, Pfeiffer WH. Biofortification: a new tool to reduce micronutrient malnutrition. Food Nutr Bull 2011;32(1_Suppl.1):S31–40.
[15] Welch RM, Graham RD. Breeding for micronutrients in staple food crops from a human nutrition perspective. J Exp Bot 2004;55(396):353–64.
[16] Shukla UN, Mishra ML. Biofortification: Golden way to save life from micronutrient deficiency-A review. Agric Rev 2018;39(3):202–9.
[17] Adu MO, Asare PA, Yawson DO, Nyarko MA, Osei-Agyeman K. Agronomic biofortification of selected underutilised *Solanaceae* vegetables for improved dietary intake of potassium (K) in Ghana. Heliyon 2018;4(8):e00750.
[18] Cakmak I, Guilherme LR, Rashid A, Hora KH, Yazici A, Savasli E, et al. Iodine biofortification of wheat, rice and maize through fertilizer strategy. Plant Soil 2017;418(1–2):319–35.
[19] Garg M, Sharma N, Sharma S, Kapoor P, Kumar A, Chunduri V, et al. Biofortified crops generated by breeding, agronomy, and transgenic approaches are improving lives of millions of people around the world. Front Nutr 2018;5:12.
[20] Ortiz-Monasterio JI, Palacios-Rojas N, Meng E, Pixley K, Trethowan R, Pena RJ. Enhancing the mineral and vitamin content of wheat and maize through plant breeding. J Cereal Sci 2007;46(3):293–307.
[21] Prasanna R, Nain L, Rana A, Shivay YS. Biofortification with microorganisms: present status and future challenges. In: Ummed Singh U, Praharaj CS, Singh SS, Singh NP, editors. Biofortification of food crops. New Delhi: Springer; 2016. p. 249–62.

[22] Ashok Kumar A, Reddy BV, Ramaiah B, Sanjana Reddy P, Sahrawat KL, Upadhyaya HD. Genetic variability and plant character association of grain Fe and Zn in selected core collection accessions of sorghum germplasm and breeding lines. J SAT Agric Res 2009;7:1–4.

[23] Doebley JF, Gaut BS, Smith BD. The molecular genetics of crop domestication. Cell 2006;127(7):1309–21.

[24] Mayer KF, Rogers J, Dolezel J, et al. A chromosome-based draft sequence of the hexaploid bread wheat (*Triticum aestivum*) genome. Science 2014;345:1251788.

[25] Alonso-Blanco C, Aarts MG, Bentsink L, Keurentjes JJ, Reymond M, Vreugdenhil D, et al. What has natural variation taught us about plant development, physiology, and adaptation? Plant Cell 2009;21(7):1877–96.

[26] Sedeek KE, Mahas A, Mahfouz M. Plant genome engineering for targeted improvement of crop traits. Front Plant Sci 2019;10:114.

[27] Henderson IR, Salt DE. Natural genetic variation and hybridization in plants. J Exp Bot 2017;68(20):5415.

[28] Gallais A, Dillmann C, Goldringer I. Quantitative genetics and breeding methods: the way ahead. In: Proceedings of the eleventh meeting of the EUCARPIA section biometrics in plant breeding. Paris, France, August 30/31-September 1, 2000. Editions Quae; 2001.

[29] Collard BC, Jahufer MZ, Brouwer JB, Pang EC. An introduction to markers, quantitative trait loci (QTL) mapping and marker-assisted selection for crop improvement: the basic concepts. Euphytica 2005;142(1–2):169–96.

[30] Hamblin MT, Buckler ES, Jannink JL. Population genetics of genomics-based crop improvement methods. Trends Genet 2011;27(3):98–106.

[31] Provart NJ, Alonso J, Assmann SM, Bergmann D, Brady SM, Brkljacic J, et al. 50 years of *Arabidopsis* research: highlights and future directions. New Phytol 2016;209(3):921–44.

[32] Arabidopsis Genome Initiative. Analysis of the genome sequence of the flowering plant *Arabidopsis thaliana*. Nature 2000;408(6814):796.

[33] Holland CK, Jez JM. *Arabidopsis*: the original plant chassis organism. Plant Cell Rep 2018;37(10):1359–66.

[34] Lamesch P, Berardini TZ, Li D, Swarbreck D, Wilks C, Sasidharan R, Muller R, et al. The Arabidopsis Information Resource (TAIR): improved gene annotation and new tools. Nucleic Acids Res 2012;40(D1):D1202–10.

[35] Krishnakumar V, Hanlon MR, Contrino S, Ferlanti ES, Karamycheva S, Kim M, et al. Araport: the Arabidopsis information portal. Nucleic Acids Res 2015;43(D1):D1003–9.

[36] Katari MS, Nowicki SD, Aceituno FF, Nero D, Kelfer J, et al. VirtualPlant: a software platform to support systems biology research. Plant Physiol 2010;152(2):500–15.

[37] Waese J, Provart NJ. The bio-analytic resource for plant biology. In: van Dijk ADJ, editor. Plant Genomics Databases. New York: Humana Press; 2017. p. 119–148.

[38] Zimmermann P, Hirsch-Hoffmann M, Hennig L, Gruissem W. GENEVESTIGATOR. Arabidopsis microarray database and analysis toolbox. Plant Physiol 2004;136(1):2621–32.

[39] Feldmann KA, Marks MD. Agrobacterium-mediated transformation of germinating seeds of *Arabidopsis thaliana*: a non-tissue culture approach. Mol Gen Genet 1987;208(1–2):1–9.

[40] Sessions A, Burke E, Presting G, Aux G, McElver J, Patton D, et al. A high-throughput *Arabidopsis* reverse genetics system. Plant Cell 2002;14(12):2985–94.

[41] Alonso JM, Stepanova AN, Leisse TJ, Kim CJ, Chen H, Shinn P, et al. Genome-wide insertional mutagenesis of *Arabidopsis thaliana*. Science 2003;301(5633):653–7.

[42] Rosso MG, Li Y, Strizhov N, Reiss B, Dekker K, Weisshaar B. An *Arabidopsis thaliana* T-DNA mutagenized population (GABI-Kat) for flanking sequence tag-based reverse genetics. Plant Mol Biol 2003;53(1–2):247–59.

[43] Woody ST, Austin-Phillips S, Amasino RM, Krysan PJ. The WiscDsLox T-DNA collection: an *Arabidopsis* community resource generated by using an improved high-throughput T-DNA sequencing pipeline. J Plant Res 2007;120(1):157–65.

[44] O'Malley RC, Barragan CC, Ecker JR. A user's guide to the *Arabidopsis* T-DNA insertion mutant collections. In: Alonso JM, Stepanova AN, editors. Plant Functional Genomics. New York: Humana Press; 2015. p. 323–342.

[45] Samson F, Brunaud V, Duchêne S, De Oliveira Y, Caboche M, Lecharny A. FLAGdb + +: a database for the functional analysis of the *Arabidopsis* genome. Nucleic Acids Res 2004;32(suppl.1):D347–50.

[46] Sundaresan V, Springer P, Volpe T, Haward S, Jones JD, Dean C, et al. Patterns of gene action in plant development revealed by enhancer trap and gene trap transposable elements. Genes Dev 1995;9(14):1797–810.

[47] Ito T, Motohashi R, Kuromori T, Mizukado S, Sakurai T, Kanahara H, et al. A new resource of locally transposed dissociation elements for screening gene-knockout lines in silico on the *Arabidopsis* genome. Plant Physiol 2002;129(4):1695–9.

[48] Tang C, Toomajian C, Sherman-Broyles S, Plagnol V, Guo YL, Hu TT, et al. The evolution of selfing in *Arabidopsis thaliana*. Science 2007;317(5841):1070–2.

[49] Bomblies K, Yant L, Laitinen RA, Kim ST, Hollister JD, Warthmann N, et al. Local-scale patterns of genetic variability, outcrossing, and spatial structure in natural stands of *Arabidopsis thaliana*. PLoS Genet 2010;6(3):e1000890.

[50] Abbott RJ, Gomes MF. Population genetic structure and outcrossing rate of *Arabidopsis thaliana* (L.) Heynh. Heredity 1989;62(3):411–8.

[51] Bergelson J, Stahl E, Dudek S, Kreitman M. Genetic variation within and among populations of *Arabidopsis thaliana*. Genetics 1998;148(3):1311–23.

[52] Bakker EG, Stahl EA, Toomajian C, Nordborg M, Kreitman M, Bergelson J. Distribution of genetic variation within and among local populations of *Arabidopsis thaliana* over its species range. Mol Ecol 2006;15(5):1405–18.

[53] Alonso-Blanco C, Andrade J, Becker C, Bemm F, Bergelson J, Borgwardt KM, et al. 1,135 genomes reveal the global pattern of polymorphism in *Arabidopsis thaliana*. Cell 2016;166(2):481–91.

[54] Fournier-Level A, Korte A, Cooper MD, Nordborg M, Schmitt J, Wilczek AM. A map of local adaptation in *Arabidopsis thaliana*. Science 2011;334(6052):86–9.

[55] Nordborg M, Hu TT, Ishino Y, Jhaveri J, Toomajian C, Zheng H, et al. The pattern of polymorphism in *Arabidopsis thaliana*. PLoS Biol 2005;3(7):1289–99.

[56] Clark RM, Schweikert G, Toomajian C, Ossowski S, Zeller G, Shinn P, et al. Common sequence polymorphisms shaping genetic diversity in *Arabidopsis thaliana*. Science 2007;317(5836):338–42.

[57] Horton MW, Hancock AM, Huang YS, Toomajian C, Atwell S, Auton A, et al. Genome-wide patterns of genetic variation in worldwide *Arabidopsis thaliana* accessions from the RegMap panel. Nat Genet 2012;44(2):212.

[58] Hoffmann MH. Biogeography of *Arabidopsis thaliana* (l.) heynh. (*Brassicaceae*). J Biogeogr 2002;29(1):125–34.

[59] Hoffmann MH. Evolution of the realized climatic niche in the genus: *Arabidopsis* (*Brassicaceae*). Evolution 2005;59(7):1425–36.

[60] Cajero-Sanchez W, Aceves-Garcia P, Fernández-Marcos M, Gutiérrez C, Rosas U, García-Ponce B, et al. Natural root cellular variation in responses to osmotic stress in *Arabidopsis thaliana* accessions. Genes 2019;10(12):983.

[61] Frachon L, Mayjonade B, Bartoli C, Hautekèete NC, Roux F. Adaptation to plant communities across the genome of *Arabidopsis thaliana*. Mol Biol Evol 2019;36(7):1442–56.

[62] Ferrero-Serrano A, Assmann SM. Phenotypic and genome-wide association with the local environment of *Arabidopsis*. Nat Ecol Evol 2019;3(2):274–85.

[63] Murren C, Kohler C, Balazs R, Bassett R, Beacham A, Cousins EA, et al. Gene type and mutation position influence responses in root traits across nutrient environments. Int J Plant Sci 2020;181(2):210–23.

[64] Kawakatsu T, Huang SS, Jupe F, Sasaki E, Schmitz RJ, Urich MA, et al. Epigenomic diversity in a global collection of *Arabidopsis thaliana* accessions. Cell 2016;166(2):492–505.

[65] Lang Z, Xie S, Zhu JK. The 1001 *Arabidopsis* DNA methylomes: an important resource for studying natural genetic, epigenetic, and phenotypic variation. Trends Plant Sci 2016;21(11):906–8.

[66] Quadrana L, Silveira AB, Mayhew GF, LeBlanc C, Martienssen RA, Jeddeloh JA, et al. The *Arabidopsis thaliana* mobilome and its impact at the species level. eLife 2016;5:e15716.

[67] Bouzid M, He F, Schmitz G, Häusler RE, Weber AP, Mettler-Altmann T, et al. *Arabidopsis* species deploy distinct strategies to cope with drought stress. Annals Bot 2019;124(1):27–40.

[68] Dietrich CC, Bilnicki K, Korzeniak U, Briese C, Nagel KA, Babst-Kostecka A. Does slow and steady win the race? Root growth dynamics of *Arabidopsis halleri* ecotypes in soils with varying trace metal element contamination. Environ Exp Bot 2019;167:103862.

[69] Nijveen H, Ligterink W, Keurentjes JJ, Loudet O, Long J, Sterken MG, et al. Ara QTL–workbench and archive for systems genetics in *Arabidopsis thaliana*. Plant J 2017;89(6):1225–35.

[70] Seren Ü, Grimm D, Fitz J, Weigel D, Nordborg M, Borgwardt K, et al. AraPheno: a public database for *Arabidopsis thaliana* phenotypes. Nucleic Acids Res 2016;45(D1):D1054–9.

[71] Togninalli M, Seren Ü, Meng D, Fitz J, Nordborg M, Weigel D, et al. The Ara GWAS Catalog: a curated and standardized *Arabidopsis thaliana* GWAS catalog. Nucleic Acids Res 2018;46(D1):D1150–6.

[72] Buniello A, MacArthur JA, Cerezo M, Harris LW, Hayhurst J, Malangone C, et al. The NHGRI-EBI GWAS Catalog of published genome-wide association studies, targeted arrays and summary statistics 2019. Nucleic Acids Res 2019;47(D1):D1005–12.

[73] Guerinot ML, Baxter I, Salt DE. From the ionome to the genome: identifying the gene networks that control the mineral content of plants. Annual Plant Rev 2018;35:290–303.

[74] Baxter I, Muthukumar B, Park HC, Buchner P, Lahner B, Danku J, et al. Variation in molybdenum content across broadly distributed populations of *Arabidopsis thaliana* is controlled by a mitochondrial molybdenum transporter (MOT1). PLoS Genet 2008;4(2):e1000004.

[75] Chen X, Yuan L, Ludewig U. Natural genetic variation of seed micronutrients of *Arabidopsis thaliana* grown in Zinc-deficient and Zinc-amended soil. Front Plant Sci 2016;7:1070.

[76] Campos AC, Kruijer W, Alexander R, Akkers RC, Danku J, Salt DE, et al. Natural variation in *Arabidopsis thaliana* reveals shoot ionome, biomass, and gene expression changes as biomarkers for zinc deficiency tolerance. J Exp Bot 2017;68(13):3643–56.

[77] Chen ZR, Kuang L, Gao YQ, Wang YL, Salt DE, Chao DY. *AtHMA4* drives natural variation in leaf Zn concentration of *Arabidopsis thaliana*. Front Plant Sci 2018;9:270.

[78] Kellermeier F, Chardon F, Amtmann A. Natural variation of *Arabidopsis* root architecture reveals complementing adaptive strategies to potassium starvation. Plant Physiol 2013;161(3):1421–32.

[79] Satbhai SB, Setzer C, Freynschlag F, Slovak R, Kerdaffrec E, Busch W. Natural allelic variation of *FRO2* modulates Arabidopsis root growth under iron deficiency. Nat Commun 2017;8(1):15603.

[80] Bouain N, Korte A, Satbhai SB, Rhee SY, Busch W, Rouached H. Systems approaches provide new insights into *Arabidopsis thaliana* root growth under mineral nutrient limitation. bioRxiv 2018;460360.

[81] Jia Z, Giehl RF, Meyer RC, Altmann T, von Wirén N. Natural variation of *BSK3* tunes brassinosteroid signaling to regulate root foraging under low nitrogen. Nat Commun 2019;10(1):2378.
[82] Lee S, I. Sergeeva L, Vreugdenhil D. Natural variation of hormone levels in *Arabidopsis* roots and correlations with complex root architecture. J Integr Plant Biol 2018;60(4):292–309.
[83] Food and Agriculture Organization of the United Nations. FAOSTAT Statistical Database. Available from: http://www.fao.org/faostat/en/#data/QC; 2020 [accessed: 30.05.2020].
[84] Xu Y, An D, Li H, Xu H. Breeding wheat for enhanced micronutrients. Can J Plant Sci 2011;91(2):231–7.
[85] Velu G, Tutus Y, Gomez-Becerra HF, Hao Y, Demir L, Kara R, et al. QTL mapping for grain zinc and iron concentrations and zinc efficiency in a tetraploid and hexaploid wheat mapping populations. Plant Soil 2017;411(1–2):81–99.
[86] Monasterio I, Graham RD. Breeding for trace minerals in wheat. Food Nutr Bull 2000;21(4):392–6.
[87] Cakmak I. Enrichment of cereal grains with zinc: agronomic or genetic biofortification? Plant Soil 2008;302(1–2):1–7.
[88] White PJ, Bradshaw JE, Finlay M, Dale B, Ramsay G, Hammond JP, et al. Relationships between yield and mineral concentrations in potato tubers. Hort Sci 2009;44(1):6–11.
[89] Bhatta M, Morgounov A, Belamkar V, Poland J, Baenziger PS. Unlocking the novel genetic diversity and population structure of synthetic hexaploid wheat. BMC Genom 2018;19(1):1–2.
[90] Morgounov A, Abugalieva A, Akan K, Akın B, Baenziger S, Bhatta M, et al. High-yielding winter synthetic hexaploid wheats resistant to multiple diseases and pests. Plant Genet Resour 2018;16(3):273–8.
[91] Bhatta M, Baenziger PS, Waters BM, Poudel R, Belamkar V, Poland J, et al. Genome-wide association study reveals novel genomic regions associated with 10 grain minerals in synthetic hexaploid wheat. Int J Mol Sci 2018;19(10):3237.
[92] Alomari DZ, Eggert K, Von Wirén N, Pillen K, Röder MS. Genome-wide association study of calcium accumulation in grains of European wheat cultivars. Front Plant Sci 2017;8:1797.
[93] Alomari DZ, Eggert K, Von Wirén N, Alqudah AM, Polley A, Plieske J, et al. Identifying candidate genes for enhancing grain Zn concentration in wheat. Front Plant Sci 2018;9:1313.
[94] Velu G, Ortiz-Monasterio I, Cakmak I, Hao Y, Singh RP. Biofortification strategies to increase grain zinc and iron concentrations in wheat. J Cereal Sci 2014;59(3):365–72.
[95] Velu G, Singh RP, Huerta-Espino J, Pena RJ, Arun B, Mahendru-Singh A, et al. Performance of biofortified spring wheat genotypes in target environments for grain zinc and iron concentrations. Field Crops Res 2012;137:261–7.
[96] Velu G., Singh R.P. Biofortified wheat, sustainable agricultural approach to mitigate micronutrient malnutrition. In: Abstract of the international hidden congress. Stuttgart, Germany; 2013.
[97] Velu G, Singh RG, Balasubramaniam A, Mishra VK, Chand R, Tiwari C, et al. Reaching out to farmers with high zinc wheat varieties through public-private partnerships: an experience from eastern-gangetic plains of India. Adv J Food Sci Technol 2015;1(3):73–5.
[98] Meher D, Samantara K, Behera A, Mohapatra SR. Folk rice: Genetic storehouse for Biofortification: A review. Int J Chem Stud 2020;8(2):1813–6.
[99] Murgia I, De Gara L, Grusak MA. Biofortification: how can we exploit plant science and biotechnology to reduce micronutrient deficiencies? Front Plant Sci 2013;4:429.
[100] Pinson SR, Tarpley L, Yan W, Yeater K, Lahner B, Yakubova E, et al. Worldwide genetic diversity for mineral element concentrations in rice grain. Crop Sci 2015;55(1):294–311.

[101] Ng'uni D, Geleta M, Hofvander P, Fatih M, Bryngelsson T. Comparative genetic diversity and nutritional quality variation among some important Southern African *Sorghum* accessions ['*Sorghum bicolor*'(L.) Moench]. Aust J Crop Sci 2012;6(1):56.
[102] Reddy BV, Ramesh S, Longvah T. Prospects of breeding for micronutrients and b-carotene-dense sorghums. Int Sorghum Millets Newslett 2005;46:10–4.
[103] Sserumaga JP, Makumbi D, Warburton ML, Opiyo SO, Asea G, Muwonge A, et al. Genetic diversity among tropical provitamin A maize inbred lines and implications for a biofortification program. Cereal Res Commun 2019;47(1):134–44.
[104] Ortiz-Monasterio JI, Palacios-Rojas N, Meng E, Pixley K, Trethowan R, Pena RJ. Enhancing the mineral and vitamin content of wheat and maize through plant breeding. J Cereal Sci 2007;46(3):293–307.
[105] Harjes CE, Rocheford TR, Bai L, Brutnell TP, Kandianis CB, Sowinski SG, et al. Natural genetic variation in lycopene epsilon cyclase tapped for maize biofortification. Science 2008;319(5861):330–3.
[106] Menkir A, Rocheford T, Maziya-Dixon B, Tanumihardjo S. Exploiting natural variation in exotic germplasm for increasing provitamin-A carotenoids in tropical maize. Euphytica 2015;205(1):203–17.
[107] Pixley K, Rojas NP, Babu R, Mutale R, Surles R, Simpungwe E. Biofortification of maize with provitamin A carotenoids. In: Tanumihardjo SA, editor. Carotenoids and human health. New York: Springer Science and Business Media; 2013. p. 271–92.
[108] Queiroz VAV, Guimarães PEDO, Queiroz LR, Guedes EDO, Vasconcelos VDB, Guimarães LJ, et al. Iron and zinc availability in maize lines. Food Sci Technol 2011;31(3):577–83.
[109] Lindhout WH, Schotte TP, Visser RG, Van Eck HJ, Hutten RC, inventors; AGVENTURE BV, assignee. Hybrid seed potato breeding. United States patent US 10,524,436; 2020 Jan 7.
[110] O'Neill KR. Statistical derivation of nutrition label for raw potatoes from 2001-2002 United States Department of Agriculture (USDA) Agricultural Research Service (ARS) Nutrient Data Laboratory (NDL). Data for raw russet, white, and red potatoes by weighting by market share. Center for Food Safety and Applied Nutrition, FDA; 2005.
[111] Singh BK, Singh B. Genotypic and breeding potential to improve mineral content of vegetable crops: an overview. Int J Veg Sci 2019;25(5):441–56.
[112] Andre CM, Ghislain M, Bertin P, Oufir M, del Rosario Herrera M, Hoffmann L, et al. Andean potato cultivars (*Solanum tuberosum* L.) as a source of antioxidant and mineral micronutrients. J Agric Food Chem 2007;55(2):366–78.
[113] Burgos G, Amoros W, Morote M, Stangoulis J, Bonierbale M. Iron and zinc concentration of native Andean potato cultivars from a human nutrition perspective. J Sci Food Agric 2007;87(4):668–75.
[114] Bonierbale M, De Haan S, Burgos G, Thiele G, Liria R, Meisner CA. Nutritious subsistence food systems. Adv Agron 2007;92:1–74.
[115] Paget M, Amoros W, Salas E, Eyzaguirre R, Alspach P, Apiolaza L, et al. Genetic evaluation of micronutrient traits in diploid potato from a base population of Andean landrace cultivars. Crop Sci 2014;54(5):1949–59.
[116] Haynes KG, Yencho GC, Clough ME, Henninger MR, Sterrett SB. Genetic variation for potato tuber micronutrient content and implications for biofortification of potatoes to reduce micronutrient malnutrition. Am Potato J 2012;89(3):192–8.
[117] Andersson MS, Saltzman A, Virk PS, Pfeiffer WH. Progress update: crop development of biofortified staple food crops under HarvestPlus. Afr J Food Agric Nutr Dev 2017;17(2):11905–35.

[118] Chipungu F, Changadeya W, Ambali A, Saka J, Mahungu N, Mkumbira J. Analysis of micronutrients variations among sweet potato (*Ipomoea batatas* [L.] Lam) genotypes in Malawi. J Agric Biotech Sustain Dev 2017;9(4):22–35.
[119] Neela S, Fanta SW. Review on nutritional composition of orange-fleshed sweet potato and its role in management of vitamin A deficiency. Food Sci Nutr 2019;7(6):1920–45.
[120] Komakech A, Ipolto OU, Odongo JC. The contributions of orange fleshed sweet potato to household Vitamin A intake. J Food Nutr Sci 2019;7(5):66–72.
[121] Hillocks RJ, Madata CS, Chirwa R, Minja EM, Msolla S. Phaseolus bean improvement in Tanzania, 1959-2005. Euphytica 2006;150(1–2):215–31.
[122] Tryphone GM, Nchimbi-Msolla S. Diversity of common bean (*Phaseolus vulgaris* L.) genotypes in iron and zinc contents under screen house conditions. African J Agri Res 2010;5(8):738–47.
[123] Guzmán-Maldonado SH, Acosta-Gallegos J, Paredes-López O. Protein and mineral content of a novel collection of wild and weedy common bean (*Phaseolus vulgaris* L). J Sci Food Agri 2000;80(13):1874–81.
[124] Haas JD, Luna SV, Lung'aho MG, Wenger MJ, Murray-Kolb LE, Beebe S, et al. Consuming iron biofortified beans increases iron status in Rwandan women after 128 days in a randomized controlled feeding trial. J Nutr 2016;146(8):1586–92.
[125] Vaiknoras KA, Larochelle C. The impact of biofortified iron bean adoption on productivity, and bean consumption, purchases and sales. In: Agricultural & Applied Economics Association Annual Meeting. Washington, D.C.; 2018.
[126] Boukar O, Massawe F, Muranaka S, Franco J, Maziya-Dixon B, Singh B, et al. Evaluation of cowpea germplasm lines for protein and mineral concentrations in grains. Plant Genet Resour 2011;9(4):515–22.
[127] Muranaka S, Shono M, Myoda T, Takeuchi J, Franco J, Nakazawa Y, et al. Genetic diversity of physical, nutritional and functional properties of cowpea grain and relationships among the traits. Plant Genet Resour 2016;14(1):67–76.
[128] Thavarajah D, Ruszkowski J, Vandenberg A. High potential for selenium biofortification of lentils (*Lens culinaris* L.). J Agri Food Chem 2008;56(22):10747–53.
[129] Thavarajah P, Sarker A, Materne M, Vandemark G, Shrestha R, Idrissi O, et al. A global survey of effects of genotype and environment on selenium concentration in lentils (*Lens culinaris* L.): Implications for nutritional fortification strategies. Food Chem 2011;125(1):72–6.
[130] Deja-Muylle A, Parizot B, Motte H, Beeckman T. Exploiting natural variation in root system architecture via genome-wide association studies. J Exp Bot 2020;71(8):2379–89.
[131] Thomas CL, Alcock TD, Graham NS, Hayden R, Matterson S, Wilson L, et al. Root morphology and seed and leaf ionomic traits in a *Brassica napus* L. diversity panel show wide phenotypic variation and are characteristic of crop habit. BMC Plant Biol 2016;16(1):214.
[132] Schaefer RJ, Michno JM, Jeffers J, Hoekenga O, Dilkes B, Baxter I, et al. Integrating co-expression networks with GWAS to prioritize causal genes in maize. Plant Cell 2018;30(12):2922–42.
[133] Ziegler G, Nelson R, Granada S, Krishnan HB, Gillman JD, Baxter I. Genome-wide association study of ionomic traits on diverse soybean populations from germplasm collections. Plant Direct 2018;2(1):e00033.
[134] Yang M, Lu K, Zhao FJ, Xie W, Ramakrishna P, Wang G, et al. Genome-wide association studies reveal the genetic basis of ionomic variation in rice. Plant Cell 2018;30(11):2720–40.

[135] Liu S, Zhong H, Meng X, Sun T, Li Y, Pinson SR, et al. Genome-wide association studies of ionomic and agronomic traits in USDA mini core collection of rice and comparative analyses of different mapping methods. Res Square 2020;. PPR: PPR158878.

[136] Zhang H, Wang ML, Schaefer R, Dang P, Jiang T, Chen C. GWAS and co-expression network reveal ionomic variation in cultivated peanut. J Agri Food Chem 2019;67(43):12026–36.

[137] Ye Z, Zeng J, Ye L, Long L, Zhang G. Genome-wide association analysis of potassium uptake and translocation rates under low K stress in Tibetan wild barley. Euphytica 2020;216(3):35.

Chapter 8

Untapped Soybeans: A Genetic Reservoir for its Improvement

Muhammad Hammad Nadeem Tahir[a] and Humera Razzaq[b]
[a]*Institute of Plant Breeding and Biotechnology, Muhammad Nawaz Shareef University of Agriculture, Multan, Pakistan;* [b]*Department of Plant Breeding and Genetics, University of Agriculture, Faisalabad, Pakistan*

8.1 Introduction

Soybean (*Glycine max* L.), also known as "soya bean" and "sojabean," is an annual legume member of the family *Fabaceae* (*Leguminosae*). The name "Glycine" is derived from Greek word "glyks" means sweet. Economically it is the most important bean in the world and a source of plant protein for millions of people. There are more than 10,000 different products made from soybean, such as tempe, tofu, tamari, miso, bean sprouts, soy milk, and soy sauce. Soybean meal is an important constituent of poultry and fish feed. Its demand is increasing due to its good nutritive profile, health benefits, and unique composition and oil. It contains ~20% oil and ~40% proteins [1]. Soybean is grown almost in 35 countries as a major oilseed crop commercially [2]. The main producers and exporters of soybean globally are USA, Brazil, and Argentina exporting 44, 33, and 11% of the total export of the world, respectively, and the largest importer is China importing 38% of the total import of the world [3].

Soybean is an erect branching plant with a 2 m high stem, self-fertilizing plant, bearing white or purple flowers. Pod may contain 1–4 seeds of yellow, green, brown, or bicolored having length of 3–7 cm [2]. Leaves are trifoliate of oval shape mostly 3–10 cm in length [4].

Fats and proteins are the major components of cotyledon, which covers the 90% part of soybean seed whereas anti-nutritional factors and carbohydrates are also part of it. Soybean seed contains 40–48% protein, 20% fat, 6% ash, 5–11% water, 10–14% neutral detergent fiber (NDF), 9–11% acid detergent fiber (ADF), and 32% carbohydrate, on dry matter basic [5,6]. Soybean protein contains best amino acid composition containing high amount of lysine (6.2 g/16 gN), isoleucine, threonine, tryptophan, and valine but cytosine and methionine limits its value. Trypsin and chymotrypsin are the most important protein inhibitors in soybean [7].

Wild Germplasm for Genetic Improvement in Crop Plants. http://dx.doi.org/10.1016/B978-0-12-822137-2.00008-4

Soybean is sown in two seasons, autumn and spring. Seeds are planted at a depth of 3.2–4.5 cm maintaining row to row distance of 30 cm. No-till is preferred for soybean as it reduces the erosion and conserves moisture. For proper growth and seed setting, 6–7 irrigations are recommended. More numbers of irrigation are required in sandy soil. Nitrogen is applied @ 25 kg/acre, phosphorus @ 60 kg/acre, and potassium @ 40 kg/acre.

It is a short-day plant showing better production in tropics. Optimum temperature of 28°C–30°C is required for better growth whereas temperature more than 32°C and less than 21°C causes reduction in flowering. It is propagated through seed and as it is a nitrogen fixing crop so to enhance the efficiency of growth, soybean seeds are inoculated with *Rhizobia* species.

8.2 Origin and domestication

Soybean was introduced in the Eastern Zhou Dynasty from north-eastern China 2510 BP, leading to agricultural revolution. Maximum genetic diversity of soybean is found around the Huang He river [8,9]. Based on phylogenetic analysis, Southern China has also been proposed as an area of origin of soybean but archeological analysis showed no evidence to support this claim [10]. Sequenced genomic regions of *G. soja*, *G. max*, and *G. gracilis* have classified these as wild and landraces of soybean [11]. *G. gracilis* has been either the result of hybridization of *G. soja* and *G. max* or preliminary evolutionary product. Recent analysis of gene flow and genetic difference shows that *G. gracilis* is a transitional species and *G. soja* is progenitor of both *G. max* and *G. gracilis*. In central China, Huang-Huai valley showed greater genetic introduction than any other region for *G. soja* [11]. Migration analysis has also shown gene flow to northern and southern China from Huang-Huai valley. Divergence studies of *G. soja* and *G. max* has revealed that *G. max–G. soja* complex has diverged from *S. soja* [12]. These evidences propose the possibility that *G. gracilis* represents *G. max–G. soja* complex. Despite all this, single origin hypothesis of soybean is widely accepted. Re-sequencing of improved varieties, wild relatives, and landraces suggests that all domesticated soybeans are derived from single origin that is wild *G. soja* [13,14]. Comparative phylogenetic studies of cultivated and wild soybeans suggest multiple origins in East Asia contrary to single origin hypothesis [15]. Independent domestication has occurred in Japanese and Korean regions having different gene pools [16]. Archeological analysis has shown larger seeds of soybean during 5000–3000 BP in Japan and Korea as compared to China [10]. Prolonged period of semi cultivation of wild soybean preceded the events of domestication at multiple locations [17,18]. Wild species exhibit great genetic diversity, but allelic frequency changes and selection has reduced genetic diversity during domestication. During domestication and introduction of landraces to North America and Asia, genetic bottleneck occurred. From *G. soja* to landrace, half of the genetic diversity has been lost during domestication. Only 19 landraces contributed around 85% of the genes to North American gene pool [13,19]. North American introduction bottleneck is weak as compared to domestication bottleneck. There are three possibilities for mitigating the bottleneck effect among elite cultivars, that is, (1) introduction of landrace or *G. soja* genome, (2) pin-pointing small regions of genome containing favorable traits, and (3) balanced selection [20].

In Europe, soybean was first reported in the year 1737 where it was grown for ornamental purpose in England and France. In North America, soybean was introduced from China in 1765 by seed dealers, scientists, and merchants [21]. The records indicate that soybean has been grown in Yugoslavia in 1804 for animal feed. In 1917, it was discovered that heated meal of soybean can be used as livestock feed. Henry Ford started industrial and nutritional uses of soybean on commercial basis. Due to its popularity and versatility, soybean has been now cultivated around the world and a big number of soybean products is available in the market.

8.3 Wild and cultivated species

The genus "*Glycine*" comprises of 28 soybean species (Table 8.1), out of which 2 annual species *G. max* and *G. soja* are used directly and indirectly for feed and food. *G. soja* is a wild species whereas *G. max* is cultivated annual species of soybean. The other 26 species are perennial and wild. Perennial species contain different genes conferring resistance against abiotic and biotic stresses which are not available in annual cultivated and wild species. Morphological differences occur between *G. max* and *G. soja* for many agronomic traits including seed related traits, pod shattering, and determinate growth habit due to artificial selection in *G. max* [22,23]. Genomic, cytological, and proteomic studies provide the evidence for wild *G. soja* as the progenitor of cultivated *G. max* [14]. These species are Asian in origin and are distributed in Far East Asia and Southeast Asia including Russia, China, Japan, Korea, and Taiwan. Wild soybean grows on wastelands, riversides, roadsides, and fertile valleys [24]. *G. max* and *G. soja,* both are cross compatible and produce fertile hybrids. Three species *G. canescens*, *G. argyrea*, and *G. tomentella* can successfully be hybridized with cultivated soybean and embryo culture is used to rescue F_1 hybrids. Cultivars of *G. max* have narrow genetic base due to progressive domestic bottleneck. Due to undesirable traits, that is, lodging, shattering, small seed associated with *G. soja,* genetic diversity in this species has rarely been exploited [19].

Wild soybean is important genetic resource containing gene families and genes responsible for biotic and abiotic stress resistance, higher protein and oil contents. Flow of characteristics from wild soybeans to cultivated species is evident due to their ability to produce fertile offspring, same ploidy level, and annual growth habit [25]. Due to clearance of land for industrial and agricultural purposes, wild soybean resources have decreased. During past few decades, genetic diversity has been reduced due to soybean domestication. Furthermore,

TABLE 8.1 List of soybean species (members of genus *Glycine*).

Species	Growth habit	Chromosome number
G. max	Annual, cultivated	2n = 40 [29]
G. soja	Annual, wild	2n = 40 [30]
G. latrobeana	Perennial, wild	2n = 40 [31]
G. clandestine	Perennial, wild	2n = 40 [32]
G. falcate	Perennial, wild	2n = 40 [32]
G. tomentella	Perennial, wild	2n= 38, 40, 78, and 80 [25])
G. tabacina	Perennial, wild	2n = 40 Modified
G. cyrtoloba	Perennial, wild	2n = 40 [32]
G. argyrea	Perennial, wild	2n = 40 [32]
G. latifolia	Perennial, wild	2n = 40 [32]
G. albicans	Perennial, wild	2n = 40 [32]
G. microphylla	Perennial, wild	2n = 40 [32]
G. arenaria	Perennial, wild	2n = 40 [32]
G. curvata	Perennial, wild	2n = 40 [32]
G. pindanica	Perennial, wild	2n = 40 [32]
G. dolichocarpa	Perennial, wild	2n = 40 [32]
G. lactovirens	Perennial, wild	2n = 40 [32]
G. hirticaulis	Perennial, wild	2n = 80 [32]
G. syndetika	Perennial, wild	2n = 40 [32]
G. pescadrensis	Perennial, wild	2n = 40 [32]
G. montis-douglas	Perennial, wild	2n = 40 [32]
G. gracei	Perennial, wild	2n = 40 [32]
G. pullenii	Perennial, wild	2n = 40 [32]
G. aphyonota	Perennial, wild	2n = 40 [32]
G. rubiginosa	Perennial, wild	2n = 40 [32]
G. peratosa	Perennial, wild	2n = 40 [32]
G. stenophita	Perennial, wild	2n = 40 [32]
G. canescens	Perennial, wild	2n = 40 [32]

selection for desirable traits such as seed size and oil content was mainly focused that caused reduction in diversity [22,23].

Sequencing of whole wild soybean genome has opened new arenas for comparative and functional genomics of soybean [26]. Half of the resistance related genes present in *G. soja* are not found in domesticated and landraces of soybeans. It is very helpful in understanding lost diversity, domestication, and bottleneck of soybean. In elite cultivars, reduced genetic diversity can be compensated through introgression of favorable variation from *G. soja* [27]. Climatic fluctuations, that is, environmental selection and genetic drift have greatly influenced natural populations of *G. soja*. Evolutionary history reveals that natural population of *G. soja*, compared to *G. max*, is constantly expanding. Development of sophisticated mechanism in *G. soja* enables it to adapt in local environment. Niche and environmental isolation plays an important role in genetic variation in *G. soja* than geographic isolation. Wild species have been exploited in many crops for revitalization of cultivated species [28]. Soybean cultivars with superior traits can be developed by using alleles/genes from wild cultivars. It is environment friendly and promising solution for improvement of soybeans.

8.4 Gene pool and germplasm conservation

Concept of primary, secondary, and tertiary gene pool was proposed by Ref. [33]. Primary gene pool consists of biological species. Hybrids between the species in this group exhibit normal chromosomal pairing, total seed fertility, and normal segregation. In primary gene pool of soybean wild annual progenitor *G. soja*, *G. max* cultivars, and landraces are included. Secondary gene pool includes all species that can be crossed with primary gene pool with some fertility in hybrids. It is suggested that soybean does not have any secondary gene pool. Tertiary gene pool includes outer limit of genetic resources. Hybrids are sterile and usually inviable. Gene transfer is not possible or can be accomplished through special techniques. Based on this concept, tertiary gene pool of soybean includes 16 wild perennial species. Wild soybean germplasm is conserved in many countries including United States of America, China, Japan, and Korea. There are many unexplored natural habitats of wild soybean so complete representative collection is yet to be achieved. To understand bio-systems of yield related traits, phylogenetic relationship, and taxonomy, wild germplasm of soybean needs to be collected, conserved, and characterized and should be made available to soybean breeders.

Seed of soybean is very short lived and is deteriorated when stored as compared to other crops like maize, wheat, rice, and sorghum etc. Therefore, it is imperative to store the soybean seed under a set of conditions of low temperature and humidity where it can be stored for a longer duration without deterioration and loss of viability. Medium and long-term storage facilities have been established in many countries of the world for conservation of soybean germplasm maintaining its health and viability. Usually a temperature of 5–7°C along with 35% relative humidity is maintained for medium-term conservation, whereas −20°C is recommended for long-term conservation of soybean germplasm. Many gene banks are working on conservation of wild germplasm of soybean. Largest soybean collection in Asia is held by China under Chinese Crop Germplasm Information System with 6172 accessions. Chung's Wild

Legume Germplasm Collection (CWLGC) ranks second holding 6012 accessions. CWLGC is primarily focused on wild soybean germplasm collections from Korea, China, Japan, and Far East Russia. In Japan, National Institute of Agro-biological Sciences maintains 1131 accessions. United States Department of Agriculture (USDA) conserves nearly 21,810 accessions. N.I. Vavilov Institute of Plant Genetic Resources has the collection of 350 soybean accession.

8.5 Stress resistance

Any such factor, in excess or deficiency, that is, a cause of reduction in crop yield crop is called stress [34]. Biotic and abiotic stresses both have impact on the yield as well as on other qualitative and quantitative traits of soybean. Genetic diversity is reduced in cultivated soybean due to change in climate, water resources, breeding activities for artificial selection for dietary requirements, and increase in demand. An opportunistic source of improving the genetic diversity in soybeans is wild germplasm [35].

8.5.1 Biotic stress

Many diseases and insect pest effect soybean causing a major harm that can be prevented by incorporating the resistant genes from wild relatives. Wild germplasm of soybean is a rich source of resistant genes (Table 8.2) against insect pest like bean leaf beetle, soybean leaf miner, soybean thrips, soybean borer, nematodes, and painted lady, and diseases including frogeye spot, brown spot, downy mildew, *Cercospora* leafspot, white mold which affect soybean production.

The perennial and the annual wild species of *Glycine* have not been properly and extensively exploited for biotic and abiotic stress resistance. Screening of perennial species was done due to curiosity, but few accessions were screened. *G. canescens* has resistance genes for powdery mildew, whereas *G. clandestine*, *G. tabacina*, *G. falcate*, and *G. tomentella* were not resistant [36]. Two accessions of *G. tabacina* and *G. clandestina* are resistant to brown spot [37]. Soybean cyst nematode (SCN) is one of the most harmful pathogen. QTLs for SCN resistance has been detected and reported in *G. soja* by using disequilibrium linkage–mapping and association mapping [38–41]. More durable resistance was achieved by accumulation of three alleles, *Rhg4*, *rhg1-a*, *and rhg1-b*, from *G. max* whereas two QTLs, cqSCN-006, and cqSCN-007 have been identified in *G. soja* for resistance against nematode [42]. *G. argyria*, *G. pescadrensis*, and *G. argyrea* also show resistance to the three types of nematodes [43].

Fungal, bacterial, oocytes, and viruses are the source of many diseases in soybeans leading to huge reduction in yield as well as deterioration in quality [44]. The number of diseases effecting soybeans has increased remarkably due to the increase in production and geographical distribution [45]. Mechanisms dealing with the resistance against these diseases at molecular level have been studied vigorously. In fungal diseases, Asian soybean rust (ASR) caused by *Phakopsora pachyrhizi* is one of the most damaging to soybean crop. Resistance to this pathogen has been detected in perennial species of *Glycine* including *G. clandestine*, *G. latifolia*, *G. tabacina*, *G. argyrea*, *G. canescens*, *G. tomentella*, and *G. microphylla* [46]. Studied of wild soybeans have identified many other potential sources for resistance against ASR, which may be exploited for improving soybeans [47].

White mold, caused by the pathogen *Scleroinia sclerotiorum* is also an important disease in many areas of the world especially in America. Many accessions of *G. tomentella and G. tabacina* have resistance to *S. sclerotiorum* [48]. A resistance loci associated with SCN has been reported in *G. lalifolia* [49]. Resistance to powdery mildew and sudden death syndrome is identified in many accessions of *G. tomentella and G. canescens* [36,48].

8.5.2 Abiotic stress

Various kinds of plant breeding activities, mainly done for the selection of desirable traits of grain legumes have resulted in the reduction of genetic diversity. The desirable characteristics present in the new varieties and land races, have been genetically diverged from the wild progenitors [13]. The wild crop relatives mostly have resistance against biotic and abiotic stresses (Table 8.2) due to their wide range of adaptation to climatic variations [40]. Therefore, the two promising options for crop improvement are introduction and genetic modification of genetic material through breeding exploiting wild relatives using conventional and modern breeding technologies [40,50,51]. *G. soja* has a broad genetic diversity as it is distributed across wide range of geographical areas making it more capable to adapt broad range of ecological conditions [52,53]. *G. soja*, because of its adaptability, has become an important source of novel and exotic alleles and genes for the improvement of soybean to meet the requirements of rapid demographic growth and environmental changes.

The wild soybeans are key genomic resource that contain unique, distinctive, and useful genetic traits that have been lost as a result of domestication of soybean [13,26,54,55]. The one example is the discovery of salt-resistant gene GmCH1X in wild soybean. In wild soybean, gene controls 80% salt tolerance compared to cultivated soybean which had insertion of retrotransposon possibly due to round of whole genome duplication [54], which strongly suggests exploring genetic diversity of wild soybean and exploited for improvement of cultivated soybeans.

G. canescens and *G. latifolia* have more drought tolerance than *G. soja* and *G. max* because of their relatively higher water content, low epidermal conductance, and higher osmotic adjustment [56]. The level of water use efficiency is higher in *G. soja* as compared to *G. max* [57] making it more drought tolerant.

Wild relatives of soybean are more tolerant to soil salinity in comparison to domesticated ones. *G. soja* reveals a higher level of tolerance to the toxicity

TABLE 8.2 Sources of resistance in wild species of *Glycine* for various characters.

Traits improved	Wild species	Cause of resistance
Biotic factors		
Disease resistance		Not known
Powdery Mildew resistance	*G. tomentella* *G. canescens*	Not known
Asian soybean rust resistance	*G. tabacina* *G. canescens* *G. tomentella* *G. argyrea* *G. clandestine* *G. latifolia*	Not known
Sclerotina stem rot resistance	*G. tomentella*	Not known
Soybean rust	*G. tabacina*	Not known
Soybean brown spot	*G. canescens*	Not known
Alfalfa mosaic virus	*G. soja*	Not known
Bean pod mottle virus	*G. latifolia*	Not known
	G. tomentella	Not known Not known Not known Not known Not known Not known
Insect/pest resistance		
Soybean cyst nematode resistance	*G. argyrea* *G. soja* *G. pescadrensis* *G. tomentella*	Not known cqSCN-006 and cqSCN-007 Not known Not known
Unique nodulation	*G. tabacina*	Not known
Abiotic factors		
Chloride tolerance	*G. argyrea*	Not known
Drought tolerance	*G. soja*	High RCW, high WUE, low epidermal conductance, high osmotic adjustment
Salt tolerance	*G. canescens*	Not known
	G. latifolia	Not known
	G. soja	Not known
		Induction of ROS, synthesis of compatible solutes, lower cell membrane permeability
Other traits		
Tolerance to 2, 4-D	*G. tomentella*	Not known
Bowman-Brink protease inhibitor	*G. tomentella*	Not known

of chloride ions (Cl^-) in leaves than that of *G. max*, and *G. max* is also more sensitive to accumulation of sodium ions (Na^+) in leaves [58]. The salinity tolerance of *G. soja* is attributed to the release of Na^+ which in turn minimizes the accumulation at toxic concentrations in different plant organs. It is believed that the observed salinity tolerance of *G. soja* is controlled by the hormonal induction, modifications of cell membrane, reactive oxygen species, and synthesis of compatible solutes [59]. A single dominant gene reported to control salinity tolerance was derived from a wild soybean accession [60]. The gene Gs-WRKY20 has shown ability to increase drought and salt tolerance of sensitive alfalfa after transformation. This gene was isolated from *G. soja* and it encodes a WRKY- type transcription factor. The transgenic alfalfa also showed relatively lower monodialdehyde and lower membrane permeability along with higher accumulation of soluble sugars and free proline as compared to plants of wild type under water deficit and high salinity conditions [61]. In *Arabidopsis*, the over-expression of the gene GsJAZ2, derived from *G. soja,* has been reported to increase tolerance to salinity and alkali stress [62].

8.6 Miscellaneous traits

Along with resistance to different types of stresses, genetic diversity for other traits like morphological characters, yield, and quality traits has also been eroded due to consistent artificial selection for desirable traits thus leading to loss of many desirable genes and/or alleles. Genes related to yield, quality, flowering, maturity, and shattering are present in the wild species of *Glycine* that can be exploited and incorporated in the cultivated varieties leading to the genetic diversity as well as conservation of genetic makeup. Flower color in soybean is a multi-allelic trait, genes for white and pinkish white flowers have been discovered in wild species of *Glycine*. Several loci such as W1, W3, W4, w1-Ip, w1-p2, w1-s1, w1-s2 responsible for pinkish-white and white flowers have been discovered in the wild species of soybean. GmMATE 1,2,4 genes have been discovered that are related to phenolics, antioxidants, and flavonoids. Saponins are the cause of astringent taste in soybean. The gene Glyma15g39090, related to saponins reduction, has been detected in CWS5095 that is a natural wild mutant of soybean [35].

8.7 Prospects and limitations

G. soja has the greatest potential to impart novel alleles and genes to cultivated soybeans and various other legume species for their improvement. The use of biotechnological tools (such as genome editing) and high throughput sequencing technology can remarkably accelerate the discovery of novel genes and transfer of beneficial genes to soybeans. The problem of linkage drag may appear as a result of introgression of *G. soja* × *G. max*, which can be resolved by the rapid advances in biotechnology such as genome editing and improved genetic transformations. The advanced technologies of breeding including genomic selection and MABC (Marker-Assisted Backcrossing) can also be used to improve the efficiency.

The genetic variation in *G. soja* has not been explored to its full potential yet. The information about various regions of genomes will help with selective arches in *G. soja* is useful for understanding the evolutionary mechanisms that changed and produced the characteristics in *G. max* as a result of domestication. The genetic variations explored in *G. soja* will facilitate to improve soybean. There is a dire need of a systematic screening by collecting ecotypes from several natural populations by focusing the association between *G. soja* origins and SCN resistance. The non-availability of high-throughput technology results in several challenges for the research of genotype-phenotype interactions that act as a major hurdle in screening.

The use of wild soybean for crop improvement has proved very progressing but still it lacks few promising areas. At present very less research has been done on the SMV (Soybean Mosaic Virus). To develop SMV-resistant soybean varieties, much dedicated research and hard work is required. Moreover, in wild soybeans the observed variations in early vigor traits have a much potential to improve.

As the world has become a global village and the breeders now have access to a larger range of genetic diversity of soybean through networking of national and international resources. The information regarding the list of accessions with their origin and characterization is available on the websites of many genetic repositories and can be accessed very easily. Therefore, the wild soybean breeding material should be exploited to overcome the problem of narrow genetic base and breeding new varieties with a wider genetic base making the soybean more adapted to diverse environmental conditions and resistant to biotic and abiotic factors.

References

[1] Qiu L, Chang R, Jianying S, Xianghua L, Zhanglin C, Zenglu L. The history and use of primitive varieties in Chinese soybean breeding. In: Proceedings of World Soybean Research Conference VI. Chicago, IL, USA; 1999. p 165–172.

[2] Banaszkiewicz T. Nutritional value of soybean meal. In: El-Shemy H, editor. Soybean and nutrition. IntechOpen; 2020.

[3] Rynek R. Stan I prespektywy. IERiGZ 2010;1231–69.

[4] Martin WJ. Completely regular designs. J Combi Design 1998;6(4):261–73.

[5] Ensminger ME, Oldfield JE, Heinemann WW. Feeds and nutrition. Clovis, California: The Ensminger Publishing Company; 1990.

[6] Poultry Feeding Standards, editor. The Kielanowski Institute of animal physiology and nutrition. Polish Academy of Sciences; 2005. p. 7–9.

[7] Moroz, LA, Young WH. Kunitz soybean trypsin inhibitor. The New England J Med 1980;32:1126–1128.

[8] Ho Pt. Cradle of the East: an enquiry into the indigenous origins of techniques and ideas of neolithic and early historic China, 5000-1000 B. C. Hong Kong, China: Chinese University of Hong Kong; 1975.

[9] Li YH, Li W, Zhang C, Yang L, Chang RZ, Gaut BS, et al. Genetic diversity in domesticated soybean (*Glycine max*) and its wild progenitor (*Glycine soja*) for simple sequence repeat and single-nucleotide polymorphism loci. New Phytol 2010;(188):242–53.
[10] Lee GA, Crawford GW, Liu L, Sasaki Y, Chen X. Archaeological soybean (*Glycine max*) in East Asia: does size matter? PLoS ONE 2011;6:e26720.
[11] Han YP, Zhao X, Liu DY, Li YH, Lightfoot DA, Yang ZJ, et al. Domestication footprints anchor genomic regions of agronomic importance in soybeans. New Phytol 2016;(209):871–84.
[12] Kim MY, Lee S, Van K, Kim TH, Jeong SC, Choi IY, et al. Whole genome sequencing and intensive analysis of the undomesticated soybean (*Glycine soja* Sieb. and Zucc.) genome. Proc Natl Acad Sci U S A 2010;(107):22032–7.
[13] Zhou Z, Jiang Y, Wang Z, Gou Z, Lyu J, Li W, et al. Resequencing 302 wild and cultivated accessions identifies genes related to domestication and improvement in soybean. Nat Biotechnol 2015;(33):408–14.
[14] Guo J, Wang Y, Song C, Zhou J, Qiu L, Huang H, et al. A single origin and moderate bottleneck during domestication of soybean (*Glycine max*): implications from microsatellites and nucleotide sequences. Ann Bot 2010;(106):505–14.
[15] Abe J, Xu D, Suzuki Y, Kanazawa A, Shimamoto Y. Soybean germplasm pools in Asia revealed by nuclear SSRs. Theo App Genet 2003;(106):445–53.
[16] Lee Y, Park T. Origin of legumes cultivation in Korean Peninsula by viewpoint of excavated grain remains and genetic diversity of legumes. Korean Agric History Assoc 2006;5:1–31.
[17] Obata H. Jomon agriculture and paleoethnobotany in Northeast Asia. Tokyo, Japan: Douseisya; 2011.
[18] Obata H, Manabe A. Issues on the early agriculture in Korea and Japan, based on recent archaeobotanical studies. In: Kenkyukai KJ, editor. Current research on the Neolithic period in Japan and Korea: Proceedings of the ninth conference of the Kyushu Jomon Kenkyukai; 2011.
[19] Hyten DL, Song Q, Zhu Y, Choi IY, Nelson RL, Costa TM. Impacts of genetic bottlenecks on soybean genome diversity. Proc Natl Acad Sci 2006;(103):16666–71.
[20] Valliyodan B, Dan Q, Patil G, Zeng P, Huang J, Dai L. Landscape of genomic diversity and trait discovery in soybean. Sci Rep 2016;(6):23598.
[21] Hymowitz T, Harlan JR. Introduction of soybean to North America by Samuel Bowen in 1765. Econ Bot 1983;37(4):371–9.
[22] Tian ZX, Wang XB, Lee R, Li YH, Specht JE, Nelson RL. Artificial selection for determinate growth habit in soybean. Proc Natl Acad Sci U S A 2010;(107):8563–8.
[23] Dong Y, Yang X, Liu J, Wang BH, Liu BL, Wang YZ. Pod shattering resistance associated with domestication is mediated by a NAC gene in soybean. Natl Commun 2014;(5):3352.
[24] Ohara M, Shimamoto Y. Importance of genetic characterization and conservation of plant genetic resources: The breeding system and genetic diversity of wild soybean (*Glycine soja*). Plant Species Biol 2002;(17):51–8.
[25] Chung G, Singh RJ. Broadening the genetic base of soybean: a multidisciplinary approach. Crit Rev Plant Sci 2008;(27):295–341.
[26] Lam HM, Xu X, Liu X, Chen W, Yang G, Wong FL. Resequencing of 31 wild and cultivated soybean genomes identifies patters of genetic diversity and selection. Nat Genet 2015;(12):1053–9.
[27] Song Q, Hyten DL, Jia G, Quigley CV, Fickus EW, Nelson RL. Finger printing soybean germplasm and its utility in genomic research. Bethesda 2015;(5):1999–2006.
[28] Tylka GL, Marett CC. Distribution of the soybean cyst nematode, Heterodera glycines, in the United States, Canada, 1954 to 2014. Plant Health Prog 2014;(15):13–5.

[29] Karpechenko GD. On the chromosomes of Phaseolinae. Bull Appl Bot Genet Plant Breed Leningr 1925;(14):143–8.

[30] Fukuda Y. Cyto-genetical studies on the wild and cultivated Manchurian soybeans (*Glycine* L.). Japan J Bot 1933;(6):489–506.

[31] Singh RJ, Kim HH, Hymowitz T. Distribution of rDNA loci in the genus *Glycine* Wild. Theor Appl Genet 2001;(103):212–8.

[32] Singh RJ, Nelson RL, Chung G. Soybean [*Glycine max* (L.) Merr.]. In: Singh RJ, editor. Genetic resources, chromosome engineering, and crop improvement, vol. 4. Boca Raton, Florida: CRC Press, (Taylor & Frances Group); 2007. p. 13–50.

[33] Harlan JR, de Wet JMJ. Toward a rational classification of cultivated plants. Taxon 1971;(20):509–17.

[34] Nasir B, Razzaq H, Tahir MHN. Development of best screening method at seedling stage under drought stress in *Brassica napus* L. Big Data Agric 2019;1(1):15–8.

[35] Nawaz MA, Golokhvast KS, Rehman HM, Tsukamoto C, Kim HS, Yang SH, et al. Soyisoflavone diversity in wild soybeans (*Glycine soja* Sieb. & Zucc.) from the main centres of diversity. Biochem Syst Ecol 2018;77:16–21.

[36] Mignucci JS, Chamberlain DW. Interaction of *Microsphaera diffusa* with soybean and other legumes. Phytopathology 1978;(68):169–73.

[37] Lim SM, Hymowitz T. Reaction of perennial wild species of genus *Glycine* to *Septoria glycines*. Plant Dis 1987;(71):891–3.

[38] Winter SM, Shelp BJ, Anderson TR, Welacky TW, Rajcan I. QTL associated with horizontal resistance to soybean cyst nematode in *Glycine soja* PI464925B. Theor Appl Genet 2007;(114):461–72.

[39] Vuong TD, Sonah H, Meinhardt C, Deshmukh R, Kadam S, Nelson R. Genetic architecture of cyst nematode resistance revealed by genome-wide association study in soybean. BMC Genom 2015;(16):593.

[40] Zhang H, Mittal N, Leamy LJ, Barazani O, Song BH. Back into the wild—Apply untapped genetic diversity of wild relatives for crop improvement. Evol Appl 2016;10(1):5–24.

[41] Yu N, Diers BW. Fine mapping of the SCN resistance QTL cqSCN-006 and cqSCN-007 from *Glycine soja* PI 468916. Euphytica 2017;(213):54.

[42] Kim M, Hyten DL, Niblack TL, Diers BW. Stacking resistance alleles from wild and domestic soybean sources improves soybean cyst nematode resistance. Crop Sci 2011;(51):934–43.

[43] Wen L, Yuan C, Herman T, Hartman G. Accessions of perennial glycine species with resistance to multiple types of soybean cyst nematode (*Heterodera glycines*). Plant Dis 2017;(101):1201–6.

[44] Sinclair JB, Backman P. Compendium of soybean diseases. American Phytopathological S, St. Paul; 1993.

[45] Hartman GL, West ED, Tang HK. Crops that feed the World 2. Soybean-worldwide production, use, and constraints caused by pathogens and pests. Food Secur 2011;3:5–17.

[46] Hartman GL, Wang TC, Hymowitz T. Sources of resistance to soybean rust in perennial Glycine species. Plant Dis 1992;76:396–9.

[47] Langenbach C, Campe R, Beyer SF, Mueller AN, Conrath U. Fighting Asian soybean rust. Front Plant Sci 2016;7:797.

[48] Hartman G, Gardner M, Hymowitz T, Naidoo G. Evaluation of perennial species for resistance to soybean fungal pathogens that cause *Sclerotinia* stem rot and sudden death syndrome. Crop Sci 2000;40:545–9.

[49] Chang S. Identification of loci for resistance to sclerotinia stem rot (white mold) in accessions of perennial relative of soybean (*Glycine latifolia*). Dissertation, University of Illinois at Urbana-Champaign, Urbana-Champaign, IL; 2015.

[50] Brozynska M, Furtado A, Henry RJ. Genomics of crop wild relatives: Expanding the gene pool for crop improvement. Plant Biotechnol J 2015;14:1070–85.

[51] Munoz N, Liu A, Kan L, Li MW, Lam HM. Potential uses of wild germplasms of grain legumes for crop improvement. Int J Mol Sci 2016;18(2):1–28.

[52] Qiu LJ, Xing LL, Guo Y, Wang J, Jackson SA, Chang RZ. A platform for soybean molecular breeding: the utilization of core collections for food security. Plant Mol Biol 2013;83(1–2):41–50.

[53] Sherman-Broyles S, Bombarely A, Powell AF, Doyle JL, Egan AN, Coate JE, et al. The wild side of a major crop: soybean's perennial cousins from Down Under. Am J Bot 2014;101(10):1651–65.

[54] Qi X, Li MW, Xie M, Liu X, Ni M, Shao G, et al. Identification of a novel salt tolerance gene in wild soybean by whole-genome sequencing. Nat Commun 2014;5(4340):1–11.

[55] Li YH, Zhou G, Ma J, Jiang W, Jin LG, Zhang Z, et al. De novo assembly of soybean wild relatives for pan-genome analysis of diversity and agronomic traits. Nat Biotechnol 2014;32(10):1045–52.

[56] James RA, Caemmerer SVCV, Condon S, Zwart AB, Munns R. Genetic variation in tolerance to the osmotic stress component of salinity stress in durum wheat. Funct Plant Biol 2008;35(2):111–23.

[57] Chen Y, Chen P, de los Reyes BG. Differential responses of the cultivated and wild species of soybean to dehydration stress. Crop Sci 2006;46(5):2041–6.

[58] Luo Q, Yu B, Liu Y. Differential sensitivity to chloride and sodium ions in seedlings of *Glycine max* and *G. soja* under NaCl stress. J Plant Physiol 2005;162(9):1003–12.

[59] Lu W, Bennett BD, Rabinowitz JD. Analytical strategies for LC– MS-based targeted metabolomics. J Chromatogr B Analyt Technol Biomed Life Sci 2008;871(2):236–42.

[60] Lee JD, Shannon JG, Vuong TD, Nguyen HT. Inheritance of salt tolerance in wild soybean (*Glycine soja* Sieb. and Zucc.) accession PI483463. J Hered 2009;100(6):798–801.

[61] Tang L, Cai H, Zhai H, Luo X, Wang Z, Cui L, et al. Overexpression of *Glycine soja* WRKY20 enhances both drought and salt tolerance in transgenic alfalfa (*Medicago sativa* L. Plant Cell Tissue Organ Cult 2014;118:77–86.

[62] Zhu D, Cai H, Luo X, Bai X, Deyholos MK, Chen Q, et al. Over-expression of a novel JAZ family gene from *Glycine soja*, increases salt and alkali stress tolerance. Biochem Biophys Res Commun 2012;426(2):273–9.

Chapter 9

Wild Sunflowers: The Primary Genetic Resource for Sunflower Breeding

Yalcin Kaya[a] and Roumiana Vasilevska-Ivanova[b]
[a]*Trakya University Engineering Faculty, Genetics and Bioengineering Department, Edirne, Turkey;* [b]*Institute of Plant Physiology and Genetics, Sofia, Bulgarian Academy of Sciences, Sofia, Bulgaria*

9.1 Sunflower and *Helianthus genus* genetic diversity

Sunflower produce mostly for oil in many parts of the world but mostly in Black Sea countries (Ukraine, Russia, Romania, Bulgaria, Turkey), in Eastern Europe (Hungary, Serbia), Spain, France, China, Argentina, and the United States. Sunflower production has increased in last 12 years both in planted areas as 21.000–26.000 ha from 2007 to 2019 as well as 26.000–51.000 MT seed production from 2007 to 2019s. Black sea region countries contributed more for these rises because sunflower production has tripled in the region covering over 60% of world sunflower production. With higher yielding, the problems of sunflower were changed and new traits were developed and merged into new sunflower hybrids which have over 90% in the production [1].

The genus *Helianthus* consists of 52 species, 18 annuals and 34 perennials with different ploidy level, all native to North America [2,3] (Table 9.1). Of these, only *H. annuus* L. (2n = 2x = 34) and *H. tuberosus* L. (2n = 6x = 102) are crops. The cultivated sunflower is one of the most important annual oil crops, and Jerusalem artichoke is cultivated for its tuber. The prevalent parts of *Helianthus* are perennial and annual weeds. Some annual species such as *H. argophyllus* and *H. debilis* are used as ornamental because of their attractive appearance. Wild sunflower populations occur in wide range of habitats that are characterized by variation in environmental conditions such as photoperiod, growing season, minimum/maximum temperatures, and precipitation [4,5].

Wild Germplasm for Genetic Improvement in Crop Plants. http://dx.doi.org/10.1016/B978-0-12-822137-2.00009-6

TABLE 9.1 Sections, series, and subspecies for the *Helianthus* with chromosome numbers [6–10].

Section	Series	Species	Subspecies	(2n =)
I. Agrestis		*agrestis*		34
		niveus	*niveus, tephrodes, canescens*	34
II. Annuui		*debilis*	*debilis, vestitus, tardilorus, silvestris, cucumerifolius*	34
		praecox	*praecox, runyonii, hirtus*	34
		petiolaris	*petiolaris, fallax*	34
		neglectus, annuus, argophyllus, bolanderi, anomalus, paradoxus		34
III. Ciliares	*1. Pumili*	*gracilenthus, pumilus, cusickii*		34
	2. Ciliares	*arizonensis, laciniatus*		34
		ciliaris		68
IV. Atroru-bentes	*1. Divaricati*	*mollis, divaricatus, decapetalus*		34
		occidentalis	*occidentalis, plantagineus*	34
		hirsutus, strumosus		68
		eggertii, tuberosus, strumosus		102
		rigidus	*rigidus, subrhomboideus*	102
	2. Gigantei	*giganteus, grosseserratus*		34
		nuttallii	*nuttallii, parishii, rydbergii*	34
		maximiliani, salicifolius		34
		resinosus, schweinitzii, californicus		102
	3.Microcephali	*microcephalus, glaucophyllus, smithii, longifolius*		34
		laevigatus		68
	4. Angustifolius	*angustifolius, simulans, floridanus*		34
	5. Atrorubentes	*silphioides, atrorubens, heterophyllus, radula, carnosus*		34

9.2 Genetic improvements utilizing from sunflower germplasm

Wild *Helianthus* species, that is, crop wild relatives (CWR) represents a source for disease and pest resistance, for various abiotic stresses (i.e., drought, soil salinity, and extreme temperature), early maturity, cytoplasmic male sterility (CMS), and fertility restoration (R_f) [7]. Wild germplasm can increase the allele heterogeneity, thus improving the favorable agronomic traits and yield quality. The world's largest collections of wild sunflowers are a source of valuable germplasm, which provide materials for complex research in various fields of plant science aiming to improve plant resources, and in particular sunflower genotype [11]. For sunflower, one of most important oil crops, wide hybridization (interspecic or intergeneric) is still one of the most successful strategies for improving plant production. According to Kantar and coworkers [12], the term hybridization could be defined broadly as the offspring of individuals of genetically differentiated population, and the introgression as the sexual transfer of genetic material between genetically distinguishable populations.

The narrow genetic base of cultivated sunflower varieties hinders further improvement of the sunflower crop. Both, annual and perennial wild species of *Helianthus* genus have been involved in sunflower hybrid breeding aimed at the optimal parental combinations for higher yield and quality improvements [7,13–16]. The contemporary sunflower breeding programs directed toward high yield traits combined with an enhanced biotic and abiotic stress resistance. In this sense, hybridization and introgression breeding have been of central importance in the exploration of wild *Helianthus* relatives as donors of germplasm for genetic improvement of sunflower. Concerning sunflower production, the hybrid breeding is widely used ranking second after corn. Basic requirements of the breeding objectives changes depending on a complex of factors related to various new conditions meeting economic or environmental changes (product quality, resistance to disease, changes in agricultural practices, etc.) [8]. Currently, commercial hybrid sunflower production is aimed at meeting some basic requirements such as high seed and oil yield, changed oil composition, herbicide resistance, drought tolerance, and ornamental varieties [9,10,17,18].

Development of new sunflower genotypes via interspecific hybridization is a long, time-consuming, and complex process that involves different knowledge and activities and is the result of collective efforts. Each of the aspect of plant breeding (of agronomic or food value) is focused on particular traits, each with different variation. Given the complex nature of some features, it is difficult to determine the effectiveness of the selection process and the production of a particular genotype. In a traditional sunflower breeding program, several consecutive backcrosses are required to eliminate unfavorable alleles, which are inherited simultaneously with favorable alleles due to the existing genetic correlations between the individual traits [19]. To predict breeding performance and accelerate breeding process based on favorable alleles, DNA markers for QTL analysis and marker assisted selection (MAS) was efficiently applied. In general, modern selection is associated with the application of both molecular selection methods and traditional knowledge of available genetic plant resources. Therefore, development and identification of superior sunflower genotypes could be accelerated [20].

9.3 Morphological traits

9.3.1 Plant height

Plant height is a major phenotype characteristic of a crop and important quantitative trait affecting indirectly the yield in sunflower since it significantly correlated with total leaf area and petiole length [21]. Plant height at flowering along with flowering data and seed humidity percentage at harvest time are traits usually observed in sunflower breeding programs. Beyond phenotypic characteristic, these traits are important for breeding practice because they give basic information of the sunflower habitus; in contemporary breeding early varieties and medium high plants are preferred [8]. Thus, the prevalent part of sunflower varieties is 150–180 cm in height as reduced plant stature is an advantage in higher plant density per unit area, logging resistance, and some diseases [22]. The idea to develop dwarf sunflower lines seemed to be perspective, especially for farmers who like shorter varieties but it was established that dwarf lines are low yielding and have no close relation with lodging [8]. Thus, the most research on dwarf sunflower material stopped although dwarf lines were obtained after gamma irradiation of dry seeds, and after interspecific hybridization [23]. Concerning plant height, Skoric [24] and Kaya et al., [9] indicated that there was a positive correlation with seed yield [9,24]. These findings were confirmed by the results published by Vear [8] who reported that high yield of sunflower varieties often correlated with their plant height.

9.3.2 Head diameter

In sunflower, during domestication, the phenotypic and genetic changes associated with the transformation of wild plant into a cultivated crop have led to the formation of traits that characterize only the cultivated *H. annuus.* Thus, cultivated sunflower is unbranched monocephalic plant with a single large head (capitulum) in which achenes "seeds" are formed; they are over 6 times larger than the wild sunflower seeds. On the contrary, wild sunflowers (annual and perennials) are branched with highly variation in the degree and type of branching mainly apical and basal. Such plant architecture determines the formation of numerous small inflorescences (heads) in which too small light seeds are produced [25]. Unlike cultivated sunflower whose head diameter varied from 20 to 30 cm, the head diameter of wild sunflower sufficed for 2.4–8 cm [26]. *H. annuus* plants revealed variations in the shape of head that it can be flat, concave, or convex based on the breeding requirements [25]. In addition, overall plant habit of sunflower is determined by the stem and head inclination, which vary from erect with upright head inclination, droopy stem, and semi-erect stem with horizontal head inclinations; it may consider that this characteristic is an important factor of adaptation of plants to its environmental conditions [27]. Although the morphological traits are genetically determined, they are also influenced by growing conditions (environmental and technological condition), vegetation period, number of plant per hectare, etc. [24,27].

Head diameter influences substantially the yield of sunflower since it is directly related to the number of flowers and seed set per head. Numerous researchers have found significant positive correlation between head diameter and seed yield [24,28,29]. It is supposed that breeding for high number of filled seeds/head, head diameter ranging from 20 to 25 cm, and medium plant height with more leaf area would be correct and successful selection criteria for the enhancement of the seed yield in sunflower. The finding is consistent with results reported by Skoric [24] of correlation between head diameter and yield components, that is, increasing head diameter above the optimum value led to a reduction in seed yield (g/head) and oil content in the seeds, and enhancement of husk percentage and number of empty seeds.

The sunflower crop's production (in terms of oil or protein) consisted of several distinct components as each of them is related to diameter of capitulum (head). The number of seeds *per* head also depends on vascularization of sunflower capitulum that ensure the normal development and growth of floral parts. Poor vascularization of the receptacles may cause formation of empty seeds at maturity especially in the center of the head due to insufficient nutrients. Hence, from the point of view of sunflower breeding, the development of varieties with medium head diameter would be preferable, given the limiting effect of head vascularization. Poor seed feeling might significantly reduce the sunflower yield and therefore, the breeding of varieties with a good vascularization especially in the central area, is one of the main directions in the sunflower research and breeding [10,30].

9.3.3 Male sterility

Cytoplasmic male sterility (CMS) is an extranuclear maternally transmitted trait in which a plant fails to produce viable pollen grains but maintain female fertility. CMS in sunflower was first reported by Leclercq [31], who found it in plants originated from interspecific hybridization between annual *H. petiolaris* Nutt. (2n = 2x = 34) and the cultivated sunflower *H. annuus* L. (2n = 2x = 34). Male sterility is of great interest for both academic research and plant breeding for producing hybrid seeds. At present, the extensive use of CMS system (PET1) created a situation favorable for development and production of sunflower hybrid seeds, which give a 20% higher yield than the open-pollinated cultivars [32].

The specific mitochondrial genome of all known CMS forms, due to the structural rearrangements of mitochondrial DNA entails a dysfunction of respiratory metabolism of male gamete production while the female fertility remains unaffected [33]. Restorer-of-fertility nuclear genes (Rf genes) can suppress the manifestation of mitochondrial genes responsible for CMS [34,35], and therefore Rf-genes are important in commercial hybrid seed production of sunflower. Understanding the complex interaction between the mitochondrial and nuclear genes, which provides the CMS/Rf system, is a major research focus of plant genetics and breeding [36,37]. Thus, CMS/Rf system in plants is a phenomenon, which involves quite complex interactions between mitochondrial genes causing cytoplasmic dysfunction, and nuclear genes that can suppress this dysfunction [38]. Three-line system consisting of cytoplasmic male sterility line (A-line), the maintainer line (B-line) is isogenic of A-line with a fertile cytoplasm, and the restores line (R-line, male fertile), is a main system in hybrid sunflower seed production [39].

In sunflower, the different CMS form exhibit specific genetic features [40]. The development of commercial sunflower hybrids based of the only one CMS source, CMS-PET-1 (*H. petiolaris* Nutt.) and its corresponding fertility restoration gene Rf_1 predetermine a rather narrow genetic base that is a prerequisite for a genetic vulnerability to pests and other abiotic stress factors [41,42]. The increasing genetic diversity using different cytoplasmic background in hybrid development is necessary in order to avoid extreme yield losses results from homogeny plant material. The variety of CMS forms in sunflower is considerable (as many as 72) since the wild annual and perennial relatives of the genus *Helianthus*, are the source of genes for cytoplasmic male sterility; for 34 of them, fertility restoration genes are discovered [40,43]. The genetic studies revealed that CMS could appear in the breeding lines from interspecific crossing between *H. annuus* and wild *Helianthus* species [40,44,45]. From an agricultural point of view, the exploitation of CMS/Rf system is promising for the establishment of highly efficient hybrid production systems in sunflower. The importance of identification of additional CMS/*Rf* systems is obvious since an enforcement of practical hybrid breeding based on wider germplasm use could overcome the problems associated with using a single CMS/*Rf* system [46].

9.4 Biology and phenology of the sunflower plants

9.4.1 Flower fertility and restoring systems

Sunflower (*H. annuus* L.) inflorescence is a capitulum (flower head), a specific taxonomic characteristic of the *Asteraceae* family. The capitulum is a unique and complex structure that includes receptacle bearing two types of flowers: ligulate flowers (ray florets), which formed the periphery (an outer whorl), and tubular flowers in the center (disc florets). Part of the structural elements are the bracts, small green leaves (phyllaries) that form on the underside of the developing capitulum; they are arranged in overlapping concentric circles [10,30]. The sterile ligulate flowers are usually arranged in two rows; they are zygomorphic with brightly colored petals yellow or orange, which are highly attractive to bees and other insects. Usually the number varies from 30 to 70 ray flowers per head. The actinomorphic tubular (disk) flowers occupy the entire receptacle; they are arranged in concentric circles from the center of the head. The tubular flowers are fertile and bisexual having the male (stamens) and female (carpels) reproductive organs; after pollination and fertilization they produce the one-seeded immature achenes (seeds). The number of tubular flowers varies from 30 to 3500 depending on the head diameter [47]. Although the sunflower inflorescence superficially resembles a large flower, it actually consists of many highly aggregated distinct florets with different functions [47]. Furthermore, it has been shown that the *Asteraceae* inflorescence has the characteristic of one single flower not only morphologically but also at the molecular level [48].

The flowering period is one of key phenological phase of crop development; in sunflower, it varies from 9 to 15 days depending on the plant genotype and environmental conditions [49]. The floral stage begins with the appearance of the ray florets in the sunflower head. The tubular florets gradually open in concentric rings toward the center of the head, thus increasing the flowering time. Flowering of each tubular floret lasts 3 days. An average of 1–4 rings of flowers blooming each day as each disk flower is capable of producing a seed. At the end of flowering, the ray florets wilted and dried while the disc flowers fall off just before maturity.

Sunflower is an allogamic plant, whose reproduction is associated with self-incompatibility system, that is, only compatible (non-self) pollen grains is able to effect fertilization. The pollen is relatively large with prickly appearance, which allows it to be carried by insects [24]. The pollen grains remain viable for several days. The sporophytic self-incompatibility system of sunflower is genetically determined; the two lobs of the stigma surface discriminate the incompatible (self) pollen but they can accept pollen from other florets of the same head [50]. Nowadays, the successful breeding for high self-incompatibility in sunflower resulted in developing of about 68% self-pollinating commercial hybrids [51]. Nevertheless, during flowering, honeybees visit the sunflower inflorescences and pollinate the plants, thus enhancing their yield. Therefore, the breeding of sunflower varieties attractive to pollinators is an essential part of some breeding programs [49,52].

9.4.2 Earliness, flowering, and maturity

Sunflower, a crop sown in spring and early summer is cultivated under different climatic conditions; its growth occurs mainly during the summer and includes both water shortage and high temperature [53]. Therefore, the sunflower is considered to be a drought-tolerant crop; the plants can be grown on water shortages and water-deficit soils [54]. Thus, vegetation period of sunflower is an important characteristic associated with plant productivity; obviously, a shorter growing season is a prerequisite for higher yield in some climate areas. For sunflower breeding, the earliness defined as the shortest amount of the time to produce a profitable crop, is important to avoid frost damage, soil moisture depletion, and insect and diseases infection. For example, penetration and successive developmental stages of the fungus are favored by relatively high temperature and moisture during spore germination [55]; also, the risk of end-of-cycle diseases is increased when there is a rain in summer. Earliness also has other advantage, such as allowing rotation with winter crops or extending the season for harvesting [8]. Therefore, the earliness should take advantage of the improved production efficiency through shortened growing season [53]. Earliness is affected by three main parameters-environmental conditions (photoperiod, soil temperature, precipitation decreases or frequent rainfall, low temperature either day or night, etc.), genetic potential of the genotype, and management (often more important than genotype to the maturity of the crop), and all they have substantial importance in determining productivity realized during the given season. In addition, development of new sunflower cultivars combining short vegetation period with high productivity could increase sunflower area in the north [53]. In terms of potential yield, evaluation of the interaction between duration of growing season (vegetation period) and plant productivity, which includes seed yield and oil components seems to be a key component of crop performance improvement.

Physiological maturity, that is, the stage of maturation when the seed moisture content is low enough to permit harvesting of the seeds, occur approximately 30 days after anthesis. Due to the specificity of the inflorescence, the seed filling from the outer and inner part of the head is different, as the seeds from the inner part have a shorter time for filling. Plant maturation (the time between physiological maturity and harvest) is characterized by chlorophyll degradation and the dominant manifestation of other pigments such as xanthophyll and other carotenoids presented in the chloroplasts [56]. The color of the bracts of the capitulum also changes from green to brown while the color of receptacles base turns a buttery yellow. During the period from physiological maturity to harvest, the sunflower seeds lose their moisture [56].

9.5 Yield traits

9.5.1 Seed yield

Cultivated sunflower is one of the most important annual oil crop cultivated in different regions of the world. The oil extracted from the sunflower seeds (38%–50%) is characterized by a high content of un-saturated fatty acids (about 90%), and the lack of cholesterol and toxic constituents [57,58]. Sunflower oil is used mostly for edible purposes as a cooking and salad oil, shortening and margarine as well in pharmacy, cosmetic industry, floriculture, honey production, and paints. In addition, one of the interesting applications of the sunflower lately is its use as an eco-friendly alternative to diesel fuel [10,59,60].

The main selection criteria in sunflower breeding are the high oil yield (the mathematical sum of seed yield and percent seed oil content), and oil quality. The crops yield could be considered as a complex of several yield-related parameters: (1) the number of plants *per* hectare; (2) the number of seeds *per* plant; (3) the 1000 seed weight; and (4) the oil (and protein) content of the seeds. A high seed yield ensures that farmers realize higher profit margins and high oil percentage, while improved oil quality increases demand by end-users [10].

The parameter plants *per* hectare depend on plant density, one of the most important agronomic practices. It is generally recognized that appropriate plant density is necessary for optimum yield of sunflower and affect the quality factors of oil [61]. For example, competitive relationships when the plants density is highly reduce the head size and seed weight. In addition, high density negatively affects the lodging resistance and accelerates the spread of phytopathogens [62]. It has been reported that plant density affects the seed production, the growth stage (days to 50% flowering), diameter of capitulum and stem diameter, thousand seed weight, as well as the composition of sunflower oil mainly the ratio between linoleic (18:2) and oleic (16:1) unsaturated fatty acids [63,64].

One of the main characteristics of the sunflower yield, the number of seeds *per* plant, is the result of the influence of many factors such as the plant vigor, total leaf area *prior* flowering, and the defoliation after flowering, etc. [65]. In addition, the adequate vascularization of sunflower capitulum determines the uniform nutrition and development of the seeds from the periphery toward the center. In this sense, a large capitulum diameter can be considered inappropriate for higher seed yield due to the poor vasculature that supplies nutrient to central seeds [8]. The optimal diameter for sunflower capitulum is considered to be in the range of 16–22 cm. According to Skoric [24], the smaller capitulum diameter is usually an indicator of decreased seed yield because of lower number of seeds *per* head or poor seed set in the centrum of capitulum.

The seed yield and oil content are complex characteristics with low heritability and dependent on genotype/environment interaction. It was found that all the productivity parameters which include seed yield, oil, and protein content, are highly dependent on environmental conditions (radiation, temperature, rainfall distribution, time of sowing, plant density, and planting pattern, etc.) which may act individually or collectively [9,24,51].

9.5.2 Thousand seed weight

One of the indicative parameters of seed quality in crop plants is thousand seed weight (TSW). Like other quality traits, TSW is a complex indicator that is strongly influenced by two crucial stage of plant development, that is, seed germination (viability, germination rate, percent, and energy) and seedling stage [10]. Thousand seed weight correlates with the seed size as both traits affect plant fitness and seed quality, and therefore the crop yield per unit area and harvest efficiency. The seed size is influenced by growing conditions and agronomic practices such as sowing rate and plant density during the vegetation period. The production of high quality seeds that are physiologically mature and healthy with high germination ability is one of the main goal in sunflower breeding [24].

Larger seeds were found to be superior in respect of germination, vigor of seedling, field emergence, crop growth, and biological yield. It was reported that sunflower seeds with high TSW possesses more reserve nutrients and larger embryo size providing higher germination potential of the seeds and growth acceleration, which is an advantage in unfavorable environment [28]. Seed quality is influenced by different factors that may occurred during seed formation and development in the field at flowering and pollination stages, while other occur during the harvest at different maturity stages and post-harvest. Hereditary factors crucially influence this trait, although the implementation of appropriate agronomic practices in sunflower cultivation can reduce the impact of negative factors and contribute the seed yield improvement [24].

9.5.3 Oil quality

The sunflower oil is an example of oil rich in unsaturated fatty acids; it is characterized by a predominant percentage of two-linoleic (55%–65%) and oleic (20%–30%) fatty acids. The essential polyunsaturated linoleic acid (C18:2) is the major omega-6 fatty acid, with two double bonds while oleic acid (C16:1) is a mono-unsaturated omega-9 fatty acid. The saturated fatty acids such as palmitic (16:0), stearic acid (18:0), and very small amounts of other fatty acids make up 10% of the total oil content [24,66,67]. The fatty acids composition, and in particular the ratio of unsaturated fatty acids, largely determines oil quality. Oil quality depends on the genotype and it interaction with the environment. Usually the common sunflower oil contains a high amount of linoleic acid, medium amount of oleic acid, and low amount of linolenic acid [68–70]. Oils having high percentage of oleic acid are more stable and can be helpful in minimizing the risks of cardiovascular diseases in humans due to the capability of oleic acid to decrease the triacylglycerides and cholesterol levels [70]. One preferred alternative of traditional sunflower oil is the high oleic oil (more than 75% of oleic acid), that has lighter taste, good frying characteristics, and oxidative stability, thus contributed to a healthy diet [70].

The oil content and fatty acid composition of mature sunflower seeds (achene) is dependent by the genotype and environmental conditions. The most advanced breeding program aimed to increase the oil concentration in sunflower seeds, thus improving the yield and oil content potential [67]. The oil formation in the embryo starts several days after anthesis and continues during the whole achene growth period. Studying the dynamics of the oil accumulation in the embryo revealed that during the first third of achene filling stage, very little oil was deposited. Later, lipid accumulation increases rapidly until physiological maturity and final oil concentration [71]. It should be noted that a relation between the final oil concentration in the achene and their ability to dehull was found.

Thus, sunflower seeds containing higher oil usually are difficult to dehull. In addition, the ratio between hull and kernel in sunflower achene is different when compared to the non-oilseed type sunflowers containing more hull (47%) and the oilseed-types (20%–30%) [67].

The sunflower oil content and fatty acids composition are determined by the genotype and environmental factors [72,73]. During seed filling and maturation, the sunflower plants require appropriate temperature to accumulate high quality oil [74,75]. The fatty acid composition also depends on temperature during seed filling as it has been observed that the temperature has a greater effect on the fatty acid composition than on oil content. For example, during seed development stage, each 1°C increase in temperature leads to about 2% increase in oleic acid content. At higher night temperatures, a decrease in the oleic/linoleic acid ratio has been identified [76]. Thus, the sunflower oil yield is not a constant parameter as it is largely determined by changing environmental conditions (years, location, sowing dates, etc.) [24,67].

9.5.4 Protein and other nutrient content

Dry sunflower seeds (kernel or seed endosperm) contain about 20%–40% proteins as the seed storage proteins sufficed for 85% of the total protein content [77]. The major constituents of storage proteins are salt-soluble 11S *globulins* (*helianthinin*) and water-soluble sunflower albumins (SFAs), known as the 2S albumins. The 11S *helianthinin* is an oligomeric globular protein, which consists of six spherical subunits with legumin-like structure; they contain two polypeptide chain (an acidic and a basic) linked by a single disulfide bond [78]. Taking information found in the literature, helianthinins and 2S albumins are considered the major protein fractions in sunflower kernels. It was reported that in dry sunflower seeds, the amount of helianthinins account for almost 60% of the total proteins and the 2S albumins for about 20%, respectively [79]. The 2S albumins found in seed of most dicotyledonous plants belong to a widely distributed family of seed proteins [77]. The albumin seed proteins consist of a single polypeptide chain, which possesses more sulfur containing amino acids than globulin [77].

The nutrient profile of sunflower seeds also includes a third group of proteins, the oleosins, which represent 14% of oil body's proteins. The oleosins form a thin layer around the oil bodies (intracellular organelles for storage triacylglycerols in oleaginous seeds) and prevent coalescence. Polypeptide composition of *helianthinin* and albumin fractions are qualitative and quantitative similar among *H. annuus* cultivars. It was concluded that one way of improving agronomic characteristics of sunflower cultivars or oil productivity is to include in the hybridization process the genetic resources of wild *Helianthus* species [67,80]. The studies of the genomic organization of seed storage proteins are based on the data on hereditary variation of their subunits and polypeptide composition [81]. Molecular analysis indicated that *helianthinin* heterogeneity is caused by the presence of multigene families in the sunflower genome [82].

The sunflower proteins enter in human nutrition in different forms. High quality proteins extracted from sunflower seeds are superior to most vegetable proteins in digestibility and comparable in biological valued [83]. In addition, the sunflower meal has potential as a human food because its creamy color, pleasant nutty flavor and stability, free from major allergens, and appropriate for vegan [84]. The protein isolates from industrially obtained sunflower meals are of interest as an alternative novel protein sources that can replace the current popular protein sources. Seeds also contain glutamine acid, asparagine, arginine, cysteine, and essential amino acids-phenylalanine and tyrosine, leucine, methionine, and cysteine. Overall, the sunflower seeds possess little antinutritional factors and balanced amino acid content [85]. Added to wheat-based breads, sunflower seeds improve significantly its quality due to the higher water absorption of the dough, and thus affecting its baking characteristics [86].

9.6 Other agronomic traits

9.6.1 Heavy metal tolerance

The growing sunflower has the potential to absorb and accumulate high concentrations of metal contaminant such as Ni, Cu, As, Pb, and Cd and therefore it can be classified as a plant "hyper accumulators" of heavy metals [87,88]. In general terms, such species contain from 100- to 1000-times more heavy metals in their aboveground tissues than the normal plants cultivating on field with background metal concentration [89]; at the same time these concentrations have not significant effect on the plant development and yield production. In addition, rapid plant growth and biomass accumulation is also an advantage. All this makes sunflower, a very preferred crop for remediation of heavy metal-contaminated soils as the plants have different mechanisms for adapting to the toxic effect of heavy metals [88]. Therefore, the sunflowers are the most investigated plants in remediation practices and phytoextration of contaminated sites [90]. Apart from the industrial application of sunflower for edible oil, the use of field-grown crop in phytoremediation as an alternative to the destructive physical methods seems to be an inexpensive and aesthetically pleasant approach to remediate polluted soils.

9.6.2 Mineral deficiencies and toxicities

Sunflower requires 13 elements (N, P, K, S, Ca, Fe, Mg, B, Mn, Cu, Zn, Mo, and Co) for normal functioning and development in addition to carbon (C), hydrogen (H), and oxygen (O). Plants cannot survive and grow without the presence of these elements, known as essential or nutrient elements. The essential elements in plant nutrition are required in the plant at optimal concentration; the insufficient amount of these elements can cause disturbances in the life cycle of a growing plant. Based on the plant requirements, the nutrient elements, other than C, O and H, are determined as macronutrients (major nutrients) and micronutrients (trace elements) [91]. On the other hand, the excess of some of the mineral elements needed for the plant can be toxic and lead to reduced plant growth and development. This effect of nutritional disorder is known as toxicity. In particular, aluminum (Al) in soils with low pH as well as sodium (Na) in saline and sodic soils in high concentrations may cause deleterious effects on plant development and plant growth. Different sources including air pollutants and soil application of commercial fertilizer, industrial pollution, and human action may cause heavy metal toxicity. Most often, these toxic elements are cobalt (Co), chromium (Cr), nickel (Ni), lead (Pb), thallium (Tl), and cadmium (Cd). Since plants need certain amount of nutrient elements, the excessive use of fertilizers may cause deleterious effect to plant growth. Such nutrients over concentration in the soil have toxic effect known as Mn toxicity in acid soils, B toxicity in alkaline soils, Zn and Cu toxicities [91]. Nutrient deficiency and phytotoxicity cause plant stress that often takes the form of what is recognized as mineral deficiency symptoms and/or reduced growth and yield. In addition, they cause visible symptoms that can be important markers for the diagnosis of nutritional disorders in the field conditions. Studying the effect of mineral elements and their essential and specific function in plant metabolism as well as the interaction among genetic potential of plants and mineral stresses (deficiency or toxicity) has economic relevancy since adequate sunflower nutrition is crucial for ideal growth of plants. In particular, investigation of the resistance/tolerance to mineral deficiency and mineral toxicity is an important element of sunflower breeding.

9.6.3 Cold tolerance

Cold stress is an essential climatic factor, which influences the crop yield. For surviving extreme temperatures, plants developed mechanisms of tolerance to temperature stress are common to their habitats. The optimal temperatures for plant growth and development are determined both genetically by the plant genotype and physiologically by the phase of growth and development of the given genotype [92]. Cold temperature stresses can be considered as chilling stress (0°C–15°C) or freezing stress ($<$ 0°C) depending on the tolerance of plants to different low temperatures. Cold stress affects many physiological and biological characteristics of the plants because of modification in gene expression and plant metabolism caused by low temperatures [93]. For example, plasma cell membrane disruption, changes in protein synthesis, and impaired cellular respiration, photosynthesis and carbon fixation may occur [94].

In sunflower, the cold sensitivity and its damage depend on the growth phase of the plants as the degree of negative impact depends on different factors such as crop stage, the severity, and temperature duration [92]. Sunflowers are most frost tolerant in both early growth stage—emerging and the cotyledon stage.

The plants become more sensitive to frost in the next vegetative stages (especially at flowering and during pollination) when low temperature over several days can cause branching from axillary stem buds and loss of apical dominance, multiple heads, and delay maturity; yield and quality are greatly affected [93]. A killing frost in sunflowers is considered to be at −4°C to −5°C for 6 h or more [92].

Tolerance to low temperatures arises as important issue for breeding programs [92] due to the requirements of early sowing in order to maximize the growing season and to avoid the critical water stress period [95,96]. The sunflower earliness (early sowing date) increases the importance of low temperature tolerance. To improve cold capacity and freezing tolerance of crop plants, breeding programs aiming development of cold tolerant genotypes were initiated but they have limited success [97]. However, the use of available wild sunflower germplasm in appropriate breeding programs (wide hybridization) may be an opportunity to develop cold tolerant genotypes. Such one wild species collected at low temperature areas is perennial *H. maximiliani*, which was crossed successfully with cultivated *H. annuus* [98].

9.6.4 Heat tolerance

Heat stress or exposure of the crops to temperatures higher than the optimum for plant growth and development induced negative alterations in plant productivity and its quality. The magnitude of yield reduction because of the heat stress is determined by duration, severity, and plant growth stage. High temperatures affect plants differently by either high day or night temperatures, and high aerial or soil temperatures. The optimum temperature for plant growth and development varied with plant species. The lowest base temperature at which development rate is essentially zero and seed germination is barely possible is 7.2°C [10]. The base temperature is specific since it depends on the stage of plant growth. It was established that the optimum temperature for sunflower growth is 25°C (mean daily temperature) while the optimum night temperature is 21°C [99]. It has been shown that at these optimal day/night temperatures, the sunflower plants have the highest growth rate and give maximum production [10]. The threshold mean daily temperature for sunflower growth and reproduction is 25°C–26°C. A higher temperature than this threshold causes heat stress [100]. For example, sunflower head subjected to temperature >34°C during 7 days produced smaller seeds with lower oil quality; this effect decreased with progresses of grain filling [101].

High temperatures produced an inhibition effect on all stages of plant growth and development. Heat stress might impair cellular, molecular, biochemical, physiological, and agronomic characteristics of sunflower, resulting in the restriction of plant growth. High temperature stress is associated with injuries of cellular membranes [102], reduction in leaf gas exchange (decreased rate of photosynthesis, transpiration, and stomatal conductance), and chloroplasts disruption. In addition, the detrimental effects of heat stress on pollen development probably led to a reduction of plant reproduction [103]. High temperature stress during anthesis was associated with a decrease in the pollen viability and germination ability. This type of disturbance may lead to an unsuccessful fertilization due to low germination ability of pollen grains even when large amounts of pollen are produced [104]. Likewise, high temperatures cause ovule sterility [105] and delay embryo growth [106]. Due to synthesis of specific growth regulators, heat stress led to severe changes in crop biology (phenology and plant productivity) [107].

The development of heat-tolerant sunflower varieties is an attractive option for yield improvement under high temperatures. Improving yield includes data of thermo-tolerance key traits that can be used as selection criteria for further implement heat tolerance into sunflower. The specific screening criteria for evaluation of breeding material involve clear differentiation between resistant and susceptible ecotypes under heat stress [108]. Breeding and selection of heat tolerant sunflower genotypes in the field involves a large set of strategies and depends on both genotype and environmental conditions. These strategies may include the choice of appropriate germplasm as parents, many different crosses, selection for traits that confer stress resistance and high productivity under heat stress, and testing of a large number of developed hybrid lines. The field experiments and cultivation of plants under real stress conditions are indicative of plant tolerance and their yield productivity. Based on many experimental data, two traits were identified as target of sunflower heat-resistant breeding (e.g., leaf and head orientation at the anthesis) [109]. These specific traits can be used as indicators of heat tolerant sunflower genotype. These features are simply inherited and are independent of genotype *x* environment.

9.6.5 Drought tolerance

Drought, a condition when plant is unable to meet its transpiration demands, is one of the most studied abiotic stresses [110–114]. The adverse effect of drought on growth and development of crops could affect at all stages of the plant growth; these effects are more harmful than any other abiotic stress factors (high and low temperatures, frost, heavy metals, salinity, etc.) [115,116].

In sunflower, drought is one of the most serious environmental stress that causes significant yield losses [117], even though the relative drought sensitivity of the species is defined as low to medium. Breeding for drought-resistance in sunflower aiming variety improvement, combined with theoretical research and field management is required. An approach allowing for good progress in breeding includes producing a high yielding genotype with improving drought stress tolerance [118]. A prerequisite for improving cultivated sunflower throughout wide hybridization is determining the key drought tolerant/resistant traits for *H. annuus* and wild *Helianthus* species, respectively. In addition, comparative investigations of annual and perennial relatives give more options to study in details the mechanisms responsible for drought tolerance in sunflower [119].

Wild annual and perennial *Helianthus* species are valuable genomic resources and a real reservoir of genes for resistance or tolerance to abiotic and biotic stress factors [92,120,121]. Interspecific crosses permitted to obtain sunflower varieties with improved drought resistance via introgression of drought resistant genes into *H. annuus* genome. Several wild relatives have been recognized as potential donors of drought tolerance related traits. One such species is *H. niveus* ssp. *tephrodes*, a desert-inhabiting and desert-adapted species, which is of particular interest for including in breeding [122]. Wild *H. anomalus* is also considered as drought tolerant species. This species has the largest achenes among other wild species and contains a relatively high amount of oil [123]. *H. argophyllus* (silver-leaf sunflower) has attracted interest as a possible source of genes for breeding high drought tolerant cultivars. In this species, several traits of potential interest for improving drought tolerance of *H. annuus* have been identified. For example, higher content of cuticular waxes that reduce evapotranspiration losses, stronger development of rooting system, small leaf area, and capacity to maintain net photosynthetic rate under osmotic stress [114,124]. As result, the developed interspecific lines reveal higher water use efficiency, higher harvest index, and better drought susceptibility index under drought conditions [114].

9.6.6 Salt tolerance

Soil salinity, in common with other environmental stresses, is one of the main abiotic stresses that affects the production and cultivation of crops. Salt affected soils contain high amounts of sodium (Na^+), potassium (K^+), calcium (Ca^{2+}), and magnesium (Mg^{2+}), and chlorine (Cl^-) ions [125], which result in high osmotic pressure in the root zone. It was established that salinization affects a number of processes throughout the life cycle of plants at all levels of organization (excretion, ion transport, nutrition, selectivity, and compartmentation are involved, together with growth, water use, and water use efficiency). The change in the crop-water relationship limits availability of water and nutrient, damage the water uptake ability of plants, and cause ion toxicity [126]. As a result, the growth and development in sensitive species is delayed and the physiological processes such as photosynthesis, respiration, and protein synthesis are reduced [127–131].

In terms of salinization, plants reveal different degrees of tolerance, which can be determined as sensitive, moderately sensitive, moderately tolerant, and tolerant [132]. *H. annuus* is defined as moderately tolerant to salinity [133]; seed productivity is unaffected by salinity up to EC 4.8 dS m^{-1} (50 mM NaCl) while the increase of accumulated salt in the soil above this value led to significant decrease of seed yield [134]. It has been found that salinity reduced plant growth and development, plant productivity and is associated with adverse effects on photosynthesis and some biochemical processes in plant tissues [135–138]. In addition, salt stress might influence oil production and oil yield of sunflower [139]. Salinity tolerance of sunflower varies significantly among cultivars [135].

The wild *Helianthus* species are useful group for applying hybridization with cultivated *H. annuus*, because they may serve as wild reservoir of diversity of "row material" required for the development of novel varieties [140,141]. One species, *H. paradoxus* Heiser, the Pecos or "Puzzle sunflower" is adapted to extremely saline and alkaline soils. *H. paradoxus* is a rare endangered species, and is considered a hybrid originated from a cross between two salt sensitive species *H. annuus* × *H. petiolaris* [141]. The prospects of developing sunflower genotypes with higher level of salt tolerance are much greater if one or a few defined sunflower wild species are involved in hybridizing process. Identifying phenotypic, physiological, and biochemical traits that contribute to adaptation to salinity has practical importance to enhance stress tolerance of sunflower.

9.7 Disease resistance

Downy mildew (*Plasmopara halstedi*) is one of the most harmful and economically important diseases in cultivated sunflower for which over 40 pathotypes are determined by *Pl* to genes 1–21 as well as *PlArg* [142,143]. Most of these resistant genes were transferred from wild sunflowers (e.g., Pl5 genes were originated from *H. tuberosus*, *Pl7* from *H. praecox* ssp. *runyonii*, *Pl8* and *PlArg* genes from *H. argophyllus* as well as from wild *H. annuus*, *H. petiolaris*, *H. bolanderi*, *H. salicifolius*) [2,144–147]. These resistant genes as mostly single, race-specific major dominant were used widely in the sunflower breeding programs almost all part of the world and *Pl13* and *PlArg* genes resistant against were found resistant to many races of the sunflower [10,124,148–150].

Powdery mildew (*Erysiphe cichoracearum*), is sunflower foliar disease especially in planting of warmer areas. Resistance was determined by a dominant gene observed in wild *H. decapetalus* as well as by multiple genes exhibiting full resistance in annual wild species (*H. argophyllus*, *H. agrestis*, *H. debilis*, *H. praecox*), perennials (*H. angustifolius*, *H. atrorubens*, *H. ciliaris*, *H. debilis*, *H. laevigatus*, *H. rigidus*, *H. salicifolius*, *H. pauciflorus*, *H. resinosus*, *H. tuberosus*), and some interspecific hybrids [10,144,151]. Sunflower rust (*Puccinia helianthi*) is another foliar disease observed mostly in confectionery types in almost all sunflower planted areas in the world. Some resistant genes (*R1* and *R2*) were identified in wild *Helianthus* species and were transferred into cultivated sunflower then used widely in many sunflower breeding programs in the world [10,140,150]. However, some novel genes also determined in annual species as *H. annuus*, *H. argophyllus*, and *H. petiolaris*, and in perennial species as *H. grosseserratus*, *H. maximiliani*, *H. nuttallii*, *H. pauciflorus*, and *H. tuberosus* and they were observed immune reactions in the conducted tests [2,10,32,152,153].

Verticillium wilt (*Verticillium dahliae*), one of the important foliar diseases also in sunflower production and like other diseases, the major sources for developing resistant genes is again wild sunflower species. Wild *H. annuus, H. argophyllus, H. petiolaris*, and *H. praecox* are reported in some studies as immune for *Verticillium* disease [10,140,150,154]. For resistance to *phoma* black stem (*Phoma macdonaldii*) resistance, perennial species such as *H. decapetalus, H. eggertii, H. hirsutus, H. resinosus*, and *H. tuberosus* [8,155]; for *phomopsis* stem canker (*Diaporthe helianthi*) perennials such as *H. maximiliani, H. mollis, H. resinosus, H. pauciflorus, H. tuberosus*, and *H. hirsutus* and annuals *H. debilis, H. argophyllus, H. niveus, H. neglectus, H. petiolaris*, and *H. salicifolius* [10,150,155]. Furthermore, Christov [144] reported that some resistant interspecific hybrids also developed utilizing from *H. eggertii, H. laevigatus, H. argophyllus, H. mollis*, and *H. debilis*. *Alternaria* leaf spot (*Alternaria helianthi*) resistance was observed in perennials *H. divaricatus, H. hirsutus, H. occidentalis, H. pauciflorus, H. tuberosus, H. resinosus* as well as in some wild annual species, *H. praecox, H. debilis* ssp. *cucumerifolius*, and *H. debilis* ssp. *silvestris* [10,17,140,156].

9.8 Broomrape parasitic resistance

Broomrape (*Orobanche cumana* Wallr.), is a non-photosynthetic root parasitic weed that specifically infects the sunflower (*H. annuus* L.) plants and reduces seed yield up to 100%. *O. cumana* affects crops in Mediterranean, especially in countries around the Black See, Central and Eastern Europe, Asia, China, and Mongolia, where heavy infestation of sunflower fields is a serious problem [157]. In the last 15 years, the broomrape is considered as one of the most serious biotic factor limiting sunflower cultivation. The most important feature of this pathogen that makes it difficult to overcome is its high multiplication rate, a result of production of large amount of tiny seeds, which remain viable in the soil for a long time [158]. Broomrape affects negatively on the sunflower growth and development due to its physical and metabolic overlap with the crop. The infected sunflower plants usually have reduced total biomass; in general, they are smaller with reduced heads [159–161].

The broomrape life cycle consisted of several stages, which include seed germination, plant flowering, and seed production [162]. The germination stage of the *O. cumana*/*H. annuus* pathosystems is one of the most investigated period of the broomrape life cycle [163]. The synchronization of the life cycle of the parasite weed to that of the host species occurs in early developmental stages with specific chemical stimuli from host roots [164,165]. Two main types of plant-derived germination stimulants secreted by the host root system are known to induce *O. cumana* seed germination-strigolactones and sesquiterpene lactones [166–169]. Strigolactones, a group of plant hormones, regulate shoot branching and are exuded by sunflower roots to the rhizosphere at very low concentrations; they accelerate germination of the root of broomrape seeds [170]. The guaianolide sesquiterpene lactones just like strigolactones act as interspecific chemical signal and stimulate germination of *O. cumana* seeds [167]. The critical step in parasitism is the phase after germination when start the process of parasitism. The parasite penetrates the sunflower roots through haustorial cells developed at the tip of *O. cumana* radicle; the same cells are involved in the penetration and vascular connection between *O. cumana* and the sunflower tissues [171]. After initial vascular connection, the weed develops underground structures (tubercles) on the host roots, which uptakes and accumulates nutrients and water. Finally, after pollination, *O. cumana* plants produces many capsules yielding hundred or thousand seeds per plant, which remain viable in the soil for many years [172]. The control of parasitic weeds must include a number of approaches (chemical and biological control, prevention, hand weeding, and exploitation of crop resistance by the use of genetically modified parasite-resistant and herbicide-resistant crops) at all stages of broomrape life cycle.

Broomrape has produced many races and improved itself every 15–20 years cycles in last 1960–70s history with starting from A, B, C, D, and E (calling old races), (determined by Or 1, Or2, Or 3, Or 4, and Or 5 dominant genes), respectively [173,174]. In the last 20 years, the F race controlling the Or6 gene has appeared in Spain, Romania, Russia, and Bulgaria [10,165,174,175]. Recently the G race has emerged and resistance was found in *H. debilis ssp. tardiflorus* [176,177], controlling by dominant in a single locus then they were transferred into cultural types via interspecific crossing. Furthermore, wild species are the major source for broomrape resistant and the study reported that most of the perennial species exhibited resistance fully to F race as well as *H. anomalus* and *H. agrestis* annual species and *H. grosseserratus* and *H. divaricatus* interspecific hybrids [178–181].

9.9 Insect resistance

Although more than 150 phytophagous insect species have been found to attack cultivated sunflower, only a few of them consistently produce economic damage. Most insects are considered occasional pests and are managed with conventional or reduced risk insecticides [182]. According to the National Sunflower Association, the insect species whose damage has economic implications include various *Coleopterans* and their larvae, several *Lepidopterans* and their larvae, a few *Dipterans* and their larvae, and a single *Hemipteran* [183]. Some of the most widespread and damaging insects to cultivated sunflower are: the sunflower moth and the banded sunflower moth, the sunflower seed weevils (red and gray), the sunflower stem weevil, the head clipper weevils, the sunflower beetles, cutworms, the dectes stem borer, and the sunflower midge [184].

Regular monitoring of the infestation of sunflower ecosystem requires a field check for insect activity, determining the insect species, their population, the growth stage of the insect and the crop, and yield losses due to insect pests' attack [185]. At different developmental stage, sunflowers are attacked by many insects, which utilize every part of the plant [182]. These insects (most of them not specific of sunflowers) reduce seed yield and oil quality extracted from the seeds. Several insect pests attack the early seedling stage of sunflowers causing complete destruction of plant stands; infected sunflower seedlings lose their ability to regrow or tiller. The population of insects gradually increases in parallel with the foliage of the plants and is greatest at the beginning of flowering phase. In some cases, insects are vectors of viruses and diseases [186].

According to the part of the sunflower plant attacked, insect pest of sunflower can be grouped as follows: (1) leaf feeders, (2) stem and root feeders, and (3) head feeders [187]. Leaf feeders are considered the easiest group to control because they are easily monitored, and effective insecticidal controls can be applied when necessary. Head feeders are potentially very damaging and once they have entered the sunflower head, are difficult to control with insecticides [182,187]. Some of the pests attack the crop in the field and in the storage and cause serious economic loss [188].

Resistance of the host-plant to insect pests is a cardinal tool to limit insect damage. It includes combinations of mechanisms and traits, such as (1) chemical and physical antibiosis (if the trait affects vitality or fecundity of the pest), (2) chemical or physical antixenosis (traits that counteract destructive behaviors of a pest source; e.g., insect reactions to morphological traits such as trichrome or pubescence), and (3) tolerance (ability to recover from withstand injury). In addition, understanding the life-cycle of the insect pests is an approach to reduce damages with limited resources [189].

Screening of host-plant resistance should be based on using the existing sunflower defensive traits such as achene structure (pericarp hardness), diterpenoid acids, phytomelanin barriers and terpene repellants in glandular trichomes [190,191], coumarins, and some morphological and physiological characteristics supplemented by interdisciplinary approaches for efficient screening and breeding. Considering the fact that many wild *Helianthus* species are resistant to the major insect pest of *H. annuus*, introgression in which herbivore resistance traits (alleles) has transferred from wild *Helianthus* relatives to the genome of sunflower seems to be very helpful in the development of hybrids with increased adaptation. It was supposed that such resistance is result of toxic and antifeedants effect of sesquiterpene lactones and diterpenes synthesized in glandular hairs of *Helianthus* species [192]. Thus, breeding for higher density of glandular hairs on floral parts surrounding the pollen can be successful approach to enhance resistance to sunflower moth predation [192]. In support of this hypothesis are the results obtained by Whitney et al., [193]. The authors found that a hybrid derived after cross *H. annuus* x *H. debilis* named *H. annuus texanus* was more resistant to several insects such as seed midges *Neolasioptera helianthis* and to receptacle/seed feeding *Lepidoptera*.

9.10 Herbicide resistance

Applications combining agricultural processes to reduce the costs in crop production have increased recently and preferred by farmers frequently in the world. Therefore, herbicide resistant crops (mostly GMOs) have used commonly in many important crops such as soybean, corn, wheat, etc. However, based on banning GMO products in some countries as well as in European Union and production of non-GMO herbicides resistant genes have become also popular in this context. Under the name CLEARFIELD (Clean Field) commercial brand by BASF, IMI (Imidazolinone) group herbicides (Imazamox, Imazapyr, etc.) with IMI herbicide resistant varieties are widely used in the world in some crops such as sunflower, soybean, corn, wheat, canola, and paddy rice. First IMI herbicide resistance was discovered in sunflower wild population (*H. annuus* L.—ANN-PUR) in a planted 7-years soybean field in Kansas, United States, by Al-Khatib and coworkers [194]. Applying embryo culture, these resistant genes were transferred into the *H. annuus* parental lines by backcrosses. The populations thus developed are called IMISUN populations [195–197]. IMI resistance is controlling by two genes as one major semi-dominant gene Imr1 is allelic variant of the AHAS1 locus and modified gene Imr2 [198,199]. Lately, some IMI resistance genes were found in some wild *H. annuus* population in Canada border.

In addition to this resistant sources, a new IMI herbicide resistance gene (CLHA-Plus) was developed by BASF in Argentina via chemical mutation [200,201]. However, both sources are not dominant genes so both parental lines should have IMI genes to develop IMI sunflower hybrids and now both IMI resistant ones are selling sunflower in the market in the world. Currently, IMI herbicides (mostly Imazamox) are selling in the market widely and using as post emergence when sunflower plants have 4–8 leaves, both for efficient *Orobanche* and large leaves (*Xanthium strumarium* Wallr., *Avena sterilis* L., *Chenopodium album* L., *Echinochloa cruss-galli*, *Sinapis arvensis* L., *Amaranthus spp.*, *Solanum nigrum* L., *Datura stramonium* L.) weed control [202–207].

On the other hand, another herbicide resistance as Sulfonyl Urea (SU) was found firstly in same wild sunflower population for IMI herbicides Kansas [194]. Lately, SU genetic resistant plants were discovered in *H. annuus* and *H. petiolaris* populations collected from Colorado, Kansas, Nebraska, South and North Dakota [208,209]. With similar way as IMI resistance, first SU resistant lines were developed via classical backcrossing method from wild types in Kansas [196]. Currently SU tolerant sunflower hybrids produced commercially and exists widely in sunflower production almost all parts of the world [205].

9.11 Interspecific and intergeneric hybridization

While there is no enough genetic variation among current germplasm in breeding tool, interspecific or intergeneric hybridization is widely preferred for introgressing new genes. However, it is hard procedure due to high-level crossing barriers for introgression and wide hybridization to obtain successful F1 hybrids [208]. Although it is not easy process due to chromosomal differences, many successful crosses were obtained among *Helianthus* species both with perennials and annuals and with other genus. Nevertheless, there are some intergeneric crosses obtained between *Helianthus* × *Verbesina* as well as *H. annuus* × *Tithonia rotundifolia* and *H. annuus* × *Echinacea purpurea* [209,210]. To obtain successful hybridization, Christov and Vassilevska-Ivanova [211] and Christov and Panayotov [212] proposed to use sunflower female parent in *H. annuus* (HA-89) × *V. encelioides* as well as *H. annuus (HA89)* × *Tithonia rotundifolia* crosses.

9.12 Future directions

Wild *Helianthus* sunflower is wide genus growing different type of environments in large areas so it has many valuable undiscovered genes for sunflower breeders. Until today, some genes were transferred via interspecific hybridization mostly with annual ones. However, due to chromosomal barriers and hard crossing process, the most of the perennial species stay untouched and keeps still valuable mysteries because most of them prove completely resistant to some important sunflower diseases as well as broomrape. With new developments in modern technologies in plant breeding and biotechnology, they will screen largely both classically and molecularly as well as at genome level via molecular mapping and then they will provide many resistance genes as well as other important yield traits in the future. Therefore, sunflower breeders will get benefits extensively from the wild species collections, which preserve the genetic diversity of the genus *Helianthus*.

References

[1] Kaya Y. Sunflower production in Blacksea region: The situation and problems. Int J Innov Appr Agric Res 2020;4(1):147–55.
[2] Seiler G, Marek LF. Germplasm resources for increasing the genetic diversity of global cultivated sunflower. Helia 2011;34:1–20.
[3] Kane NC, Burke JM, Marek L, Seiler G, Vear F, Baute G, et al. Sunflower genetic, genomic and ecological resources. Mol Ecol Resour 2013;13:10–20.
[4] Christov M, Vassilevska-Ivanova R. Intergeneric hybrids in *Compositae* (*Asteraceae*). Hybridization between cultivated sunflower *H. annuus* L. and species of *Compositae* genera. Helia 1999;22:13–22.
[5] Terzic S, Boniface MC, Marek L, Alvarez D, Baumann K, Gavrilova V, et al. Gene banks for wild and cultivated sunflower genetic resources. OCL 2020;27:9.
[6] Fernandez MJ, Perez-Vich B, Velasco L. Sunflower. In: Vollmann J, Rajcan I, editors. Oil crops, vol. 4, Handbook of plant breedingNew York: Springer; 2009. p. 155–232.
[7] Breton C, Gil A, Wargnier J, Seryes H, Berville A. Transfer of architectural traits from perennial *Helianthus mollis* Lam. to sunflower (*H. annuus* L.) and localization of introgression. Euphytica 2012;186(2):557–72.
[8] Vear F. Changes in sunflower breeding over the last fifty years. OCL 2016;23(2):17–23.
[9] In: Kaya Y, Jocic S, Miladinovic D, Sunflower SK, Gupta, editors. Technological innovations in major world oil crops, vol. 1. Springer Press; 2012. p. 85–129.
[10] In: Kaya Y, Sunflower A, Pratap, editors. Alien gene transfer in crop plants, vol. 2. Springer Press; 2014. p. 281–315.

[11] Christov M, Panayotov L. Hybrids between the genera *Helianthus* and *Tithonia* and their study. Helia 1991;14:27–34.

[12] Kantar MB, Sosa CC, Khoury CK, Castañeda-Álvarez NP, Achicanoy HA, Bernau V, et al. Ecogeography and utility to plant breeding of the crop wild relatives of sunflower (*H. annuus* L.). Front Plant Sci 2015;6:841.

[13] Davey MR, Masood J. Sunflower (*Helianthus annuus* L.): Genetic improvement using conventional and in vitro technologies. J Crop Improv 2010;24(4):349–91.

[14] Zhang ZW, Ma G, Zhao J, Qi I. Discovery and introgression of the wild sunflower- derived novel downy mildew resistance gene Pl_{19} in confection sunflower (*Helianthus annuus* L.). Theor Appl Gen 2017;130(1):29–39.

[15] Vassilevska-Ivanova R, Kraptchev B, Shtereva L. An intergeneric hybrid line produced after cross *Helianthus annuus x Echinacea purpurea*. Genet Recour Crop Evol 2015;62(6):829–36.

[16] Vassilevska-Ivanova R, Shtereva L, Stancheva I, Geneva M. Drought stress responses of sunflower germplasms developed after wide hybridization. Turk J Agric 2016;4(10):859–66.

[17] Seiler GJ, Jan CC. Genetic resources. In: Hu J, Seiler G, Kole C, editors. Genetics, genomics and breeding of sunflowerBoca Raton, FL: CRC Science; 2010. p. 1–51.

[18] Jocic S, Miladinović D, Kaya Y. Breeding and genetics of sunflower. In: Martinez-Force E, Dunford NT, Salas JJ, editors. Sunflower, chemistry, production, processing, utilization. Urbana, IL: AOCS; 2015. p. 1–26.

[19] Breseghello F, Coelho A. Traditional and modern plant breeding methods with examples in rice (*Oryza sativa* L.). J Agric Food Chem 2013;61(35). 8277-8276.

[20] Dimitrijevich A, Horn R. Sunflower hybrid breeding: from markers to genomic selection. Front Plant Sci 2017;8(1):2238–50.

[21] Josković M, Jocić S, Marinković R, Prodanović S, Čanak P, Ćirić M, et al. Heritability of plant height and head diameter in sunflower (*Helianthus annuus* L.). Ratar Povrt 2013;50(2):62–6.

[22] Gulya TJ, Marek LF, Gavrilova V. Disease resistance in cultivated sunflower derived from public germplasm collections. In: Proceedings of international symposium "Sunflower Breeding on Resistance to Diseases." Krasnodar, Russia; 2010. p. 7–18.

[23] Vassilevska-Ivanova R, Tcekova Z. Agronomic characteristic of a dwarf germplasm sunflower line. Helia 2005;28(42):51–6.

[24] Skoric D. The genetics of sunflower. In: Kovacevic Z, Skoric D, Sakac Z, editors. Sunflower genetics and breedingSerbian Acad Sci, Serbia: International Monogram; 2012. p. 1–125.

[25] Burke JM, Knapp SJ, Riesenberg LH. Genetic consequences of selection during the evolution of cultivated sunflower. Genetics 2005;171(4):1933–40.

[26] Rogers CE, Thompson TE, Seiler G. Sunflower species of the United States. North Dakota: National Sunflower Association, Bismarck; 1982. p. 75.

[27] Marinković R, Dozet B, Vasić D. Sunflower breeding. Školska knjiga: Novi Sad, Serbia; 2003. p. 1–367. (In Serbian).

[28] Hladni N, Jocić S, Miklič V, Mijić A, Saftić-Panković D. Direct and indirect effects of morphophysiological traits on seed yield of sunflower (*Helianthus annuus* L.). In: Proceeding of the seventeenth international sunflower conference, vol. 1. Cordoba, Spain; 2008. p. 393–397.

[29] Hladni N, Jocić S, Miklič V, Mijić A, Saftić-Panković D, Škoric´ D. Effect of morphological and physiological traits on seed yield and oil content in sunflower. Helia 2010;33(53):101–15.

[30] Seiler GJ. Anatomy and morphology of sunflower. In: Schneiter AA, editor. Sunflower technology and production. Agronomy monograph, 35. US: CSA Press; 1997. p. 67–111.

[31] Leclercq P. Une sterilite male cytoplasmique chez le tournesol. Ann Amel Plantes 1969;19:99–106.

[32] Qi LL, Seiler G, Vick BA, Gulya TJ. Genetics and mapping of the R_{11} gene conferring resistance to recently emerged rust races, tightly linked to male fertility restoration, in sunflower (*Helianthus annuus* L.). Theor Appl Genet 2012;125(5):921–32.
[33] Köhler RH, Horn R, Lössl A, Zetsche K. Cytoplasmic male sterility in sunflower is correlated with the co-trascription of a new open reading frame with the atpA gene. Mol Gen Genet 1991;227:369–76.
[34] Riesenberg LH, Van Fosse C, Arias D, Carter RL. Cytoplasmic male sterility in sunflower: origin, inheritance, and frequency in natural production. J Hered 1994;85(3):233–8.
[35] Horn R. Molecular diversity of male sterility inducing and male-fertile cytoplasms in the genus *Helianthus*. Theor Appl Genet 2002;104:562–70.
[36] Hanson MR, Bentolila S. Interactions of mitochondrial and nuclear genes that affect male gametophyte development. Plant Cell 2004;16:154–69.
[37] Chase C, Babay-Laughnan S. Cytoplasmic male sterility and fertility restoration by nuclear genes. In: Daniell H, Chase C, editors. Molecular biology and biotechnology of plant organellesDordrecht, The Netherlands: Kluwer Academic Publishers; 2004. p. 593–622.
[38] Eckardt N. Cytoplasmic male sterility and fertility restoration. Plant Cell 2006;18(3):515–7.
[39] Bohra A, Iha UC, Adhimoolam P, Bisht D, Singh NP. Cytoplasmic male sterility (CMS) in hybrid breeding in field crops. Plant Cell Rep 2016;35(5):967–93.
[40] Serieys H. Identification, study, utilization in breeding programs of new CMS sources in the FAO subnetwork. In: Proceedings of the sunflower subnetwork progress report. Rome: FAO; 2005. p. 47–53.
[41] Haevekes FWJ, Miller JF, Jan CC. Diversity among sources of cytoplasmic male sterility in sunflower (*Helianthus annuus* L.). Euphytica 1991;5:125–9.
[42] Horn R, Friedt W. Fertility restoration of new CMS sources in sunflower (*Helianthus annuus* L.). Plant Breed 2006;116(4):317–22.
[43] Jan CC, Vick BA. Inheritance and allelic relationships of fertility restoration genes for seven new sources of male- sterile cytoplasm in sunflower. Plant Breed 2007;126:213–7.
[44] Jan CC. Cytoplasmic male sterility in two wild *Helianthus annuus* L. accessions and their fertility restoration. Crop Sci 2000;40:1535–8.
[45] Encheveria MM, Salaberry MT, Rodriguez RH. Characterization for agronomic use of cytoplasmic male-sterility in sunflower (*Helianthus annuus* L.) introduced from *H. resinosus* Small. Plant Breed 2003;122(4):357–61.
[46] Liu Z, Wang D, Feng J, Seiler G, Cai X, Jan CC. Diversifying sunflower germplasm by integration and mapping of a novel male fertility restoration gene. Genetics 2013;193(3):727–37.
[47] Elomaa P, Zhao Y, Zhang T. Flower heads in *Asteraceae*-recruitment of conserved developmental regulators to control the flower-like inflorescence architecture. Hort Res 2018;36(5):317–31.
[48] Zhao Y, Zhang T, Broholm K, Tähtiharju S, Mouhu K, Albert V, et al. Evolutionary co-option of floral meristem identity genes for pattering of the flower-like *Asteraceae* inflorescence. Plant Physiol 2016;172(1):284–96.
[49] Miklić V, Sakać Z, Dusanić N, Miladinović D. Effects of genotype, growing conditions, and several parameters of sunflower attractiveness for bee visitation. In: Proceeding of the sixteenth international sunflower conference. Fargo, ND, USA; 2004. p. 871–876.
[50] Newcomb W. The development of the embryosac of sunflower (*Helianthus annuus*) after fertilization. Can J Bot 1973;51(5):879–90.
[51] Fick G, Miller J. Sunflower breeding. Agronomy monograph, sunflower technology and production, 35. USA: CSA, ASA Press; 1997. p. 395–439.

[52] Montilla F, Gomez-Arnau J, Duhing P. Bee-attractiveness and self-compatibility of some inbred lines and their hybrids. In: Proceeding of the twelfth international sunflower conference, vol. 1. Novi Sad, Yugoslavia; 1988. p. 423–428.
[53] Debaeke F, Casadebaig P, Flenet F, Langlade N. Sunflower crop and climate change: vulnerability, adaptation, and mitigation potential from case-studies in Europe. OCL 2017;24(1):201–26.
[54] Garcia-Vila M, Fereres F, Prieto M, Ruz C, Soriano M. Sunflower. Crop yield response to waterFAO, Irrigation and Drainage; 2012. p. 66.
[55] Chervet B, Vear F. Study of the relationship between sunflower earliness and yield, oil content, development and morphology. Agronomy 1990;10(1):51–6.
[56] Hernandez LZ, Larsen AO, Lindstrom LI, Iriarte LB. Physiological maturity in sunflower. Correspondence between the quantitative and the visual definition. In: Proceedings of the seventeenth international sunflower conference. Cordoba, Spain; 2008. p. 337–340.
[57] Leland EF. Salinity effect on four sunflower hybrids. Agron J 1996;88:215–9.
[58] De la Vega, Hall AJ. Does subdivision of the sunflower growing region of Argentina increase the response to selection?. In: Proceedings of the sixteenth international sunflower conference, vol. 2. Fargo, ND, USA; 2004; p. 459–464.
[59] Antolin G, Tinaut FV, Briceno Y, Castano V, Perez C, Ramirez AI. Optimization of biodiesel production by sunflower oil transesterification. Bioresour Technol 2002;83:111–4.
[60] Barnwal BK, Sharma MP. Prospects of biodiesel production from vegetable oils in India. Renew Sustain Energy Rev 2005;9:363–78.
[61] Zubriski JC, Zimmerman DC. Effect of nitrogen, phosphorus, and plant density on sunflower. Agron J 1974;66(6). 798-780.
[62] Andrade FH. Analysis of growth and yield of maize, sunflower and soybean grown at Balcarce, Argentina. Field Crop Res 1995;41:1–12.
[63] Xiao S, Chen SY, Zhao LQ, Wan G. Density effects on plant height growth and inequality in sunflower population. JIPB 2006;48(5):513–9.
[64] El-Satar M, Ahmed A, Hassan TH. Response of seed yield and fatty acid composition for sime sunflower genotypes to plant spacing and nitrogen fertilization. Inf Process Agric 2017;4(3):241–52.
[65] Iona V, Dicub G, Basaa A, Dumbravaa M, Temocicoa G, Epurea L, et al. Sunflower yield and yield components under different sowing conditions. Agric Agric Sci Proc 2015;6:44–51.
[66] Skoric D, Jocic S, Sakac Z, Lecic N. Genetic possibilities for altering sunflower oil quality to obtain novel oils. Can J Physiol Pharmacol 2008;86(4):215–21.
[67] Rauf S, Jamil N, Ali Tariq S, Khan M, Kausar M, Kaya Y. Progress in modification of sunflower oil to expand its industrial value. J Sci Food Agric 2017;97:1997–2006.
[68] Fernandez-Martinez JM, Velasco L, Perez-Vich B. Progress in the genetic modification of sunflower oil quality. In: Proceedings of the sixteenth international sunflower conference, vol. 2. Fargo, ND, USA; 2004. p. 1–14.
[69] Fernandez-Martinez JM, Perez-Vich B, Velasco L, Dominguez J. Breeding for specialty oil types in sunflower. Helia 2007;30(46):75–84.
[70] Burton JW, Miller JF, Vick BA, Scarth R, Holbrook CC. Altering fatty acid composition in oil seed crops. Adv Agron 2004;84(1):273–306.
[71] Mantese A, Medan D, Hall A. Achene structure, development and lipid accumulation in sunflower cultivars differing in oil content at maturity. Ann Bot 2006;97(6):999–1010.
[72] Connor DJ, Hall AJ. Sunflower physiology. In: Schneiter AA, editor. Sunflower technology and production. Agronomy monograph, 35. Madison, Wisconsin, US: ASA, CSSA, SSSA; 1997. p. 113–82.

[73] Roche J, Bouniols A, Mouloungui Z, Barranco T, Cerny M. Management of environmental crop conditions to produce useful sunflower oil components. Eur J Lipid Sci Technol 2006;108:287–97.
[74] Harris HC, McWilliam JR, Mason WK. Influence of temperature on oil content and composition of sunflower seed. Aust J Agric Res 1978;29:1203–12.
[75] Anastasi U, Cammarata M, Abbate V. Yield potential and oil quality of sunflower (oleic and standard) grown between autumn and summer. Ital J Agron 2000;4:23–36.
[76] Silver JG, Rochester CP, Bishop DG, Harris HC. Unsaturated fatty acid synthesis during the development of isolated sunflower (*Helianthus annuus* L.) seeds. J Exp Bot 1984;35:1507–14.
[77] Shewry PR, Pandya MJ. The 2S albumins storage proteins. In: Shewry PR, Casey R, editors. Seed proteinsDordrecht: Kluwer Academic Publishers; 1999. p. 619–64.
[78] Casey R. Distribution and some properties of seed globulins. In: Shewry PR, Casey R, editors. Seed ProteinsThe Netherlands: Klewer Academic Publishers; 1998. p. 159–69.
[79] Kortt AA, Caldwell JB. Low molecular weight albumins from sunflower seed: identification of a methionine-rich albumin. Phytochemistry 1990;29(9):2805–2810.
[80] Raymond J, Robin M, Azanza J. 11S seed storage proteins from *Helianthus* species (*Compositae*): biochemical, size and charge heterogeneity. Plant Syst Evol 1995;198:195–208.
[81] Anisimova I, Gavrilova V, Loskutov A, Rozhkova V, Tolmachev V. Polymorphism and inheritance of seed storage protein in sunflower. Russ J Plant Genetics 2004;40(9):1215–23.
[82] Haar V, Allen RA, Cohen RD. Organization of the sunflower 11S storage protein gene family. Gene 1988;74:433–43.
[83] Dorrell DG, Vick BA. Properties and processing of oilseed sunflower. In: Schneiter AA, editor. Sunflower technology and production. Agronomy monograph, 35. Madison, WI, USA: ASA, CSSA and SSSA; 1997. p. 709–45.
[84] Betschalt A, Lyon CK, Kohler N. Sunflower, safflower, sesame and castor protein. In: Pirie NW, editor. Food protein sourcesCambidge Univ Press; 1975. p. 79–88.
[85] Vassilevska-Ivanova R, Kraptchev B, Stancheva I, Geneva M. A compact sunflower line produced after cross *Helinathus annuus x Verbesina encelioides*. Centr Eur J Biol 2013;8(5):492–8.
[86] Skrbic B, Filipcev B. Nutritional and sensory evaluation of wheat breads supplemented with oleic-rich sunflower seed. Food Chem 2008;108(1):119–29.
[87] Lin C, Liu J, Liu L, Zhu T, Sheng L, Wang D. Soil amendment application frequency contributes to phytoextraction of lead by sunflower at different nutrient levels. Environ Exp Bot 2009;65(203):410–6.
[88] Rizvan M, Ali S, Rizvi H, Rinklebe J, Tsang D. Phytomanagement of heavy metals in contaminated soils using sunflower: A review. Crit Rev Environ Sci Technol 2016;46(18):1498–528.
[89] McGrath SP, Zhao FJ, Lombi E. Phytoremediation of metals, metalloids, and radionuclides. Adv Agron 2002;75:1–56.
[90] Niu ZX, Sun LN, Sun TH, Li YS, Wang H. Evaluation of phytoextracting cadmium and lead by sunflower, ricinus, alfalfa and mustard in hydroponic culture. J Environ Sci 2007;19:961–7.
[91] Blamey FPC, Edwards SDG, Asher CJ. Nutritional disorders of sunflower. Queensland, Australia: Department of Agriculture, University of Queensland St Lucia; 1987.
[92] Skoric D. Sunflower breeding for resistance to abiotic stresses. Helia 2009;32(40):1–16.
[93] Sanghera GS, Wani SH, Hussain W, Singh NB. Engineering cold stress tolerance in crop plants. Curr Genom 2011;12(1):30–43.
[94] Duman JG, Wisniewski M. The use of antifreeze proteins for frost protection in sensitive crop plants. Environ Exp Bot 2014;106:60–9.
[95] Soriano MA, Orgaz F, Villalobos FJ, Fereres E. Efficiency of water use of early plantings of sunflower. Eur J Agron 2004;21:465–76.

[96] Allinne C, Maury P, Sarrafi A, Grien P. Genetic control of physiological traits associated to low temperature growth in sunflower under early sowing conditions. Plant Sci 2009;177(4):349–59.
[97] Jha UC, Bohra A, Jha R. Breeding approaches and genomics technologies to increase crop yield under low-temperature stress. Plant Cell Rep 2017;36(1):1–35.
[98] Tetreault H, Kawakami T, Ungerer M, Levy C. Low temperature tolerance in the perennial sunflower *Helianthus maximiliani*. Am Midland Naturalist 2016;175(1):91–102.
[99] Manunta P, Kirkham MB. Respiration and growth of sorghum and sunflower under predicted night temperatures. J Agron Crop Sci 1996;176(4):267–74.
[100] Rondanini D, Mantese A, Savin R, Hall A. Responses of sunflower yield and grain quality to alternating day/night high temperature regimes during grain filling: Effects of timing, duration and intensity of exposure to stress. Field Crop Res 2006;96:48–62.
[101] Rondanini D, Savin R, Hall A. Dynamics of fruit growth and oil quality of sunflower (*Helianthus annuus* L.) exposed to brief intervals of high temperature during grain filling. Field Crops Res 2003;83:79–90.
[102] Corbineau F, Gay-Mathieu C, Vinel D, Côme D. Decrease in sunflower (*Helianthus annuus*) seed viability caused by high temperature as related to energy metabolism, membrane damage and lipid composition. Physiol Plant 2002;16(4):489–96.
[103] Zinn KE, Tunc-Ozdemir M, Harper JF. Temperature stress and plant sexual reproduction: uncovering the weakest links. J Exp Bot 2010;61:1959–68.
[104] Herrero MP, Johnson RR. High temperature stress and pollen viability in maize. Crop Sci 1980;20:796–800.
[105] Hatfield JL, Prueger JH. Temperature extremes: Effect on plant growth and development. Weather Clim Extremes 2015;10(A):4–10.
[106] Chimenti CA, Hall AJ. Grain number responses to temperature during floret differentiation in sunflower. Field Crops Res 2001;72:177–84.
[107] Moriondo MC, Giannakopoulos C, Bindi M. Climate change impact assessment: the role of climate extremes in crop yield simulation. Clim Change 2011;104:679–701.
[108] Souza MA, Pimentel AJB, Ribeiro G. Breeding for heat-stress tolerance. In: Fritsche-Neto R, Borém A, editors. Plant breeding for abiotic stress toleranceBerlin: Springer; 2012. p. 137–56.
[109] Kalyara T, Saed R, Silvab JA, Shahzad M. Handling sunflower (*Helianthus annuus* L.) population under heat stress. Arch Agron Soil Sci 2014;60(5):655–72.
[110] Levitt J. Responses of plants to environmental stresses. New York: Academic Press; 1972.
[111] Levitt J. Stress terminology. In: Turner NC, Kramer PJ, editors. Adaptation of plants to water and high temperature stressNew York, USA: Wiley-Interscience; 1982. p. 437–9.
[112] Yordanov I, Velikova V, Tsonev T. Plant responses to drought, acclimation, and stress tolerance. Photosynthetica 2000;38:171–86.
[113] Rauf S, Al-Khayri JM, Zaharieva M, Monneveux P, Khalil F. Breeding strategies to enhance drought tolerance in crops. In: Al-Khayri JM, Jain SM, Johnson DV, editors. Advances in plant breeding strategies: Agronomic, abiotic and biotic stress traitsSpringer; 2015. p. 397–445.
[114] Hussain MM, Rauf S, Riaz MA, Al-Khayri J, Monneveux P. Determination of drought tolerance related traits in *Helianthus argophyllus*, *Helianthus annuus* and their hybrids. Breed Sci 2017;67(3):257–67.
[115] Athar HR, Ashraf M. Photosynthesis under drought stress. In: Pessarakli M, editor. Handbook of photosynthesisNew York, NY, USA: CRC Press; 2005. p. 793–804.
[116] Rauf S. Breeding sunflower (*Helianthus annuus* L.) for drought tolerance. Commun Biom Crop Sci 2008;3:29–44.

[117] Dragović S, Maksimović L. Drought phenomenon and impact on crop yields in the Vojvodina Province, Yugoslavia. Proceedings of the international workshop on drought in the Carpathian Region. Budapest; 1995. p. 207–217.
[118] Tardieu F, Parent B, Caldeira C, Welcker C. Genetic and physiological controls of growth under water deficit. Plant Physiol 2014;164(4):1628–35.
[119] Seiler GJ, Qi LL, Marek LF. Utilization of sunflower crop wild relatives for cultivated sunflower improvement. Crop Sci 2017;57:1083–101.
[120] Seiler GJ, Gulya TJ, Marek L. Exploration for wild *Helianthus* species from the desert southwestern USA for potential drought tolerance. Helia 2006;29:1–10.
[121] Vassilevska-Ivanova R, Shtereva L, Kraptche B, Karceva T. Response of sunflower (*Helianthus annuus* L.) genotypes to PEG-mediated water-stress. Cent Eur J Biol 2014;9:1206–14.
[122] Bowsher AW, Milton EF, Donovan LA. Comparison of desert-adapted *Helianthus niveus* (Benth.) Brandegee ssp. *tephrodes* (A. Gray) Heiser to cultivated *H. annuus* L. for putative drought avoidance traits at two ontogenetic stages. Helia 2016;39(64):1–19.
[123] Seiler GJ. Variation in agronomic and morphological characteristics of several populations of wild annual sunflower (*Helianthus annuus* L.). Helia 1984;7:29–32.
[124] Rauf S. Breeding strategies for sunflower (*Helianthus annuus* L.) genetic improvement. In: Al-Khayri J, Jain S, Johnson D, editors. Advances in plant breeding strategies: industrial and food cropsSpringer; 2019. p. 637–73.
[125] Chowdhury FMT, Halim MA, Hossain F, Akhtar N. Effects of sodium chloride on germination and seedling growth of Sunflower (*Helianthus annuus* L.). Jahangirnagar University. J Biol Sci 2018;7(1):35–44.
[126] Munns R. Comparative physiology of salt and water stress. Plant Cell Environ 2002;2:239–50.
[127] Meloni DA, Oliva MA, Martinez CA, Cambraia J. Photosynthesis and activity of superoxide dismutase, peroxidase and glutathione reductase in cotton under salt stress. Environ Exp Bot 2003;49(1):69–76.
[128] Chinnusamy V, Jagendorf A, Zhu JK. Understanding and improving salt tolerance in plants. Crop Sci 2005;45:437–48.
[129] Ashraf M, Athar HR, Harris PJC, Kwon TR. Some prospective strategies for improving crop salt tolerance. Adv Agron 2008;97:45–110.
[130] Ahmad P, Prasad MNV. Environmental adaptations and stress tolerance in plants in the era of climate change. New York: Springer Science & Business Media; 2012.
[131] Tuteja N, Gill SS, Tuteja R. Improving crop productivity in sustainable agriculture. Weinheim, Germany: Wiley-VCH Verlag GmbH & Co. KGaA; 2012.
[132] Maas EV, Hoffman GJ. Crop salt tolerance-current assessment. J Irrig Drain Div 1977;103(2):115–34.
[133] Steduto P, Albrizio R, Giorio P, Sorrentino G. Gas-exchange response and stomatal and non-stomatal limitations to carbon assimilation of sunflower under salinity. Environ Exp Bot 2000;44:243–55.
[134] Flagella Z, Giuliani MM, Rotunno T, Di Caterina R, De Caro A. Effect of saline water on oil yield and quality of a high oleic sunflower (*Helianthus annuus* L.) hybrid. Eur J Agron 2004;21(2):267–72.
[135] Ashraf M, Tufail M. Variation in salinity tolerance in sunflower (*Helianthus annuus* L.). J Agron Crop Sci 1995;174(5):351–62.
[136] Akram NA, Ashraf M. Pattern of accumulation of inorganic elements in sunflower (*Helianthus annuus* L.) plants subjected to salt stress and exogenous application of 5-aminolevulinic acid. Pak J Bot 2011;43:521–30.

[137] Shahbaz M, Ashraf M, Akram NA, Hanif A, Hameed S, Joham S, et al. Salt-induced modulation in growth, photosynthetic capacity, proline content and ion accumulation in sunflower (*Helianthus annuus* L.). Acta Physiol Plant 2011;33:1113–22.

[138] Gaballah MS, Shaaban S, Aboelleil AA. The role of antioxidants in improving sunflower cultivars' performance grown under saline conditions. Int J Acad Res 2012;4(6):33–6.

[139] Szabolcs I. Soils and salinization. In: Pessarakali M, editor. Handbook of plant and crop stressNew York: Marcel Dekker; 1994. p. 3–11.

[140] Seiler GJ. Utilization of wild *Helianthus* species in sunflower breeding. In: Škorić D, editor. Sunflower genetics and breeding international monogramNovi Sad, Serbia: Serbian Academy of Sciences and Arts; 2012. p. 355–413.

[141] Lexer C, Lai Z, Riesenberd L. Candidate gene polymorphism associated with salt tolerance in wild sunflower hybrids: implications for the origin of *Helianthus paradoxus*, a diploid hybrid species. New Phytol 2004;161(1):225–33.

[142] Hulke BS, Miller JF, Gulya TJ, Vick BA. Registration of the oilseed sunflower genetic stocks HA 458, HA 459, and HA 460 possessing genes for resistance to downy mildew. J Plant Reg 2010;4:1–5.

[143] Tabrizi MM. Identification of downy mildew resistance loci in sunflower germplasm. Notulae Sci Biol 2017;9(4):515–9.

[144] Christov M. Helianthus species in breeding research in sunflower. In: Proceedings of the seventeenth international sunflower conference. Cordoba, Spain; 2008. p. 709–714.

[145] Wieckhorst S, Bachlava E, Dußle CM, Tang MS, Gao W, Saski C, et al. Fine mapping of the sunflower resistance locus *Pl ARG* introduced from the wild species *H. argophyllus*. Theor Appl Genet 2010;12:1633–44.

[146] Qi LL, Long YM, Jan CC, Ma GJ, Gulya TJ. Pl17 is a novel gene independent of known downy mildew resistance genes in the cultivated sunflower (*Helianthus annuus* L.). Theor Appl Genet 2015;128:757–67.

[147] Qi LL, Foley ME, Cai XW, Gulya TJ. Genetics and mapping of a novel downy mildew resistance gene, Pl18, introgressed from wild *Helianthus argophyllus* into cultivated sunflower (*Helianthus annuus* L.). Theor Appl Genet 2016;129:741–52.

[148] Dussle CM, Hahn V, Knapp SJ, Bauer E. *Plarg* from *H. argophyllus* unlinked to other known downy mildew resistance genes in sunflower. Theor Appl Genet 2004;109:1083–6.

[149] Mulpuri S, Liu Z, Feng J, Gulya TJ, Jan CC. Inheritance and molecular mapping of a downy mildew resistance gene, Pl_{13} in cultivated sunflower (*Helianthus annuus* L.). Theor Appl Genet 2009;119:795–803.

[150] Seiler GJ. Utilization of wild *Helianthus* species in breeding for disease resistance. In: Lucomets V, editor. Proceedings of the international sunflower association (ISA) symposium: Sunflower breeding on resistance to diseases, 23-24 June, All-Russia Research Institute of Oil Crops by V.S. Pustovoit (VNIIMK). Krasnodar, Russia; 2010. p. 36–50.

[151] Reddy KP, Sankaraneni CR, Pulugurtha BK, Mulpuri S. Development of a scoring scale for powdery mildew *G. cichoracearum* disease and identification of resistance sources in cultivated and wild sunflowers. Euphytica 2013;190:385–99.

[152] Qi LL, Gulya TJ, Seiler GJ, Hulke BS, Vick BA. Identification of resistance to new virulent races of rust in sunflowers and validation of DNA markers in the gene pool. Phytopathology 2011;101:241–9.

[153] Gong L, Hulke BS, Gulya TJ, Markell SG, Qi LL. Molecular tagging of a novel rust resistance gene R12 in sunflower (*Helianthus annuus* L.). Theor Appl Genet 2013;126:93–9.

[154] Quresh Z, Jan CC, Gulya TG. Resistance of sunflower rust and its inheritance in wild sunflower species. Plant Breed 1993;110:297–306.

[155] Encheva J, Christov M, Shindrova P, Drumeva M, Encheva V. New sunflower restorer lines developed by direct organogenesis method from interspecific cross *Helianthus annuus* L. (cv. Albena), *Helianthus salicifolius* L.—Disease resistance, combining ability. Helia 2006;29:95–106.

[156] Sujatha M, Prakabaran AJ. Ploidy manipulation and introgression of resistance to *Alternaria helianthi* from wild hexaploid *Helianthus* species to cultivated sunflower (*Helianthus annuus* L.) aided by another culture. Euphytica 2006;152:201–15.

[157] Molinero-Ruiz L, Delavault P, Perez-Vich B, Pacureanu M, Bulos M, Altieri E, et al. History of the race structure of *Orobanche cumana* and the breeding of sunflower for resistance to this parasitic weed: A review. Spanish J Agric Res 2015;13(4):1–19.

[158] Abbes Z, Kharrat M, Delavault P, Simier P, Chaibi W. Field evaluation of the resistance of some faba bean (*Vicia faba* L.) genotypes to the parasitic weed *Orobanche foetida* Poire. Crop Prot 2007;26(12). 1777-1774.

[159] Alcántara E, Morales-García M, Díaz-Sánchez JD. Effects of broomrape parasitism on sunflower plants: growth, development and mineral nutrition. J. Plant Nutr 2006;29:1199–206.

[160] Lins RD, Colquhoun JB, Mallory-Smith C. Effect of small broomrape (*Orobanche minor*) on red clover growth and dry matter partitioning. Weed Sci 2007;55:517–20.

[161] Duca M. Historical aspects of sunflower researches in the republic of Moldova. Helia 2015;38:79–92.

[162] Yang Z, Wafula EK, Honaas LA, Zhang H, Das M, Fernandez-Aparicio M. Comparative transcriptome analyses reveal core parasitism genes and suggest gene duplication and repurposing as sources of structural novelty. Mol Biol Evol 2015;32:767–90.

[163] Louarn J, Boniface MC, Pouilly N, Velasco L, Perez-Vich B, Vincourt P, et al. Sunflower resistance to broomrape (*Orobanche cumana*) is controlled by specific QTLs for different parasitism stages. Front Plant Sci 2016;7:590–9.

[164] Smith CE, Dudley MW, Lynn DG. Vegetative—parasitic transition control and plasticity in *Striga* development. Plant Physiol 1990;93. 208-22.

[165] Fernandez-Aparicio M, Reboud X, Gibot-Leclerc S. Broomrape weeds. Underground mechanisms of parasitism and associated strategies for their control: a review. Front Plant Sci 2016;7:135.

[166] Cook CE, Whichard LP, Wall E, Egley GH, Coggon P, Luhan PA, et al. Germination stimulants. II. The structure of strigol, a potent seed germination stimulant for witch weed (*Striga lutea* Lour.). J Am Chem Soc 1972;94:6198–9.

[167] Joel D, Chaudhuri S, Ziadna H, Steffens J. Dehydrocostus lactone is exuded from sunflower roots and stimulates germination of the root parasite *Orobanche cumana*. Phytochemistry 2011;72(7):624–34.

[168] Yoneyama K, Awad A, Xie X, Yoneyama K, Takeuchi Y. Strigolactones as germination stimulants for root parasitic plants. Plant Cell Physiol 2010;51:1095–103.

[169] Raupp FM, Spring O. New sesquiterpene lactones from sunflower root exudate as germination stimulants for *Orobanche cumana*. J Agric Food Chem 2013;61:10481–7.

[170] Umehara M, Hanada A, Yoshida S, Akiyama K, Arite T, Takeda N, et al. Inhibition of shoot branching by new terpenoid plant hormones. Nature 2008;455:195–201.

[171] Hassan E, Al-Akkad S, Moustafa S, El-Awadi M. Histochemical aspects of penetration and vascular connection of broomrape haustoria in the host root, and the possible implication of phenylpropanoids. Int J Agri Biol 2004;6(3):430–4.

[172] Foy C, Jain R, Jacobsohn R. Recent approaches for chemical control of broomrape (*Orobanche* ssp.). Rev Weed Sci 1989;4:123–52.

[173] Vrânceanu AV, Tudor VA, Stoenescu FM, Pirvu N. Virulence groups of *Orobanche cumana* Wallr., differential hosts and resistance sources and genes in sunflower. In: Proceedings of the ninth international sunflower conference. Torremolinos, Spain; 1980. p. 74–80.

[174] Fernandez-Martinez JM, Dominguez J, Perez-Vich B, Velasco L. Update on breeding for resistance to sunflower broomrape. Helia 2008;31(48):73–84.

[175] Kaya Y, Evci G, Pekcan V, Yilmaz MI. Broomrape resistance breeding in sunflower: a case study in Turkey. In: Proceedings of the third international symposium on broomrape (*Orobanche* spp.) in sunflower. Cordoba, Spain; 2014. p. 194–199.

[176] Höniges A, Wegmann K, Ardelean A. *Orobanche* resistance in sunflower. Helia 2008;31(49):1–12.

[177] Velasco L, Pérez-Vich B, Yassein AM, Jan CC, Fernández-Martínez JM. Inheritance of resistance to broomrape (*Orobanche cumana* Wallr.) in an interspecific cross between *Helianthus annuus* and *Helianthus debilis* ssp. tardiflorus. Plant Breed 2012;131:220–1.

[178] Fernández-Martínez JM, Melero-Vara JM, Muñoz-Ruz J, Ruso J, Domínguez J. Selection of wild and cultivated sunflower for resistance to a new broomrape race that overcomes resistance to *Or*5 gene. Crop Sci 2000;40:550–5.

[179] Fernández-Martínez JM, Velasco L, Pérez-Vich B. Progress in research on breeding for resistance to sunflower broomrape. Helia 2012;57:47–56.

[180] Velasco L, Perez-Vich B, Jan CC, Fernandez-Martinez JM. Inheritance of resistance to broomrape (*Orobanche cumana* Wallr.) race F in a sunflower line derived from wild sunflower species. Plant Breed 2007;126(1):67–71.

[181] Cvejic S, Radanović A, Dedić B, Jocković M, Jocić S, Miladinović D. Genetic and genomic tools in sunflower breeding for broomrape resistance. Genes 2020;11:152.

[182] Rogers CE. Insect pests and strategies for their management in cultivated sunflower. Field Crop Res 1992;30:301–32.

[183] Sikora D. Evaluation of host plant resistance against sunflower moth, *Homoeosoma electellum* (Hulst) in cultivated sunflower in western Nebraska. Thesis MS, Univ of Nebraska, Entomology; 2017.

[184] Schultz JT. Insect pests. In: Carter JF, editor. Sunflower science and technology. Agronomy 19Madison, WI: American Society of Agronomy, Crop Science Society of America, and Soil Science Society of America; 1978. p. 169–223.

[185] El-Wakeil N. Insect economic levels in relation to crop production. Arch Phytopathol Plant Prot 2010;43(17):1710–44.

[186] Akinkunmi OY, Akintoye HA, Umeh VC, AdeOluwa O. Influence of spacing on the feeding activities of major pests of sunflower and their associated damage. Agric Biol J N Am 2012;3(6):233–6.

[187] Arthur AP. Insect problems in sunflower. J Am Oil Chem Soc 1983;60(7):1251–9.

[188] Irum M. Sunflower disease and insect pest in Pakistan: A review. Afric Crop Sci J 2010;17(2):109–18.

[189] Charlet LD, Gavloski J. Insects of sunflower in the Northern Great Plains of North America. In: Arthropods of Canadian grasslands, vol. 2. Inhabitants of a changing landscape, Floate KD, editor. Biological Survey of Canada; 2011. p. 159–178.

[190] Rossiter M, Gershenzon J, Mabry TJ. Behavioral and growth responses of specialist herbivore, *Homoeosoma electellum*, to major terpenoid of its host, *Helianthus* spp. J Chem Ecol 1985;12(6):1505–21.

[191] Rogers CE, Gershenzon J, Ohno N, Mabry TJ, Stipanovic RD, Kreitner GL. Terpenes of wild sunflowers (*Helianthus*): An effective mechanism against seed predation by larvae of

the sunflower moth, *Homoeosoma electellum* (Lepidoptera: *Pyralidae*). Environ Entomol 1987;16(3):586–92.

[192] Gershenzon J, Rossiter MC, Mabry TJ, Rogers CE, Blust MH, Hopkins TL. Insect antifeedant terpenoids in wild sunflower—a possible source of resistance to the sunflower moth. In: Hedin PH, editor. Bioregulators for pest controlWashington, D.C.: ACS Symposium Ser. 276, American Chemical Society; 1985. p. 433–46.

[193] Whitney KD, Randell RA, Riesenberg LH. Adaptive introgression of herbivore resistance traits in the weedy sunflower *Helianthus annuus*. Am Nat 2006;167(6):794–807.

[194] Al-Khatib K, Baumgartner JR, Peterson DE, Currie RS. Imazetherapy resistance in common sunflower (*Helianthus annuus*). Weed Sci 1998;46:403–7.

[195] Al-Khatib K, Miller JF. Registration of four genetic stocks of sunflower resistant to imidazolinone herbicides. Crop Sci 2000;40:869–70.

[196] Miller JF, Al-Khatib K. Registration of two oilseed sunflower genetic stocks, SURES-1 and SURES-2 resistant to Tribenuron herbicide. Crop Sci 2004;39:301–2.

[197] Malidza G, Skoric D, Jocic S. Imidazolinone–resistant sunflower (*Helianthus annuus*): Inheritance of resistance and response towards selected sulfonyrurea herbicides. In: Proceedings of the fifteenth international sunflower conference. Toulouse (France); 2000. p. O42–O47.

[198] Bruniard JM, Miller JF. Inheritance of imidazolinone-herbicide resistance in sunflower. Helia 2001;24(35):11–6.

[199] Kolkman JM, Slabaugh MB, Bruniard JM, Berry S, Bushman SB, Olungu C, et al. Acetohydroxyacid synthase mutations conferring resistance to imidazolinone or sulfonylurea herbicides in sunflower. Theor Appl Genet 2004;109:1147–59.

[200] Sala CA, Bulos M, Echarte AM, Whitt S, Budziszewski G, Howie W, et al. Development of CLHA-Plus: a novel herbicide tolerance trait in sunflower conferring superior *Imidazolinone* tolerance and ease of breeding. In: Proceedings of the seventeenth international sunflower conference. Cordoba, Spain; 2008. p. 489–494.

[201] Sala CA, Bulos M, Altieri E, Weston B. Response to imazapyr and dominance relationships of two *Imidazolinone*-tolerant alleles at the Ahasl1 locus of sunflower. Theor Appl Genet 2012;124:385–96.

[202] Soare T, Soare G, Calcagno F, Calcagno F, Bozzini A, Calcagno L. Sunflower breeding for alternative genetic resistance to the parasite *Orobanche cumana* Wall. (*Helianthus annuus* L.). J Genet Breed 2005;59(3–4):269–75.

[203] Kaya Y, Evci G. Herbicide resistance in sunflower (*Helianthus annuus* L.). International Research Conference, Plant Genetic Stocks—The Basis of Agriculture of Today, June 13-14, vol. 2. Plovdiv. Bulgaria; 2007. p. 45–47.

[204] Demirci M, Kaya Y. Status of *Orobanche cernua* Loefl. and weeds in sunflower production in Turkey. Helia 2009;32(51):153–60.

[205] Evci G, Pekcan V, Yılmaz MI, Kaya Y. The Resistance breeding to IMI and SU herbicides in sunflower. Plant Sci 2012;49:6–11.

[206] Evci G, Sezer N, Pekcan V, Yılmaz MI, Kaya Y. Chemical control of broomrape and weeds with *Imidazolinone* herbicide and resistant hybrids in sunflower production in Turkey. J Acad Sci Moldova 2011;2(314):118–24.

[207] Elezovic I, Datta A, Vrbnicanin S, Glamoclija D, Simic M, Malidza G, et al. Yield and yield components of imidazolinone-resistant sunflower (*Helianthus annuus* L.) are influenced by pre-emergence herbicide and time of post-emergence weed removal. Field Crops Res 2012;128:137–46.

[208] Christov M. Contribution of interspecific hybridization to sunflower breeding. Helia 2012;35:37–46.

[209] White AD, Owen MDK, Hartzler RG, Cardina J. Common sunflower resistance to acetolactate synthase-inhibiting herbicides. Weed Sci 2002;50(4):432–7.

[210] Olson BLS, Zollinger RK, Thompson CR, Peterson DE, Jenks B, Moechnig M, et al. Pyroxasulfone with and without sulfentrazone in sunflower (*Helianthus annuus*). Weed Technol 2011;25:217–21.

[211] Vassilevska-Ivanova R, Naidenova N. Intergeneric hybridization between cultivated sunflower and Asteraceae genera: *Helianthus annuus* × *Echinacea purpurea*. C R Acad Bulg Sci 2005;58:729–34.

[212] Vassilevska-Ivanova R. A short-cycle sunflower line derived from intergeneric hybridization *Helianthus* × *Verbesina*. Helia 2005;28:55–60.

Chapter 10

Brassicas: A Complete Guide to the Potential of Their Wild Relatives

Humera Razzaq[a], Sue J. Armstrong[b] and Hina Saleem[a]
[a]*Department of Plant Breeding and Genetics, University of Agriculture, Faisalabad, Pakistan;*
[b]*School of Bioscience, University of Birmingham, Birmingham, United Kingdom*

10.1 Introduction of the genus

A great fascinating and biological journey through the evolutionary time period is required in order to understand the genus "*Brassica.*" Witnessing the population of wild plants interbreeding and producing stable hybrids, this genus has always been under the limelight since the beginning of agriculture as mankind used both its wild, cultivating and hybrid types and improved them by further selection and combination. Flexibility and genetic diversity are the main features of all the members of *Brassicaceae* (*Cruciferae*) family. Due to the presence of these traits their domestication was encouraged during the Neolithic era. Records indicate that ancient Romans, Chinese, Greeks, and Indians all treasures this family and use them widely. The etymology of the term *Brassica* has always under great debate, H. Boerhaave in 1727 suggested that the word is evolved from Greek word while Henslow, 1908 debated that it is derived from the Latin word. But the most acceptable suggestion is that it s derived from the Latin word which means "to cut off early" since it leaves were harvested in early august and winter for food and fodder. While the term *Brassicaceae* comes from *Brassica*, the additional—*aceae*, is a standardized suffix approved by international scientific vocabulary, for the names of plant family. It belongs to the 10 most economically important plant families. This is so because mankind has derived the greatest diversified products from the single genus. As all part of the plants of this genus are consumed. Therefore, this genus is important for consumption either in terms of edible oil or direct consumption of its leaves as a vegetable source or for fodder purposes. It is rich in proteins, used as lubricants and illumination, condiments: that includes herbs, mustards, and flavorings. Apparently new uses of *Brassica* are opening new avenues, which involve emphasis on its wild relatives too. Its wild crop species are also reported

Wild Germplasm for Genetic Improvement in Crop Plants. http://dx.doi.org/10.1016/B978-0-12-822137-2.00010-2

TABLE 10.1 Classification of the genus *Brassica*.

- Kingdom Plantae—Plants
  - Subkingdom Tracheobionta—Vascular plants
    - Superdivision Spermatophyta—Seed plants
      - Division Magnoliophyta—Flowering plants
        - Class Magnoliopsida—Dicotyledons
          - Subclass Dilleniidae
            - Order Capparales
              - Family Brassicaceae—Mustard family
                - Genus Brassica

in ancient civilization such as in Asia and the Mediterranean region [1]. Middle ages clearly revealed various types of *Brassica* and Dutch paintings of 16 and 17 centuries also shows various examples of *Brassicas*.

The family *Brassicaceae* contains around 338 genera and 3709 species. Taxonomic studies of *Brassicas* began around 1700's by Tournefort and were extended by Linnaeus [2–4] who classified and grouped the plants based on characteristics of carpels. The classification of the genus is as shown in Table 10.1:

The *Brassicaceae* family consists of 25 tribes with about 5 additional tribes which are under study [5]. Tribe *Brassiceae*, that contains genus *Brassica* and its wild relatives, consists of approximately 240 species and 40 genera [6,7].

10.2 Importance

Worldwide *Brassicas* are consumed with various modifications [8]. Such modifications include the stem or root to form a bulb, the compacting of the leaves to form a head, the seed to provide both condiment and oil [9,10]. *Brassica* species include a great number of taxonomically related but morphologically diverse plants which have been under cultivation since centuries and presently have been utilized by extensive crossing and hybridization [11].

Brassicas are capable of surviving and growing at relatively lower temperatures, hence are among the edible oil sources, they can be successfully grown and breed at cool temperate regions of the world. This is the reason they are well adapted even at higher elevations and as well as winter crop in the subtropical regions. As a matter of fact *Brassicas* provide 15% of the world vegetable edible oil and is therefore considered to be an important commodity.

Recently Fast Plants, molecular biology are producing important research and educational tools, and more importantly *Brassicas* are seen as "functional foods" having long-term roles in the fight against coronary diseases and cancer.

The long breeding program aims at the understanding about the genetic potential of wild relatives of *Brassicas*. Moreover, it is obvious that many of the wild *Brassicas* have great potential value as a source for new crops. From host systems in molecular breeding to the use as condiments and industrial oils, the wild *Brassicas* have emerged as a great source possessing numerous useful agronomic related traits that could be incorporated in breeding program. Wild *Brassicas* are gaining a lot of importance recently for the projects that includes: intermediate C3-C4 photosynthetic activity; resistance to insect pests and insect and diseases; tolerance to chilling heat, salinity, and drought condition; and in cytoplasmic male sterility program.

10.3 Taxonomy

A most comprehend study on the taxonomy of *Brassicas* indicates that it contains 242 species with 48 genera. Out of which 20 are monotypic. From geographical point of view it is centered in Mediterranean region specifically in Spain, Morocco, and Algeria. This region contains 40 genera that exhibit the fact that either these are native or this region is their primary center of origin. This range extends to eastern regions that include Pakistan, India, and China, and having an extremely small or negligible representation in the new world.

The detail about the names of genera is given in Table 10.2.

Intra-generic association among the wild and cultivated genras is the definite proof of their random mating fruit character, which is generally used as an

TABLE 10.2 Wild genera.

• Ammosperma • Brassica • Cakile • Carrichtera • Ceratocnemum Chalcanthus • Coincya • Conringia • Cordylocarpus Crambe • Crambella • Didesmus • Diplotaxis • Douepea • Enarthrocarpus Eremophyton • Eruca • Erucaria • Erucastrum • Fezia • Foleyola • Fortuynia • Guiraoa • Hemicrambe	• Henophyton • Kremeriella • Moricandia • Morisia • Muricaria • Orychophragmus • Otocarpus • Physorhynchus • Pseuderucaria • Pseudofortuynia • Psychine • Quezeliantha • Raffenaldia • Raphanus • Rapistrum • Rytidocarpus • Savignya • Schouwia • Sinapidendron • Sinapis • Succowia • Trachystoma • Vella • Zilla

identification mark for differentiation between genera and species. Taxonomic classification of wild and cultivated types is under debate since long. But recent advent in science includes placement of wild generas as Dolichorhynchus in Douepea [12]; Hirschfeldia in Sinapis [13]; Euzomodendron in Vella and Coincya[13,14]; Quidproquo in Raphanus and of Hirschfeldia in Sinapis [13]; of the monotypic genera Dolichorhynchus in Douepea [14] and of Quidproquo in Raphanus [15].

10.4 Wild *Brassicas*

Wild diploid *Brassicas* and their related amphidiploids hybrids have evolved naturally in unpleasant places with abilities to withstand drought, heat, and salt stresses. A range of wild diploid *Brassica* still holds its survival in unpleasant habitats and they are actually indicators of the natural diversity of this genus. This diversity has spread too many folds in domestication and the service of mankind.

10.5 Wild *Brassicas* history

In the year 1925, Blairinghem founded *B. oleraceae* perennial form growing at the cliffs of French coast and named it as var. *suffruticosa.* The collection of such seeds from coastal areas began from the year 1901 but had a long fate to go. They were accidently destroyed in the area they were sown. The following years data revealed that it was self sterile. Meanwhile inland cultures resulted in full fertile plants. This specie resembles a lot to the sub specie of *B. oleraceae* var. *sylvestris.* This indicated that these were the wild form of the same species.

B. oleracea subsp. *oleracea* [16–19] is the perennial form of wild cabbage which is the closest wild relative and is found in the Atlantic coast of Europe. A study was carried on the glucosinolate multiphyletic origin which indicated that cultivated forms are derived from different races of wild [20]. *B. oleracea* is a good source where a both cultivated and wild race shows distinct pathways. With wild ones mainly originating from Mediterranean and spreading to several islands and mainly found on cliffs.

Brassica carinata (n = 17), the hybrid was formed by the cross between *B. oleracea* (n = 9) and *B. nigra* (n = 8). The specie formed has slow growth like *B. oleracea* and oil properties of *B. nigra.* Its wild form is not known but their cultivated types are found in Kenya and Ethiopia. This specie might have originated from the wild form of *B. oleracea.*

The third hybrid, *B. napus* (n = 19), has its wild forms in Denmark and Sweden, United Kingdom and The Netherlands. It is developed from a cross between *B. rapa* (n = 10) and *B. oleracea* (n = 9). *B. rapa* extended from the Irano Turanian regions. Comparatively, *B. napus* may have Mediterranean origins or, as seems likely, there were several centers of evolution. The wild populations of *B. napus* have gained major scientific importance recently as they present a potential determinant of gene flow in the modified cultivars of oilseed *rapa*. The association between hybrids and amphidiploids were well documented by in U-triangle and indicates how wild could have been originated from them.

10.6 History of Nagaharu U triangle

Woo Jang-Choon was a Korean Japanese agriculturist and botanist who was well noted for his work in breeding and genetics. In the year 1935, synthetic hybrids were made by Woo by crossing diploid and tetraploids in *Brassica* and examines how the chromosomes pairing results in triploid. He also presented the theory about the evolution and genetics of the species of *Brassica* involved in his triangle for breeding. And that triangle was given the name Nagaharu U, here "U" is an estimated phonetic rendering of the Korean perusing of his family name (禹,우), while "Nagaharu" is a Japanese perusing of the Chinese characters in his Korean given name (長春).

The triangle of U explains the evolutionary relationship between the species of *Brassica* showing how the genome of three ancestral species actually gave rise to the new combinations which were stable and were used as common oilseed source and vegetable. This triangle as shown in Fig. 10.1 is a demonstration of how six species are interrelated and how new species originated from three ancestral species.

The letter set n indicates the number of chromosomes in every genome. *B. rapa* is indicated by AA and 19 sets of chromosome. This demonstrates each

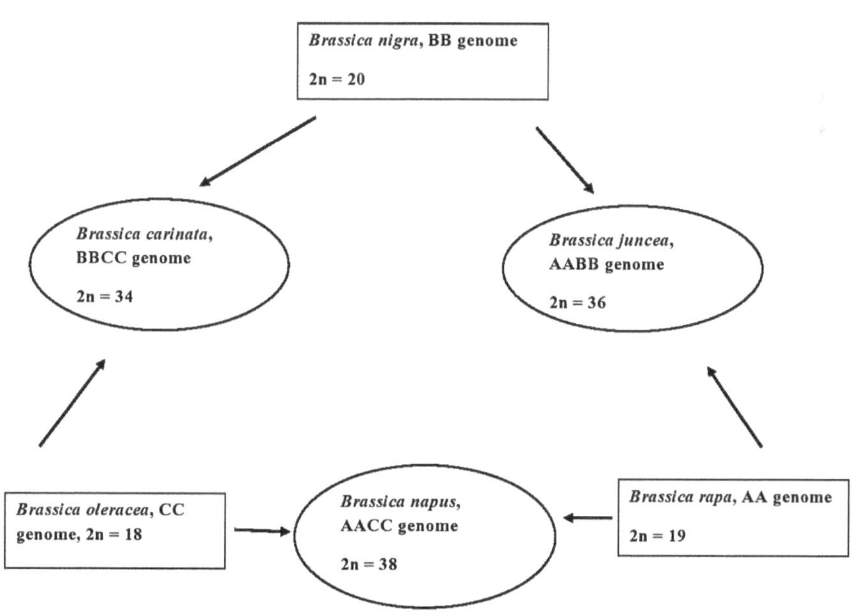

FIGURE 10.1 **Nagaharu U triangle.**

somatic cell of the plant contains two complete genome duplicates (diploid) and every genome has 10 chromosomes. Every cell will contain 20 chromosomes. *B. rapa*, *B oleracea*, and *B napus* are isolated species, yet it was feasible for them to interbreed on the grounds that they are firmly related. This interspecific reproducing led to the creation of three new species of tetraploid *Brassica* AABB (*B. juncea*), AACC (*B. napus*), and BBCC (*B. carinata*).

Brassica species have cultivated as well as weedy types which exhibit expansive phenotypic plasticity. This is shown by contrasts between root vegetables, for example, turnip, the leafy ones, for example, Chinese cabbage, and oilseeds. With the exception of economic point of view, there are various reason for the studying of this genome along with related enhancements in vegetable and oilseed crops, they presents an amazing model to study the impact of polyploidization as well as have dynamic morphological characters which aids in studying plant species.

Even though many plant species are utilized for food and horticulture purposes, but just 30 crops represent the 95% of the food that is being produced around the world. Crops that are domesticated have least diversity in comparison to their crop wild relatives (CWRs). This "bottleneck" of genetics has brought about a great loss of important alleles. During the domestication process, the addition of wild alleles add genetic variability to the existing germplasm.

As wild "progenitors" of crops keep on developing under abiotic and biotic conditions, it is essential to preserve this subsequent genetic biodiversity, which can be valuable for agricultural purposes specially (in situ conservation). Seed banks or collection of germplasm are additionally essential to preserve as another source for agribusiness (ex situ preservation). The complete genome sequencing of CWRs might be utilized first to characterize wild populations and inform about the characters that might be useful for future program. Moreover the genetic information would be quiet useful for the crops to be introgressed by the wild alleles as it would help to improve the food security, climate adaptation, and nutritional aspects.

The genus *Brassica* is listed in Annex I of the International Treaty on Plant Genetic Resources for Food and Agriculture. *B. oleracea* (or wild cabbage), a species native to southern and western Europe, has become established as an important human food crop plant because of its large reserves stored over the winter in its leaves.

A few weedy races reaching at E. Asia. Evolution based on amphidiploid advancement is the most important character for the genus. The diploid species *B. nigra* (L.) Koch (n = 8), *B. rapa* L. improve. Metzg. (n = 10, syn.: *B. campestris* L.), *and B. oleracea* L. (n = 9) all show a rich variety under the process of domestication. From the amphidiploids *B. juncea* (L.) Czern. (n = 18), *B. napus* L. improve. Metzg. (n = 19), and the uncommon *B. carinata* A. Braun (n = 17) some vegetable races are derived. The man-made *B harmsiana* O.E. Schulz (*B oleracea* × *B. rapa*, n = 29, n = 39) hybrids have serves for the improvement of new vegetables. *B. tournefortii Gouan* (n = 10) from another

Brassica-cytodeme, not the same as the *B. rapa* group, is used as a vegetable in India. *Brassica* has created two hotspots under its cultivation regions, in the Mediterranean zone and in E. Asia.

10.7 Genome A

This genome contains the turnip greens, rape-turnip Chinese cabbage, turnip, and sarson crops, which are assigned on the basis of morphology, separately, to the verdant, rapifera and oleifera types. The Chinese cabbage is of immense importance and hold great significance in Asia as it is economically important in this region and widely used for salad purposes where as sarson turnip is a minor crop utilized primarily in the regions of Europe and in New Zealand where it is used for edible purposes; moreover turnip tops and turnip greens are exceptionally utilized in Portugal and northern part of Spain for culinary purposes. Rape-Turnip is widely used as an oilseed crop in the Northern America. *B. campestris* has been renamed as *B. rapa* as indicated by the International Code of Botanical Nomenclature. This variant is more widespread hence the name is renewed.

10.8 Genome B

The genome-B is predominately represented by black mustard, which is these days considered in Europe as a weed yet was once notable in the Middle Ages in Europe as topping. Over 2,000 years back, the plant was utilized as a topping; it was referenced by the Roman creator Columella in the first century A.D. The plant leaves were additionally used by immersing in vinegar. In France, in the 13th century, the seeds were ground and utilized. They were blended in with incompletely matured grape juice to make a sour wine (or consuming must). This turned out to be later what is now called as mustard in English.

It produces a very sharp flavor that is commonly produced using ground seeds of the plant, with the seed coats separated. The little (1 mm) seeds are thick and ranges in color from muddy-brown colored to dark. They are tasty, despite the fact that they have basically no smell. The seeds are ordinarily utilized in Indian cuisine for instance in curry, where it is known as rai. The seeds are normally tossed into hot oil or ghee, after which they pop, discharging a trademark nutty flavor. The seeds have a lot of greasy oil, fundamentally oleic acid. This oil is utilized frequently as cooking oil in India, it is classified "sarson ka tel" in local language.

10.9 Genome C

This genome is represented by *B. oleracea,* and broadening itself into various botanical varieties and related varieties by process of domestication, for example, var. acephala, var. botrytis, var. capitata, var. gemmifera, var. gongylodes, var. italica, and var. sabauda, which are known to, separately, as by kale, cauliflower, cabbage, Brussel's sprout, kohlrabi, broccoli furthermore, Savoy cabbage crops. In certain circles, kale has become super famous. This was seen in ancient Roman and Greek culture where it was artificially selected to present day leafy form indicated the fact that it ruled that era.

10.10 Diversity of amphidiploids of U-triangle

Great amount of genetic diversity is present among the three amphidiploids species. Based upon studies, it is indicated that *B. napus* is the most ancient amphidiploids. Two major factors that are quiet possible for the reasons of the diversity among amphidiploids are: (1) polyploidization and genome modification. (2) Multiple hybridization with different diploid parents.

Artificial synthesis, re-synthesis, and hybridization indicated how important wild relatives are in production of new or improved cultivar. Genome changes include either the loss or gain of parental fragments. This rapid genome changes results that shows evolutionary lineage of the amphidiploids *Brassicas*.

10.11 Morphology and physiology of wild and cultivated *Brassicas*

It will be clear at this point that a large number of the wild *Brassica* spp. and their close relatives inhibit arid, dry coastal, desert type areas. These wild types have less chlorophyll, thicker leaves, and much more cell wall compartments as compared to the cultivated ones. Well developed xylem structures that can hold or conserves more water is the best feature of all in the wild ones. Wild relatives have high photosynthetic rate per leaf unit area in dry conditions where as cultivated types have low photosynthetic rate per leaf unit area.

10.12 Historical findings of comparative taxonomy of *Brassicas* in different eras

Brassica seeds are often found in ancient archeological sites. As one could see the utilization of *B. oleracea* in Neolithic or Bronze Age, however a reliable proof of this is missing at present, in spite of the fact that records for *B. rapa* L. also, *B. nigra* (L.) are better reported and documented [23,24]. The challenges in the distinguishing proof of archaeo-natural *Brassica* records all around the world are well documented [25], who pinpointed the fact that there are no promising variation exists between various *Brassica* species in terms of seed size and shape necessary for the identification of the of *Brassica* fossil material. The online database of Helmut Kroll on "Writing on archeological survives from developed plants 1981–2004" additionally shows that no specific proof could be found of *B. oleracea* archeological stays before Roman occasions.

The consumption of coles in Egypt is recorded for Greco-Roman occasions by Greek, Demotic, and Coptic writings. For instance, the Hibeh Papyri (1.121), written in Greek around 250 B.C.E., contain a private record, likely rendered by a worker to his lord, of consumption for different purposes, including krambe, which cost 4 oboloi. The Oxyrhynchus Papyri (12.1479), written in Alexandria somewhere in the range of 25 and 1 B.C.E. discusses a decoction of krambe (kramben). A demotic clinical papyrus (Wien D 6257, IV, 6) of the II century C.E. portrays a formula including coles (grmb) to treat a malady of the throat [26].

Theophrastus offers some different insights regarding *Rhaphanos* and its development. Plants, for example, *Rhaphanos* and lament are given by Theophrastus as guides to show that it is so hard to build up obvious groupings, since they have just one stem as though they were trees, therefore some call these "tree-herbs" (PPH 1.3.4) and they can be named under-bushes, because of their woodiness (PPH 6.1.2). *Rhaphanos* is likewise characterized as having a solitary root (PPH 1.6.6), being evergreen (PPH 1.9.4), having leaves (PPH 1.10.4), and bearing organic products both on the top and along the edges (PPH 1.14.2). Time of planting is after the late spring, in the long stretch of Metageitniōn (July) (PPH 7.1.2–3). It is propagated by a piece detached yet it is important to take a piece which has roots joined to it (PPH 7.2.1). The plant shoots again if the stem is broken and the re-developed stem is better, given the leaves are peeled off before the plant deteriorates (PPH 7.2.4). This indicates the fact that these were commonly used and domesticated in Greek Roman era. This information collected clearly indicates the use of *Brassicas* as early as possible also highlights its importance in those eras.

Kale and collard greens, which are a part of similar subspecies, were made by making the predecessor plant's leaves greater, and were tamed in Europe at some point before 300 B.C.E. Red, green, and savoy cabbages were made from a kale cultivar (likely the European collard greens) during the 1200s by choosing for a huge terminal bud the developing end at the highest point of the plant. The leaves are firmly twisted around a short, wide stem (the cabbage's center).

Kohlrabi was created by selecting for a thicker stalk in a kale plant around the 1400s.Chinese broccoli, another subspecies of *B. oleracea,* was domesticated in China and is distinct from broccoli, which was created from a kale predecessor in the 1500s. Broccoli was developed by selecting for the larger flower clusters, which are then harvested before they bloom. Cauliflower was developed from one of hundreds of broccoli varieties.

The amazing evolution of *B. oleracea* goes to show that humans have been tinkering with the genetics of our food creating what are now known as genetically modified foods, or GMOs for a very long time. New lab techniques just let us do that in a more precise and directed way.

10.13 Conservation of wild *Brassicas*

In the previous decades, a significant loss of common genetic diversity of variety of numerous crops has been seen due to many factors, for example, the widespread of new F_1, dry seasons, changes in food priorities and horticultural practices, and human exercises, for example, deforestation and relocation to urban regions from rural region. This procedure is known as genetic erosion. The loss of genetic constituents, precisely to the loss of wild germplasm as well as the loss of developed landraces coming about because of the genotype into environment interaction that results in selection of certain genes present in both wild and developed populaces. The Brassica family has not been a special case and, specifically, preservation of wild *B. oleracea* species has been a high need. During the 1970s, wild germplasm of Brassica was widely collected and cytogenetic examinations were begun. Concentrated endeavors were made in the most recent decades to look and gather this material that would be irreversibly lost if not preserved because it presents the highest risk [21].

After 1970s, the presentation of the idea of biodiversity was a solid help for some upgrades in ex situ and in situ preservation systems. Seed banks were made to keep up the genetic variety of numerous crops, to limit the genetic drift and erosion, moreover to preserve the seeds of various crops that are near to extinct or to create a large database which could help in taxonomy of various crops and would provide wild germplasm for improvement in present breeding programs. The majority of the *Brassicas* are preserved in the form of seeds and, in general, they are preserved under long haul storage system conditions to keep up seed viability for a long time. The main special case inside *Brassica* crops is a kale (*B. oleracea* L. var. ramose DC) that can be produced by vegetative proliferation because of the loss of its ability to bloom [22].

During a series of IPGRI missions, with the basic objective of collecting germplasm of wild family members to cultivated *B. oleracea,* around 200 territories all through the whole widespread territory have been visited. Besides from seed gathering, additional information was documented. Accordingly, the field overviews included data about related species, environmental affiliations, population structure, danger of extinction, and other information of related to preservation were collected.

10.14 Ex situ

Huge efforts has been made for protection and conservation of *Brassica* germplasm, for the most is primarily initiated by IPGRI (earlier IBPGR, International Board for Plant Genetic Resources), have been made in the course of the most recent decade. A great amount of seeds of all the species is presently accessible for reproducing, breeding, and research purposes at UPM, run by Professor C. Gomez-Campo, Universidad Politecnica, Madrid. A database of all collected information has additionally been set up. Some of the wild species are shown in Fig. 10.2

10.15 In situ

Possibilities for in situ preservation of the (n = 9) Mediterranean *Brassica* species have been discussed and debated about by [27]. During a conversation inside the ECP/Gr *Brassica* group of IPGRI they have focused on various perspectives.

FIGURE 10.2 Plant morphological diversity of Brassica wild species. *(Branca F, Cartea E.* Brassica. *In: Kole C, editor. Wild crop relatives: genomic and breeding resources. 2011.)*

To preserve the (n = 9) *Brassica* species is not a particularly hard assignment in comparison with other plant preservation issues. A considerable amount of sites are now protected either by their own structures (cliffs) or in light of the fact that they are either placed in generally remote territories (massifs, capes, islands). Legal security of however many *Brassica* sites as could be allowed is an initial step. In any case, *Brassicas* protection and conservation should be enhanced as completely passive type of conservation is not the solution. Various actions have to be taken for the diverse germplasm conservation. Legal actions are necessary to maintain the conservation. It is an opinion by Gustafsson that a major portion of important wild *Brassicas* are seen in Sicily.

10.16 Conclusion

Various authors have worked on the collection and importance and history of *Brassicas*. For this, several authors have laid different hypothesis. Evidences supporting the fact that these were cultivated in early eras as certain excavation sites revealed the facts. Moreover, the insight on the wild relatives of *Brassicas* is also very important because these wild relatives would help in improvement in future breeding programs. As population is expected to rise by 9 billion by 2020 and impact of climate change that might brought with it various insect pest and diseases, these wild *Brassicas* will provide the opportunity to improve the crops in future.

References

[1] Branca F, Cartea E. *Brassica*. In: Kole C, editor. Wild crop relatives, genomic and breeding resources, vol. Oilseeds. Heidelberg, Dordrecht, London, New York: Springer; 2011. p. 17–36. Doi: 10.1007/978-3-642-14871-2_2.

[2] Linnaeus C. Species plantarum, vol. 2. Stockholm, Sweden; 1753. p. 561.

[3] De Candolle AP. Cruciferae. Syst Nat 1821;2:139–700.

[4] Hooker JD. In: Bentham G, Hooker JD, editors. Genera plantarum, vol. 1. London: Lovell Reed; 1862. p. 57–102.

[5] Baillon HE. Cruciferes. Historie des plantes (Paris) 1871;3:188–95. 248.

[6] Prantl K. Cruciferae. In: Engler A, Prantl K, editors. Die Natürlichen Pflanzenfamilien. Leipzig: Wilhelm Englmann; 1891. p. 145–208.

[7] Schulz OE. *Cruciferae–Brassicaceae*. Part I: *Brassicinae* and *Raphaninae*. In: Engler A, editor. Das Pflanzenreich, vol. 68–70. Leipzig: Wilhelm Engelmann; 1919. p. 1–290.

[8] Schulz OE. Cruciferae. In: Engler A, Prantl P, editors. Die Natürlichen Pflanzenfamilien. Leipzig: Wilhelm Engelmann; 1936. p. 227–658.

[9] Beilstein MA, Al-Shehbaz IA, Kellog EA. *Brassicaceae* phylogeny and trichome evolution. Am J Bot 2006;93:607–19.

[10] Al-Shehbaz IA, Beilstein MA, Kellogg EA. Systematics and phylogeny of the *Brassicaceae* (*Cruciferae*): an overview. Plant Syst Evol 2006;259:89–120. https://doi.org/10.1007/s00606-006-0415-z.

[11] Warwick SI, Hall JC. Phylogeny of *Brassica* and wild relatives. In: Gupta, SK, editor. Biology and breeding of crucifers. Boca Raton: CRC Press; 2009.

[12] Appel O, Al-Shehbaz IA. Cruciferae. In: Kubitzki K, Bayer C, editors. The families and genera of vascular plants. Berlin: Springer; 2003. p. 75–174.

[13] Warwick SI, Francis A, Al-Shehbaz IA. *Brassicaceae*: species checklist and database on CD-Rom. Pl Syst Evol 2006;259:249–58.

[14] Al-Shehbaz IA, Warwick SI. The generic disposition of *Quidproquo confusum* and *Sinapis aucheri* (*Brassicaceae*). Novon 1997;7:219–20.

[15] Appel O, Al-Shehbaz IA. Cruciferae. In: Kubitzki K, editor. Families and genera of vascular plants, vol. 5. Berlin: Springer; 2003. p. 75–174.

[16] Snogerup S, Gustafsson M, Bothmer von R. *Brassica* sect. *Brassica* (*Brassicaceae*). I. Taxonomy and variation. Willdenowia 1990;19:271–365.

[17] Song KM, Osborn TC, Williams PH. Brassica taxonomy based on RFLPs. 3. Genome relationships in *Brassica* and related genera and the origin of *B. oleracea* and *B. rapa*. Theor Appl Genet 1990;79: 497–506.

[18] Hodgkin T. Cabbages, kales, etc. In: Small J, Simmonds NW, editors. Evolution of crop plants. Harlow, UK: Longman Group; 1995. p. 76–82.

[19] Gustafsson M, Lanner-Herrera C. Diversity in natural populations of wild cabbage (*Brassica oleracea* L.). Bocconea 1997;7:95–102.

[20] Mithen RF, Lewis BG, Heaney RK, Fenwick GR. Resistance of *Brassica* species to *Leptosphaeria maculans*. Trans Br Mycol Soc 1987;88:525–31. 10.1016/S0007-1536(87)80036-0.

[21] Gomez-Campo C, Aguinagalde I, Ceresuela J, Lazaro A, Martınez-Laborde J. Erosion of genetic resources within seed Genebanks: the role of seed containers. Seed Sci Res 2006;16:291–4.

[22] Gomez-Campo C, Aguinagalde I, Arús P, Jimenez-Aguilar C, Lázaro A, Martın-Clemente JP, et al. Geographical distribution and conservation status of *Brassica montana* in NE Spain. Cruciferae Newslett 2007;27:32–4.

[23] Schlichterle H. Cruciferen als Nutzpflanzen in neolitischen Ufersiedlungen Südwestdeutschlands und der Schweiz. ZArchaol 1981;15:113–24.

[24] Korber-Grohne U. Nutzpflanzen in Deutschland von der Vorgeschichte bis heute. Stuttgart: Konrad Theiss; 1995.

[25] Tomlinson P, Hall AR. A review of the archaeological evidence for food plants from the British Isles: an example of the use of the Archaeobotanical Computer Database (ABCD), Internet Archaeology, (1). Council for British Archaeology; 1996. doi:10.11141/ia.1.5. http://intarch.ac.uk/journal/issue1/tomlinson_index.html.

[26] Reymond EAE, editor. A medical book from Crocodilopolis: P. Vindob. D. 6257. From the contents of the libraries from the Suchos Temples in the Fayyum. Part I. Vienna: Brüder Hollinek; 1976.

[27] Gomez-Campo, Gustafsson. Report of a working group on Brassica. Second meeting, 13-15 November 1994, Lisbon, Portugal European Cooperative Programme for Crop Genetic Resources Networks (ECP/GR). International Plant Genetic Resources Institute; Rome: 1994.

Chapter 11

Wild Germplasm: Shaping Future Tomato Breeding

Asif Saeed and Noreen Fatima
Department of Plant Breeding and Genetics, University of Agriculture, Faisalabad, Pakistan

11.1 Introduction

Domesticated tomato (*Solanum lycopersicum* L.) is an important edible plant species that is also used as model organism for scientific research. It belongs to family *Solanaceae* that comprises 98 genera. It is botanically a fruit but consumed as a vegetable. Cultivated tomato is consumed in almost every country. Tomato is a rich source of nutrients and consumed as salad, soup, ketchup dessert, beverages, and in variety of ways. China and India are among the top tomato producers worldwide [1]. Being a nutritious vegetable, tomato contains the following mineral nutrients and vitamins (Table 11.1) as key constituents of tomato [2] Tomato is rich in vitamin A in the form of β-carotene. Its red color is due to high amount of lycopene which is useful for the human health because of its anti-oxidative properties and potential to reduce the chances of the cancer and cardiovascular diseases. Moderate consumption of tomato reduces the chances of chronic heart diseases because it makes DNA resistant to oxidative damage, but some studies reveal that these beneficial effects are not only due to lycopene rather some other compounds are also involved [3].

11.2 Wild germplasm of tomato

Importance of tomato for scientific research has been reported more than 100 years ago. It was mapped for the morphological attributes in the start of 20th century [4]. Early scientific study of its clade of almost 13 species shows that cultivated tomato has less genetic variation while wild relatives of domesticated tomato are rich source of natural variation that can be easily crossable with domesticated tomato to increase the diversity and thus population can further be used in genetic mapping.

Wild Germplasm for Genetic Improvement in Crop Plants. http://dx.doi.org/10.1016/B978-0-12-822137-2.00011-4

TABLE 11.1 Nutritional components in tomato.

		Raw tomato (100 g)	Canned tomato paste (100 g)
Minerals	Calories (kcal)	22	82
	Proteins (g)	1	3.4
	Fats	0.2	0.4
	Fiber (g)	0.5	0.9
	Carbohydrates (g)	4.7	18.6
	Ca (mg)	13	27
	P (mg)	27	70
	Na (mg)	3	790
	Mg (mg)	17.7	20
	K (mg)	244	888
	Fe (mg)	0.5	3.5
	Zn (mg)	0.2	0.8
	Cu (mg)	0.01	0.59
Vitamins	Vitamins A (IU)	900	3300
	Vitamins D (IU)	0	
	Vitamins E (mg)	0.4	
	Vitamins C (mg)	23	49
	Thiamin (mg)	0.06	0.2
	Riboflavin (mg)	0.04	0.12
	Niacin (mg)	0.7	3.1
	Pantothenic acid (mg)	0.33	0.44
	Vitamin B6 (mg)	0.1	0.38
	Folic acid (mg)	39	
	Biotin (mg)	4	

From Ref. [2].

11.2.1 Plant defense

Wild species of tomato have resistance genes called R genes which provide resistance against various pathogens by the production of specific compounds. R genes can be divided into four classes: CNL, RLP/RLK, TNL, and miscellaneous, all these are represented in tomato genome. These classes have many genes that provide the resistance to tomato against bacterial, fungal, and viral pathogens. For example, *Bs4* gene from the class TNL provides resistance against *Xanthomonas compestris* while the *Prf* gene from the CNl class provide resistance against *P. syringae*. One of the first R genes (resistance gene) was isolated from wild species and cloned in domesticated tomato [5].

11.2.2 Speciation and mating system

Tomato species have mating systems that range from autogamy to allogamy. Molecular studies help in understanding the mating systems in tomato by using molecular markers. Various researches on this aspect consider the floral attributes which are involved in evolution of the mating system. Some researchers have focused on QTL mapping and isolating of loci that contribute to floral attributes associated with change from the allogamous to autogamous mating system [6]. Use of interspecific crosses for the creation of mapping lines in tomato provides the ideal plant material which can be further used to examine the speciation process and genetics of species by isolating barriers.

11.2.3 Divergence and adaptation

Tomato is used as a model plant to study the divergence and adaptation. It shows a vast morphological and physiological diversity among wild species which make it possible to have different adaptation mechanisms that can be helpful in future tomato improvement programs. These species also show tolerance to biotic and abiotic stresses that is another useful factor in future tomato breeding [7]. Tomato species have allogamous to autogamous mating systems and have various levels of genetic variations. Due to these attributes, botanists consider this clade for population studies to explore the association between mating system, genetic variation, recombination, and demographic processes that affect the diversification of population and evolutionary studies.

11.3 Evolution and taxonomy

Spaniards introduced the tomatoes into Europe in the start of 16th century [8]. Early botanists of Europe named the tomatoes as *S. pomiferum* because they found the association of tomatoes with the genus *Solanum*. Tournefort named the cultivated tomato as the Lycopersicon in 17th century. Then in 18th century Linnaeus suggested the botanical names of tomato as *Solanum peruvianum* and *S. lycopersicum*. Later in 18th century Miller was the first person that described the genus *Lycopersicon*.

Monograph of wild tomatoes with their wild relatives was identified [9]. He also showed the 13 species of wild and cultivated tomatoes that were part of tomato clade (Tables 11.2 and 11.3). The age of *Solanum* genus is approximately 12 million years on the basis of genetic markers. Arcanum group has bright carotenoid content in fruit. It is difficult to know the evolutionary relationship within its clade because of diversification in group and low amount of genetic variation in allogamous species. Different genetic markers have been used to study the association of species with its clade [10].

TABLE 11.2 Species list for *Solanum* sect. *Lycopersicoides*, sect. Juglandifolia and sect. Lycopersicon with equivalent names in Lycopersicon.

Names in genus *Solanum* [9]	Equivalent to *Lycopersicon*
Solanum neorickii	*Lycopersicon parviflorum*
Lycopersicon juglandifolium	*Lycopersicon juglandifolium*
Solanum peruvianum L.	*Lycopersicon peruvianum* (L.)
Solanum pennellii	*Lycopersicon pennellii*
Solanum chilense	*Lycopersicon chilense*
Lycopersicon lycopersicoides	*Lycopersicon lycopersicoides* (Dunal)
Solanum habrochaites	*Lycopersicon hirsulum*
Solanum corneliomuelleri	Part of *Lycopersicon peruvianum* (L.) also known as *L. glandulosum*
Solanum huaylasense	Part of *Lycopersicon peruvianum* (L.)
Solanum chmielewskii	*Lycopersicon chmeilewskii*
Solanum pimpinellifolium L.	*Lycopersicon pimpinellifolium* (L.)
Solanum ochranthum	*Lycopersicon ochranthum* (Dunal)
Solanum cheesmaniae (L. Riley)	*Lycopersicon cheesmaniae* L. Riley
Solanum galapagense	Part of *Lycopersicon cheesmaniae* L. Riley
Lycopersicon sitiens	*Lycopersicon sitiens*
Solanum lycopersicum L.	*Lycopersicon esculentum*
Solanum arcanum (4 geographic races: 'humifusum', lomas, Marañon, Chotano- Yamaluc)	Part of *Lycopersicon peruvianum* (L.)

Wild tomatoes of section *lycopersicon* with its close relatives' sections *lycopersicoides* and *Juglandifolia,* are the monophyletic. Sections *lycopersicoides and Juglandifolia* are also monophyletic but similar to *lycopersicon* section. In this section, *S. pennellii* is considered as the base of the phyletic tree where *S. habrocaties*, *S. chmeilewskii*, and *S. neorickii* are closely associated to *S. arcanum* which produces brightly red colored tomatoes.

Solanum galapagense, *S. cheesmaniae*, *S. pimpinellifolium*, and *S. lycopersicon* species produces orange or yellow, typical red color fruits which reveals that they belong to monophyletic group. Based on molecular markers, morphological markers, distribution of species on geographic area, phylogeny, and mating system, the new classification of wild and domestic relatives of tomatoes have been published [9] and described in Table 11.3.

Both biotic and abiotic factors affected the evolutionary process of wild tomatoes [8]. Evolutionary studies reveal that putative allogamous species of tomatoes occupy a wide area of central Peru. The environmental pressure due to

TABLE 11.3 Taxonomic treatments of wild and wild relatives of tomatoes.

Section Lycopersicon	Lycopersicon group	*S. lycopersicon*
		S. pimpinellifolium
		S. cheesmaniae
		S. galapagense
	Neolycopersicon group	*S. pennellii*
	Eriopersicon group	*S. habrochaites*
		S. chilense
		S. peruvianum
		S. corneliomulleri
		S. huaylasense
	Arcanum group	*S. Arcanum*
		S. chmielewskii
		S. neorickii
Section Lycopersicoides		*S. lycopersicoides*
		S. sitiens
Section Juglandifolia		*S. juglandifolium*
		S. ochranthum

From Ref. [9].

severe climate change and desert formation of Peruvian-Chilean region increases the process of adaptation to the new environment that leads to speciation.

11.4 Evaluation and utilization of genetic resources

The tomato germplasm has been used by many breeders and screened for various objectives. Evaluation has been done for various biochemical, physiological, and genetically related characteristics [11]. First evaluation of tomato was done in its native habitat and variation among wild species was observed [12]. Plants showed the resistance to various biotic and abiotic stresses that indicated the presence of beneficial attributes for breeding programs.

Second most important part during evaluation is measurement of variation in the germplasm. Evaluation is easier if plant populations have uniform and homozygous lines. If there is higher rate of outcrossing, then large sample is recommended to test. Mostly, evaluation is done based on morphological traits but biochemical analysis gives more accurate results. At cellular level, chromosomal evaluation has been done in tomato to comprehend the compatibility in species. The tomato germplasm has also been evaluated for various diseases and insect pests.

By the utilization of wild and semi-wild germplasm, improvement of dry matter content, ascorbic acid, carotene content, and sugar in tomato has been successfully attained. Ascorbic acid and carotene content have been improved by crossing our cultivated germplasm with *L. peruvianum.* F_1 hybrids developed by the cross between cultivated tomato and wild (*L. minutum*) that yield ascorbic acid up to 59 mg/100 g of fruit. Two wild species; *S. pennellii* and *L. cheesmanii* var. *minor* have high amount of carotene content. Wild specie *L. chilense* have been used in tomato breeding that increased the ascorbic acid content (upto 49.45 mg/100 g) and the β-carotene content (upto 885 mg/100 g). While *L. hirsutum, L. esculentum* var, *cerasiforme*, and *L. chmielewskii* are rich in carotene and total soluble salts respectively [13].

Wild germplasm of tomato also shows the resistance to abiotic factors like cold, drought, and heat, etc. *L. cheesmanii* and *L. pennellii* shows the resistance to salinity and drought, respectively while *L. hirsutum* have the resistance for cold and drought. Now stress resistance in these species have been used in tomato breeding programs.

11.5 Variation and its use in future

Different levels of isozyme polymorphism using well-developed maps of isozyme show a little genetic variation in *L. esculentum.* Various mutants were evaluated for specific morphological attributes and less genetic variation was found among these mutants. Experiment was conducted for the evaluation of 10,000 diploid plants that were derived from a single haploid parent but only one heritable variant was found [14]. Later 0.02% mutation rate was found [15] that affected the genes related to fruit formation. Less level of isozyme polymorphism, morphological variation, and above-mentioned study support the genome stability and less variation within the *L. esculentum.*

11.5.1 Wide hybridization

Wide hybridization plays a pivotal role for the improvement of useful attributes that increase genetic variation in *L. esculentum* [16]. Most useful role of wide hybridization is the improvement of tomato for various biotic stresses. The recent improvement in tomato has been made by the utilization of wild tomato genotypes to develop pseudo-diploid hybrids of *S. lycopersicoides* and *L. esculentum* [17]. Monosomic alien addition lines and the diploid individuals can be developed by using the hybrid that carries the attributes inherited from the wild relatives.

Many fruitful attributes from wild species have been used and transferred into tomato via *L. esculentum* × *S. lycopersicoides. Solanum lycopersicoides* mutants comprise of characters like white anthers (Wa), bifurcate inflorescence (Bif) day length sensitivity (Dls), frilly leaves (Frl), rugose leaf surface (Rug), fimbriate leaves (Fmb), lacinate leaves (Lac), and various new variants of isozymes. Single dominant gene has been identified for high anthocyanin contents but this gene is environmentally affected, mostly by sunlight. Purple spotting in fruits is also one of its expression.

Recently, *S. rickii* have been crossed with tomato. The hybrid obtained by aforementioned pseudo-diploid hybrid is pistillate flowered. Diploid hybrid *L. esculentum* × *S. rickii* are sterile with less extent of female fertility and incompatibility with the *L. esculentum*. Amphidiploid hybrids have been developed and now are being used to transfer various *S. rickii* attributes to tomato. Progress is notable due to these two species that provide new source of variation. These two species provide not only morphological variation but also new horticultural attributes [18]. Therefore, it can be concluded that *Solanum Lycopersicon* wild species can provide useful genetic variation for tomato breeding programs.

11.5.2 Traditional mutagenesis

Explanation with historical perspective of induced and spontaneous mutation in tomato was explained by scientists [19]. Many procedures have been used to induce mutation in tomato that include treatments of radium, UV light, X-rays, neutrons, induced osmotic stress, and chemical mutagenesis.

Hans Stubbe played important role in mutagenesis by creating 200 mutants in *L. pimpinellifolium* and 300 mutants in *L. esculentum* [20]. Fruit size of *L. pimpinellifolium* and *L. esculentum* has been improved through mutagenesis. By use of mutagenesis and direct selection, fruit size has been increased but *L. esculentum* and *L. pimpinellifolium* show decrease in fruit size. Recently, seed of tomato is treated with nitroso-methyl urea which results in the chimeric plants which show segregation for plastid mutation [21]. Both plastid and nuclear gene mutation can be obtained.

So, mutation make it ensure that improvement of tomato and the mutant lines can further be used in tomato breeding programs to achieve the targeted traits.

11.5.3 Somaclonal variation

Now mutagens have been used to induce the heritable variability via tissue or cell culture. Heritable variations developed from somatic tissues are known as somaclonal variation. Many experiments have reported that mutation in single gene can be created by using somaclonal variation. Tissue culture technique has been used to improve complex attributes, so it has gained so much attention. The chemical mutagenesis (using ethyl methane sulfonate) compared with somaclonal variation according to their relative efficiency [22]. Plant breeders have found high number of mutants which were produced from the somaclonal mutations.

More heritable variations have been found in tomatoes that were treated with ionizing radiation as compared to tissue culture mutants. In future both methods can be used to produce heritable variation in tomatoes.

11.5.4 Transposon mutagenesis

Transgenic tomato plants containing the maize transposable elements have already been developed. These incorporated transposable elements have been used as induced mutagens in gene cloning and gene identification. Scientists have also used these transposable elements for identification of genes and clones in maize as well. Scientist explained the transposon *Activator* (Ac) that can be transferred into tomato plants [23]. Later study showed the maize *Activator* (Ac) can also be mobilized by using *Dissociation* (Ds). However, no tomato genes have been cloned or tagged via transposons mutagenesis till now.

11.5.5 Genetic engineering for variation induction

Genetic engineering is an advanced field to the plant breeders for increasing genetic diversity in crop plants. Transgenic tomato plants have been developed by using *Agrobacterium* mediated transformation and other methods. There are many examples of tomato, improved through using genetic engineering approaches.

11.5.5.1 Insect resistance

Insects impose serious issues in tomato production all around the world. Wild *Lycopersicon spp.* have the genes for insect resistance but such type of insect resistance is not possible in cultivated tomato yet. Transgenic tomato plant has been developed by the transfer of *Bt* gene that express in tomato. The gene obtained from the *Bacillus thurigienesis* [24] shows resistance for tobacco hornworm, tomato fruit worm, tobacco budworm, and tomato pinworm.

11.5.5.2 Herbicide resistance

Herbicide resistance availability is valuable for growers especially for large-scale crop production. The *aroA* gene, isolated from the *Salmonella typhimurium* resistance strain was introduced into tomato. This gene provides the glyphosate resistance in tomato. Other crops are responsible for developing resistance against the sulfonylureas, bromoxynil [25], bialaphos, and glufosinate. The transgenic plant having resistance gene can be used as parent in conventional tomato breeding program.

11.5.5.3 Virus resistance

Constitutive expression of tobacco mosaic virus (TMV) coat protein produces the notable resistance in tomato plants against TMV [26]. Viral coat protein production inhibits or prevents the infection of closely associated virus infections. Transgenic tomato plants were developed that showed partial resistant to various strains of TMV. Transgenic plants developed for the virus resistance may be effective for other viral groups also [27].

11.5.5.4 Fruit quality

Another use of genetic engineering in tomato is the improvement of fruit quality. Plant breeders strive for succeeding in manipulation of fruit ripening components. Fruit ripening in tomato is achieved by the manipulation of polygalacturonase (PG) enzyme that causes the fruit softening by the degradation of middle lamella pectin [28]. Early reports show the association between PG activity, fruit softening, and association between PG level and fruit softening in the ripening mutants *nor*, *rin*, and *Nr* [29].

The polygalacturonase cDNA has been cloned and sequenced [30]. The PG clone is used for the evaluation of developmental expression of PG. In immature green fruits, mRNA of PG is not detectable. It is only detectable in mature green fruits that increase the concentration of fruit ripens [31]. Three mutants (*nor*, *rin*, and *Nr*) showed the different accumulation patterns of PG by the analysis of mRNA level. This shows that a unique molecular mechanism contributes toward the PG pattern expression in each fruit. Expression with *rin* was especially wonderful as this mutant showed the ability to accumulate less level of PG mRNA and active protein with no elevation of ethylene level. This shows that ethylene does not regulate the PG level. Other groups are also evaluated for PG mRNA level in ripening tomato mutants [32].

11.6 Genetic stock and mutants

There are hundreds of total numbers of known mutants in tomato. The tomato gene list committee designated total 21 classes of mutant tomato based on pigmentation, leaves, trichomes morphological features, fruit, disease resistance, sterility, maturity, root characteristics, and wilting [33]. This list also includes the isozyme variants. In 1988, TGC reported a list of 94 mutants of allozyme allele variants at 30 different loci. The stock helps the researcher to improve the tomato crop. Many of these mutants are easily available to researcher. The TGRC (Tomato Genetic Resources Centre, Davis, California) have 750 monogenic mutant stock. In England, Glasshouse Crops Research Institute have the "Alisa Carig" near isogenic mutants up to 155. This institute also have the closely related mutants to "Carigella" and also have the mutants for fruit color, disease resistance, and habit modification [34].

11.6.1 Cytological variants

Many variants of tomato have been reported with different polyploidy levels like haploids, tetraploids, triploids, and some aneuploids. Euploid mutants were at one time quickly checkable at large commercial planting. These off-types were characterized as the "bull" plants because of high vegetative growth but they showed inability for fruit setting at large scale [15]. Rick assessed 9000 populations and found only 66 plants that were haploids, diploids, aneuploids, triploids, or tetraploids. Now in the presence of cultural practices including high planting densities, bull plants or off-types plant are not observable because of plant to plant competitions. Haploids are useful for tomato breeders for various investigation but they are not commonly used because it requires extra care to maintain them. Colchicine can be used for chromosome doubling that result in tetraploid tomato production. Induction of tetraploidy is carried out by the use of dissection or grafting [19] but success is less as compared to colchicine. Tetraploids are easy to maintain but their progeny shows the variability in population, but they fail in fruit setting and if fruit setting happens then few fruits with reduced size and seed numbers are produced [19]. TGSC have maintained the tetraploids of eight cultivars [16]. The triploid tomatoes were first time observed in 1925 and scientists reported the "Dwarf Aristocrat" that was a single unfruitful plant. Triploids can be made from the cross between tetraploid and diploid by the decapitation but former method is comparatively difficult. [19]. Most of the triploids showed the dark green leaves, gigantism, and stout petioles with thick leaves and sterile behavior in mostly male gametophyte. Triploids when crossed with diploids can be used to develop aneuploids. Treatment of mature pollen with ionizing radiation produces the primary and tertiary monosomic traits up to 18 chromosome number. Monosomic condition cannot be transmitted through seed because there is no proper chromosomal complementation that causes functional sterility in pollen. Tomato is true diploid because it does not transmit the deficient gametes to next generation. Extra-chromosomal mutants have been developed by many ways in tomato. Extra-chromosomal mutants have been observed in primary trisomics. Many types of extra-chromosomal mutants like primary, secondary, tertiary, and telotrisomic have been observed. These trisomic stock is available in TGRC that includes set of primary trisomic and telotrisomic of six types, compensating trisomic of three types, secondary trisomic of eight types, and tertiary trisomic of six types [16]. Trisomic stock is also being used for mapping. For example, secondary and tertiary trisomics are beneficial to identify the marker genes in arm location, linkage group orientation, and location of centromere.

11.6.2 Commercially important mutants

Natural mutants and artificial mutants both are important in breeding point of view. They show disease resistance, fruit coloring, fruit ripening, parthenocarpy, male sterility, and plant architecture.

11.6.2.1 Disease resistance

Tomato species are useful because they provide the novel sources of various resistance of diseases. For example, in tomato species, *Fusarium* wilt resistance is available that can lead to better tomato yields. Before the development of disease resistant lines, growers labeled the infected field as "wilt land" and they always found new areas for tomato production. Now, two races of *Fusarium* wilt cause the problem in fields of tomato and now farmers grow only those cultivars that show the resistance against both. The third race of *Fusarium* wilt though has been reported recently but its genetic resistance has already been found in *L. pennellii* [35].

11.6.2.2 Plant architecture

Spontaneous mutants were found in Florida in 1914 [36]. For tomato processing, these mutants showed the compact habit that is desirable for the mechanically harvested tomatoes. Modifying genes can affect the expression of spontaneous mutants. The other mutants like *u* and *ug* for uniform ripening, *j* and *j-2* for jointless pedicle, fruit color mutants, male sterile, and those mutants that lead to stronger shelf life of tomatoes. Uniform ripening gene is desirable because typical dark green shoulder is lacking that leads to uniform fruit ripening. This is fruitful character for processed as well as fresh market types. The jointless (*j*) mutants found [37] in French cultivars and he also determined that this trait is monogenic which is expressed by recessive gene. The jointless mutant was found in *L. chessmanii* accessions [38]. These mutants were used by breeders for processing and fresh tomatoes. It was believed that jointless helps in separation from fruit of calyx and stem. That leads to less damage to fruit during picking and make the hand picking easy and faster because stem don't need to be removed to protect fruit. Now, many cultivars lack this character.

11.6.2.3 Color mutants

The mutants *β* have potential to increase the amount of pro-vitamin A. This mutant converts the lycopene into *β*-carotene. Orange tomatoes have 13 times higher level of *β*-carotene as compared to normal. This mutant is useful because it has no side effect on flavor and has nutritional components [39]. This mutant is not preferred by consumers because of its high pigment (*hp*), crimson (*og*), and dark green color (*dg*). Gene *hp* is similar to *dg*, the gene *dg* increases the carotenoid content. Plant breeders do not use these both genes because they cause adverse pleiotropic effect on plant vigor so they are not preferred for commercial production. These mutants have potential to increase the fruit color but no single mutant has gained the acceptance for commercial production [40].

11.6.2.4 Fruit ripening mutants

Three mutants, non-ripening (*nor*), ripening inhibitor (*rin*), and never ripe (*Nr*) have gained the attention of consumers and growers [28]. Never ripe (*Nr*) fruit is dominant mutants and the gene located on chromosome 9 turns its fruit color in normal condition but do not develop the full pigmentation level. Due to dominant gene action and less level of carotene pigment production it did not gain much attention on commercial scale. The mutants bearing non-ripening (*nor*) and ripening inhibitor (*rin*) characters are controlled by recessive gene, are similar to *Nr* mutants and do not show the respiratory climacterics but they are responsible for high level of ethylene production [41]. Both mutants show the normal pectin-esterase activity and have trace level of polygalacturonase activity. Both mutants have less carotene level and they do not fully ripe with better flavors. The mutant *rin* fruit have yellow color and soften very slowly but with *nor* fruits turn into orange color. The mutant *nor* also express allelic similarity to the mutant Alcobaca (*ale*). *Nr* mutant was not useful for commercial production of tomato but greater interest was shown by growers in *rin* and *nor* because of fruit firmness and long shelf life. It has little potential in homozygous condition but considered as best in heterozygous condition [42].

11.6.2.5 Male sterility mutants

Till 1986, 47 male sterile mutants had been found in tomato [43]. Due to these mutant hybrid seed production cost has been greatly reduced because of less labor requirement for emasculation in the seed production companies.

11.7 Conclusion

Tomato wild relatives are good source of genetic variation for various attributes like fruit color, fruiting ripening, plant architecture, fruit quality, and pathogenic infection. The wild germplasm can be used in future breeding of tomato because cultivated tomato has considerable gap that needs research for improvement in domains of heat stress, disease resistance, and fruit quality.

References

[1] Faostat. Available from: http://faostat.fao.org/site/339/; 2015 [accessed on March, 19, 2015].

[2] Ensminger AH, Ensminger ME, Konlande JE, Robson JRK. The concise encyclopedia of foods and nutrition. Boca Raton, FL, USA: CRC Press; 1995.

[3] Laquatra I, Yeung DL, Storey M, Forshee R. Health benefits of lycopene in tomatoes—conference summary. Nutr Today 2005;40:29–36.

[4] Jones DF. Linkage in *Lycopersicum*. Am Nat 1917;51:608.

[5] Martin GB, Brommonschenkel SH, Chunwongse J, Frary A, Ganal MW, Spivey R, et al. Map-based cloning of a protein-kinase gene conferring disease resistance in tomato. Science 1993;262:1432–6.

[6] Chen KY, Cong B, Wing R, Vrebalov J, Tanksley SD. Changes in regulation of a transcription factor lead to autogamy in cultivated tomatoes. Science 2007;318:643–5.

[7] Frankel N, Hasson E, Iusem ND, Rossi MS. Adaptive evolution of the water stress- induced gene Asr2 in *Lycopersicon* species dwelling in arid habitats. Mol Biol Evol 2003;20:1955–62.

[8] Peralta IE, Spooner DM. Morphological characterization and relationships of wild tomatoes (*Solanum* L. Section Lycopersicon). In: Keating RC, Hollowell VC, Croat TB, editors. A Festschrift for William G D'Arcy: The legacy of a taxonomist. Monogr Syst Bot Missouri Bot Gard 104: 227-257.

[9] Peralta IE, Spooner DM, Knapp S. Taxonomy of wild tomatoes and their relatives (*Solanum* Sect. Lycopersicoides, Sect Juglandifolia, Sect Lycopersicon; *Solanaceae*). Syst Bot Monogr 2008;84:183.

[10] Peralta IE, Spooner DM. Granule-bound starch synthase (GBSSI) gene phylogeny of wild tomatoes (*Solanum* L. section Lycopersicon [Mill.] Wettst. subsection Lycopersicon). Am J Bot 2001;88:1888–902.

[11] Khan MA, Butt SJ, Nadeem F, Yousaf B, Javed HU. Morphological and physio-biochemical characterization of various tomato cultivars in a simplified soilless media. Ann Agric Sci 2017;62(2):139–43.
[12] Rick CM. Conservation of tomato species germplasm. Cal Agric 1977;31(9):32–3.
[13] Rick CM. High soluble solids content in large fruited tomato lines derived from a wild green fruited species. Hilgardia 1974;42:493–510.
[14] Lindstrom EW. Genetic stability of haploid, diploid and tetraploid genotypes in the tomato. Genetics 1941;26:387–97.
[15] Rick CM. A survey of cytogenetic causes of unfruitfulness in the tomato. Genetics 1945;30:347–62.
[16] Rick CM. list of miscellaneous stocks. Rep Tomato Genet Coop 1988;38:62.
[17] Rick CM, De Vema JW, Chetelat RT, Stevens MA. Meiosis in sesquidiploid hybrids of *Lycopersicon esculentum* and *Solanum lycopersicoides*. Proc Nat Acad Sci U S A 1986;83: 3580–3.
[18] Rick CM. Tomato-like nightshades: affinities, auto-ecology and breeders' opportunities. Econ Bot 1988;42:145–54.
[19] Rick CM, Butler L. Cytogenetics of the tomato. Adv Genet 1956;8:267–382.
[20] Rick CM. The tomato. In: King RC, editor. Handbook of genetics, vol. 2. New York: Plenum; 1975. p. 247–80.
[21] Hosticka LP, Hanson MR. Induction of plastid mutations in tomatoes by nitrosomethylurea. J Hered 1984;75:242–6.
[22] Gavazzi G, Tonelli C, Todesco G, Arreghini E, Raffaldi F, Vecchio F, et al. Somaclonal variation versus chemically induced mutagenesis in tomato (*Lycopersicon esculentum* L.). Theor Appl Genet 1987;74:733–8.
[23] Yoder n, Palys J, Alpert K, Lassner M. Ac transposition in transgenic tomato plants. Mol Gen Genet 1988;213:291–6.
[24] Fischhoff DA, Bowdish KS, Perlak FJ, Marrone PG, McCormick SM, Niedermeyer JG, et al. Insect tolerant trans-genic tomato plants. Biotechnology 1987;5:807–13.
[25] Stalker DM, McBride KE, Malyj LD. Herbicide resistance in transgenic plants expressing a bacterial detoxification gene. Science 1988;242:419–23.
[26] Powell Abel P, Nelson RS, De B, Hoffmann N, Rogers SG, Fraley RT, et al. Delay of disease development in transgenic plants that express the tobacco mosaic virus coat protein gene. Science 1986;232:738–43.
[27] Wilson TMA. Plant viruses: a toolbox for genetic engineering and crop protection. Bio Essays 1989;10:179–86.
[28] Grierson D, Thcker GA, Keen J, Ray J, Bird CR, Schuch W. Sequencing and identification of a cDNA clone for tomato polygalacturonase. Nucleic Acids Res 1986;14:8595–603.
[29] Brady CJ, Meldrum SK, McGlasson WB, Ali ZM. Differential accumulation of the molecular forms of polygalacturonase in tomato mutants. J Food Biochem 1983;7:7–14.
[30] Sheehy RE, Pearson J, Brady CJ, Hiatt WR. Molecular characterization of tomato fruit polygalacturonase. Mol Gen Genet 1987;208:30–6.
[31] Bennett AB, DellaPenna D. Polygalacturonase gene expression in ripening tomato fruit. In: Nevins D, Jones R, editors. Tomato biotechnologyNew York: Liss; 1987. p. 299–308.
[32] Biggs MS, Handa AK. Temporal regulation of polygalacturonase gene expression in fruits of normal, mutant and heterozygous tomato genotypes. Plant PhysioI 1989;89:117–25.
[33] Clayberg CD, Butler L, Rick CM, Kerr EA, Robinson RW. Report of the gene list committee. Rep Tomato Genet Coop 1971;21:2–10.
[34] Maxon-Smith JW, Ritchie DB. A collection of near-isogenic lines of tomato: Research tool of the future? Plant Mol Bioi Rep 1983;1:41–5.

[35] McGrath DJ, Gillespie D, Vawdrew L. Inheritance of resistance to *Fusarium oxysporum* f. sp. lycopersici races 2 and 3 in *Lycopersicon pennellii*. Aust J Agric Res 1987;38:729–33.
[36] Rick CM. Tomato mutants: freaks, anomalies and breeders' resources. Hort Sci 1986;21:918.
[37] Butler L. Inherited characters in the tomato. 2-jointless pedicel. J Hered 1936;27:25–6.
[38] Rick CM. Fruit and pedicel characters derived from Galapagos tomatoes. Econ Bot 1967;21:171–84.
[39] Tomes ML, Quackenbush FW, Kargl T. Action of the gene B in biosynthesis of carotenes in the tomato. Bot Gaz 1956;117:248–53.
[40] Wann EV, Jourdan EL, Pressey R, Lyon BG. Effect of mutant genotypes hp, oK" and dg occ on tomato fruit quality. J Am Soc Hortic Sci 1985;110:212–5.
[41] Brady CJ. Fruit ripening. Annu Rev Plant PhysioI 1987;38:155–76.
[42] Strand LL, Morris LL, Heintz CM. Taste life of *rin* and *nor* hybrids. In: Proceeding of the fourth tomato quality workshop, Veg Crops Res Rep VEC 83-1, Univ Fla, Gainesville, pp 68–77.
[43] Stevens MA, Rick CM. Genetics and breeding. In: Atherton JG, Rudich I, editors. The tomato crop: a scientific basis for improvement, London: Chapman & Hall; 1986. p. 35–109.

Chapter 12

Wild Potatoes: A Genetic Reservoir for Potato Breeding

Ikram Bashir[a], Rodrigo Nicolao[a] and Gustavo Heiden[a,b]
[a]*Postgraduate Program in Agronomy, Federal University of Pelotas, Rio Grande do Sul, Brazil;* [b]*Embrapa Temperate Agriculture, Pelotas, Rio Grande do Sul, Brazil*

12.1 Introduction

Wild potatoes are a large genetic reservoir for potato breeding with current use and potential to provide genes for novel traits, and resistances for abiotic and biotic stresses absent in commercial cultivars. The first steps to use wild relatives in modern potato breeding started almost two centuries ago. Wild species from South America had been introduced in Europe in 1824, especially *S. commersonii* (Fig. 12.1) and *S. maglia*. Before 1920 many efforts were made to cross *S. demissum* with *S. tuberosum* and by 1932 introgression of resistance genes succeed first for late blight, later for virus (1941) and potato cyst nematode [1]. Since then, many resistance genes have been introduced into modern potato cultivars.

Potatoes (*Solanum* sect. *Petota*) taxonomy is a matter of debate. Accepted wild species ranges from 107 [2] to 180 [3], and domesticated from 4 (*S. ajanhuiri, S. curtilobum, S. juzepczukii, S. tuberosum* group *Andigenum*) [2] to 7, being 4 of them hybrids (*S.* × *ajanhuir*, *S.* × *chaucha*, *S.* × *curtilobum*, *S.* × *juzepczukii*), 2 of them with 4 subspecies (*S. stenotomum* subsp. *goniocalyx*, *S. stenotomum* subsp. *stenotomum*, *S. tuberosum* subsp. *andigenum*, *S. tuberosum* subsp. *tuberosum*) and 1 with no subspecies (*S. phureja*) [3].

Wild potatoes occur from southwestern United States of America (38°N) to central Argentina and Chile (41°S) and the Juan Fernández Archipelago. Most species are from South America with the diversity peaking at 21°S and a secondary center of diversity lies around 20°N in the central Mexican highlands [4]. They occur along a wide range of habitats from deserts to rainforests and from sea level to elevations around 4,700 m above sea level in the tropical Andean mountains. Potato wild relatives grow in sunny to partially sunny areas and in preserved or anthropogenic disturbed environments, including tussocks and grazed areas and edges or glades in temperate and subtropical forests [5].

Wild Germplasm for Genetic Improvement in Crop Plants. http://dx.doi.org/10.1016/B978-0-12-822137-2.00012-6

FIGURE 12.1 *Solanum commersonii*, a potato wild relative native from Southern Brazil, Argentina, Paraguay, and Uruguay. (A) Wild potato plant growing in the pampas grasslands at Chuí, Rio Grande do Sul, Brazil. (B) Wild potato flowers. (C) Wild potato berries. (D) Wild potato tubers. *(Photos by Gustavo Heiden.)*

12.2 Current and potential uses of wild potatoes for breeding

12.2.1 Commercial traits

Potatoes are a high source of carbohydrates with high nutritional value providing vitamins and minerals in the form of cooked tubers and processed products such as chips and canned potatoes. Consumption heavily depends upon its appearance and taste, commonly known as quality traits according to consumer preference or commercial use. Due to the huge quantity of potato tuber yield used by the processing industry, commercial traits have a high importance. Breeding objectives of these traits depend upon the market use [6]. Quality traits are in two groups. The first refers to "external quality traits" and includes tuber size, shape, eye depth, skin color, and pulp color. The second refers to "internal quality traits" such as flavor, nutrients, dry matter, and starch content, glycoalkaloids and processing traits. Quality traits are genetically controlled and further classified into three major groups such as biological traits (proteins, carbohydrates, vitamins, minerals, reduced amounts of toxic glycoalkaloids), sensorial traits (flavor, texture, color), and industrial traits (tuber shape and size, dry matter content, cold-induced sweetening, oil absorption, starch quality). Breeding strategies to improve quality traits are achieved by sexual hybridization or genetic engineering [7]. For instance, *S. commersonii* has high dry matter content and through negative assistant selection methods, these traits were transferred to cultivated potato [8,9]. Hybrid populations of *S. tuberosum* and *S. commersonii* were reported to have low levels of glycoalkaloids and acceptable tuber quality [9,10]. *Solanum stoloniferum* and *S. verrucosum* have quality traits used in breeding programs and low levels of glycoalkaloids [11], while *S. chacoense* has good chipping quality traits, although high level of glycoalkaloids [12]. The wild species *S. medians, S. okadae, S. pinnatisectum, S. raphanifolium*, and *S. sogarandinum* possess traits which make them resistant to cold-induced sweetening, an important factor in processing industry which defines the chip frying quality after tuber storage [13,14]. Other species with noteworthy quality traits are *S. vernei* for starch and *S. stoloniferum* for ascorbic acid [15]. Unfortunately, there is a lack of focus and a large gap on exploring the genetic potential of important quality traits from wild relatives of potatoes by biofortification of micro- and macro-nutrients which are important to enhance the nutritional value of potatoes.

12.2.2 Abiotic stresses

Climate change is the main reason for the increase of many abiotic stresses such as heat, cold, drought, and salinity. Wild potatoes occur in several climate conditions from high temperature regions to high altitude with freezing temperatures. Plants which show resistance to cold or frost stress are also most likely found tolerant to stresses such as heat, drought, and salinity [16]. For the introgression of resistant traits, diploid potato breeding is the most widely known strategy which has been used. Nowadays, molecular techniques for the identification of resistant traits at genetic level are very robust and useful. *S. commersonii*, is an example of having high levels of abiotic resistance genes for frost as well as drought, heat, and salinity stresses [17].

Frost. Cold stress tolerance is the oldest goal in potato breeding for abiotic stress. Potatoes are a cool season crop, but chilling temperature has an adverse effect. This condition is commonly present at nights and sudden increase in temperature during the day makes the crop more susceptible to cold injury. If the temperature goes below chilling it causes crystallization of water and other minerals within the leaf veins and leads to cold injury or leaf necrosis. Due to cold injury one can lose all the crops in short time [18]. *S. commersonii, S. demissum*, and their hybrids are reported to be highly resistant to frost stress. After overcoming barriers for introgression of genes, breeders successfully transferred genes from *S. commersonii* to commercial cultivars especially for tolerance to low temperatures [4,10,19–21]. Gene Scdhn1 has been reported by applying different low temperature treatments in *S. commersonii*, which has shown to be resistant to frost stress. CBF gene clusters responsible for cold responses were also reported from *S. commersonii* and could be transferred to cultivated potato with deletion or duplication of gene for resistance to frost stress [22]. The highest frost tolerance assessed by ability to survive cell injury in frost killing temperature was found in *S. acaule* (100%), and *S. albicans* (100%), followed by *S. commersonii* (99%), *S. demissum* (92%), and *S. paucissectum* (92%) [23]. Wild potatoes as *S. boliviense* and *S. colombianum* were also reported to be highly resistant to frost, while *S. brevicaule, S. candolleanum, S. chomatophilum*, and *S. infundibuliforme* were fairly resistant [24].

Drought. Potato is a hydrophilic crop requiring water from germination or sprouting until tuberization. They are grown in various climatic conditions but are sensitive to drought due to the shallow root system [25]. Physiological parameters as photosynthetic and respiration rate and accumulation of glycoalkaloids could be affected by drought stress including reduction in tissue water content, and inhibition of cell elongation [26]. Incorporation of desirable traits from highly drought resistant *S. gandarillasii* to *S. tuberosum* by phenotyping seedling growth under drought stress was useful for initial selection of wild genotypes due to different growth behaviors [27]. Other wild relative's species as *S. acaule, S. boliviense, S. bulbocastanum, S. chacoense, S. iopetalum, S. kurtzianum, S. polyadenium*, and *S. raphanifolium* are somewhat resistant for drought stress while *S. jamesii* and *S. okadae* have been reported highly resistant [27]. *S. ajanhuiri, S. curtilobum*, and *S. demissum* also were somewhat resistant for drought [24]. Hybrids of *S. tuberosum* and *S. commersonii* with CBF1 (ScCBF1) gene driven by 35S promoter were evaluated for drought tolerance and results were promising [25].

Heat. Potato is a crop which requires an average 20°C throughout its life cycle. Increased temperature causes heat stress which affect potato crop to a large extent. Many biotic stresses are also triggered by heat stress such as pests and diseases [28]. Sprout development requires optimum 6°C with optimum stem elongation at 18°C and best temperature for tuber initiation at 20°C [29,30]. A difference of 5°C in temperature causes 1–3 weeks delay in tuberization [31]. The effects of heat stress affect tuber quality, shape, color, size, and maturity and tuber physiological characters resulting in bitter taste due to accumulation of glycoalkaloid, necrosis, brown spot, and changes in hormones for dormancy. Moreover, heat stress intervals during growth stages ultimately decrease shelf life of tuber and profit margins. Negative effects of high temperature are inhabitation of vegetative growth, reduction in photosynthesis, and increasing in respiration, delay in tuber initiation, reduction in tuber growth, tuber disorders, shortened period of tuber dormancy, reduction in tuber dry matter content, and increased levels of glycoalkaloids [32,33]. F_1 interspecific hybrids of *S. phureja* and *S. chacoense, S. berthaultii*, and *S. microdontum* [34] crossed with dihaploid *S. tuberosum* subsp. *andigenum* produced diploid hybrids later crossed with *S. tuberosum* resulting in tetraploid hybrids resistant for heat stress [35]. Other wild potatoes such as *S. boliviense, S. chacoense, S. iopetalum, S. kurtzianum, S. polyadenium*, and *S. raphanifolium* were somewhat resistant too [24]. Ongoing studies, prospecting accessions of *S. chacoense, S. commersonii*, and *S. malmeanum* from Embrapa Clima Temperado Potato Genebank [36], aiming the introgression of resistance to heat stress, has shown that some accessions of *S. chacoense* performed better than control (*S. tuberosum*) in a heat stress environment based on photosynthesis rate, transpiration rate, chlorophyll content, and tuber related traits.

Salinity. Reduction in agriculture land and shortage of fresh water for irrigation due to urbanization and climate change are possibly the main reasons to grow the potato crop on salty soils or irrigated with saline water leading to salinity stress. Although potatoes are moderately sensitive to salt and can bear salinity level approximately 2.0 dS/m [37], their tolerance depends on the soil type and saline water quality. The effects of salinity stress affects germination, causes non-uniform growth, reduction in tuber yield, and decrease quality reported up to 50% [38]. Salinity also affects leaves by accumulation of toxic compounds resulting in injury and death [39]. It also affects other physiological processes important for the tuber development and negatively affect the relative water content, leaf stomatal conductance, transpiration rate, and changes chloroplast structure causing serious damage to plants and crop loss [40,41]. Abundance of salt ions in irrigation water restrict the plants from water and mineral uptake and affect the soil structure for many years followed by inhibition of soil oxygen supply to roots [42]. *S. chacoense* showed resistance against salinity stress for plant survival and dry matter content under saline conditions [43]. *S. curtilobum, S. juzepczuckii*, and *S. kurtzianum*, were also found to be tolerant to salinity stress [44,45].

12.2.3 Biotic stresses

Potatoes and their pathogens have an intimate relationship since the two major events of dispersal outside South America occurred after the 16th century. Few genotypes were brought to Europe and originated the modern breeding programs. These materials were grown for a long period to the late 18th and early 19th centuries [46,47]. However, the narrow genetic variability facilitates genetic erosion and resistance to be overcome [48], since most pathogens can mutate and acquire resistance [49]. This situation is well illustrated by the Irish Famine occurred in the late 19th century (1845–49) and caused by a severe outbreak of late blight [50]. The use of potato wild relatives in breeding programs aims to introgress genetic variability into the gene pool of the new elite potato germplasm. In the beginning of the 20th century, wild potatoes have been introduced in Europe toward the restoration of potato crop, especially with the introgression of late blight resistance from Mexican *S. demissum* [1]. The potential uses of wild potatoes has been reported by Bukasov in 1933 as "The revolution in potato breeding" [51]. However, potatoes are still vulnerable to pathogens such as bacterial, fungi, oomycete, viruses, and pests, all together responsible for 22% of yield losses [15]. Currently, potatoes are one of the most dependent crop of chemical treatment, which increases costs and causes negative impacts on agroecosystems and human food chain [52]. Cultivar resistance is always the most effective solution, but due to tetrasomic inheritance, broadening the genetic diversity base of cultivated gene pool with standard techniques progresses slowly.

12.2.3.1 Bacterial diseases

Control of bacterial diseases in potato fields is challenging. Chemical antibiotics have been widely employed to reduce damage, but the prolonged use led to bacterial adaptation and concerns on human health. Once a disease is disseminated into the fields, it may persist for long periods and spread. Depending on the environmental condition the damages could be severe, and growers are obligate to leave the infected areas for a quarantine period. The most widely recommended and sustainable management is planting resistant cultivars, followed by the adoption of good agronomic practices [53].

Bacterial wilt or brown rot (*Ralstonia solanacearum*). This disease is caused by an endophyte soil-borne organism, distributed mainly in tropical and subtropical regions [54]. It infects roots from natural skin openings or lenticels and spread through the vascular system causing disfunction, wilting, and plant death [55]. Susceptible cultivars suffer losses that can exceed 50% [56]. It has a large genetic and phenotypic diversity and an ability to survive season by season [54]. The strains are classified in four broad phylotypes (I–IV), based on phylogenetic relationships and geographic origin, and in biovars [57]. Wild potatoes carrying resistance are *S. acaule, S. andreanum, S. brevicaule, S. bulbocastanum, S. candolleanum, S. cardiophyllum, S. chacoense, S. clarum, S. commersonii, S. jamesii, S. malmeanum, S. microdontum, S. pinnatisectum*, and *S. sparsipillum* [19,24,58–66]. *S. candolleanum, S. chacoense*, and *S. sparsipillum*, reported as resistant to bacterial wilt, when hybridized with *S. tuberosum*, revealed undesirable high glycoalkaloid contents in the offspring [60]. Ploidy manipulation to incorporate resistance from *S. commersonii* into tetraploid *S. tuberosum* resulted in resistant clones already released [58,63].

Bacterial ring rot (*Clavibacter michiganensis* subsp. *sepedonicus*). Longevity in the soil, latency, and low cell levels, that remains undetected for up to three or four generations, make this disease difficult to control [67,68]. Tubers are infected through natural skin openings or lenticels [69]. The infection has a slow spreading into vascular tissues at early stages, then spread increases with plant growth. Symptoms are wilting, interveinal leaf chlorosis, and upward leaf curl [70]. Regulatory agencies classify it as a quarantine phytopathology [71]. Some documented sources of resistance are found in *S. acaule* and *S. stoloniferum* [72]. The defense has been attributed to the accumulation of the glycoalkaloid aglycone and other secondary metabolites [73]. Resistant tetraploid clones derived from backcrosses between somatic hybrids of *S. acaule* and dihaploid *S. tuberosum* were already obtained [73].

Blackleg (*Pectobacterium* spp.) and tuber soft rot (*Dickeya* spp.). Blackleg of stems (*P. atrosepticum*) and tuber soft rot diseases (*D. dadantii* and *P. carotovorum*) infect natural openings in the tuber skin or stems via lenticels [74]. The diseases start in the field, but the symptoms are more evident in the post-harvest [74]. Wild potatoes reported for resistance are *S. acaule, S. acroscopium, S. berthaultii, S. brevicaule, S. brevidens, S. boliviense, S. bulbocastanum, S. candolleanum, S. chacoense, S. commersonii, S. demissum, S. hjertingii, S. kurtzianum, S. microdontum, S. paucijugum, S. pinnatisectum*, and *S. stoloniferum* [3,19,24,59,66,75–77]. Resistant clones obtained from somatic hybrids between *S. brevidense* and dihaploid *S. tuberosum*, are alternatives for breeding of new soft rot resistant cultivars [75].

Scab (*Streptomyces scabies*) and powdery scab (*Spongospora subterranea*). These soil-borne gram-positive bacterial diseases affect tubers [78]. The pathogenicity is mainly attributed to the production of phytotoxin thaxtomin A [79], which reduces seedling growth [80]. Scab infects roots and induces root galling [81]. Crust lesions reduce the quality of tubers and seed tubers and losses are higher at post-harvest, due to unmarketable tubers [81,82]. Wild potatoes with some resistance are *S. boliviense, S. candolleanum, S. commersonii, S. chacoense*, and *S. jamesii* [24,66,81,83,84]. Resistance has been introgressed from *S. chacoense* into tetraploid hybrids with *S. tuberosum* via unilateral sexual polyploidization [85].

Zebra chip (*Candidatus Liberibacter solanacearum*). This disease is caused by a non-culturable phloem-limited gram-negative agent transmitted by the phloem-feeding psyllid pest *Bactericera cockerelli* [86]. No resistant cultivar is known and the development of cultivars against potato psyllid could allow cost-effective control. So, although a bacterial disease, an alternative strategy is to focus on wild potatoes reported to be resistant to the disease vector such as *S. bulbocastanum, S. guerreroense, S. habrochaites, S. verrucosum*, and somatic hybrids of *S. tuberosum* × *S. berthaultii* [87–90].

12.2.3.2 Fungal diseases

Fungi are responsible for causing significant losses of productivity from sowing to post-harvest. Pathogenic fungi are responsible for the demand of high rates of application of fungicides and higher costs of production. Resistant cultivars are recommended, but there are no cultivars fully resistant to the main diseases and any containment via chemical fungicides, could become ineffective after prolonged use due to acquired resistance [91].

Early blight (*Alternaria solani*). This is one of the most severe leaf-spotting and defoliation agent. Damage is severe when high humidity and 20°C–25°C temperatures are combined. Early cultivars are less affected than late ones, but all cultivars are susceptible and if no strategy of control is adopted, the disease could destroy a potato field in 6 weeks [92,93]. Some resistant level to early blight are found in *S. acaule, S. acroscopium, S. berthaultii, S. brevicaule, S. bulbocastanum, S. cardiophyllum, S. chacoense, S. commersonii, S. malmeanum, S. neorossii*, and *S. raphanifolium* [66,77,94–97]. Some hybrids between *S. raphanifolium* and haploid *S. tuberosum* demonstrated high level of resistance in field trials [97–99].

Fusarium wilt (*Fusarium* spp., especially *F. oxysporum* complex). This disease has been considered as one of the most yield limiting [100], causing losses between 15% and 70%, if no effective control is adopted [101]. Potential sources of resistance are *S. acaule, S. brevicaule*, and *S. kurtzianum* [66].

Late blight (*Phytophthora infestans*). This is the main fungal disease affecting potatoes and the protagonist of the Irish Famine which caused hunger and deaths between 1845 and 1849, obligating more than 1 million people to migrate [50]. Introductions of *S. demissum, S.* × *edinense* and their hybrids with *S. tuberosum* presented levels of resistance that enabled the restoration of the crop in Europe [1]. Until the late 20th century, the A1 mating type was the predominant race outside Mexico. After 1980, a new population of the A2 mating type race has been reported to emerge and spread from Mexico to many regions worldwide. The specific resistance was not durable since the two mating-types races are able to recombine and originate new virulence strains [49,102]. Wild potatoes with different levels of resistance are *S. acroglossum, S. cardiophyllum, S. chacoense, S. demissum, S. dolichocremastrum, S.* × *edinense, S. flahaultii, S. guerreroense, S. hougasii, S. iopetalum, S. jamesii, S. laxissimum, S. huancabambense, S. malmeanum, S. microdontum, S. morelliforme, S. oxycarpum, S. polyadenium, S. stipuloideum, S. tarnii, S. venturi*, and *S. verneii* [24,66,77,103–111]. The *R*-genes confer race-specific resistance to pathogen race and are classified from *R1* to *R11* and confer race-specific resistance to pathogen race [112,113]. *R*-genes are present in *S. brevicaule* [114], *S. berthaultii* [115], *S. bulbocastanum* [24,116–118], *S. cajamarquense* [119], *S. cardiophyllum, S. clarum, S. colombianum, S. polyadenium* [24], *S. chiquidenum* [120], *S. microdontum* [121], *S. mochiquense* [122], *S. papilla, S. stoloniferum*, [123], *S. pinnatisectum* [124], and *S. verrucosum* [125,126]. *S. cajamarquense* seems to have unique late blight resistance genes [119] as well as *S. huancabambense* which resist the AVR2 effector [103]. Pyramiding of *R*-genes by race non-specific (horizontal resistance) using major *Rpi*-genes, that are different alleles of one gene, or the same alleles (allele-dosage) into a single cultivar (multi-line) seems promising. Studies are mapping *R*-genes as quantitative trait loci (QTLs) and developing molecular-marker for field resistance [127]. Pyramiding *R*-genes by classic breeding is difficult and time-consuming as well as marker-assisted selection. Protoplast fusion was successful to introgress late blight resistance from *S. tarnii* to the cultivar "Delikat" [111]. Cisgenesis mediated by *Agrobacterium tumefaciens* resulted in many *R*-genes cloned into elite cultivars, improving durability, and broad-spectrum resistance [121,128].

Potato wart (*Synchytrium endobioticum*). The disease is caused by a soil-borne biotrophic fungus [129], which can persist more than 30 years and justifies the classification as an A2 quarantine disease [130,131]. It is an obligate parasite, which infect tubers and develop galls, turning them unmarketable. Chemical control is not effective and only quarantine and phytosanitary measures, as well as cultivation of resistant cultivars, are efficient prevention methods. Wild potatoes reported for resistance are *S. acaule*, *S. acroscopicum*, *S. berthaultii*, *S. boliviense*, *S. brevicaule*, *S. bulbocastanum*, *S. cardiophyllum*, *S. chacoense*, *S. demissum*, *S. endobioticum*, *S. infundibuliforme*, *S. iopetalum*, *S. jamesii*, *S. kurtzianum*, *S. microdontum*, *S. pinnatisectum*, *S. polyadenium*, *S. raphanifolium*, *S. stoloniferum*, and *S. vernei* [24,66].

Verticillium wilt (*Verticillium* spp., especially *V. dahliae*). This is one of the most important soil-borne diseases [132]. Control is difficult due the survival as microsclerotia on the soil or as mycelia in the vascular tissue or tuber remaining in the field [133]. It infects through the roots and hyphae spread efficiently in all vascular tissues obstructing them, causing wilting of leaves, chlorosis, necrosis, and premature senescence [134]. Wild potatoes with high levels of resistance are *S. berthaultii*, *S. brevicaule*, *S. candolleanum*, *S. chacoense*, *S. malmeanum*, *S. raphanifolium*, and *S. verneii* [77,135–137]. A single dominant gene in *S. chacoense* is responsible for resistance [138]. Hybrids from reciprocal crosses between *S. brevicaule*, and *S. chacoense* evaluated under greenhouse conditions, reported resistance [139].

12.2.3.3 Pest insects

Insects affect yield and tuber quality with global losses achieving 10%–16%. The damage can be direct, by feeding on leaves or tubers, or indirect through the transmission of viruses [140]. Trichome-mediated insect-resistance is related to glandular trichomes-types on leaves and high levels of glycoalkaloids protecting from herbivory [141,142]. The glandular trichomes can be of two types: A-type has a short stalk and a four-lobed head, which ruptures when touched releasing a phenolic fluid and polyphenol-oxidase (PPO) [143], while B-type exudes a sticky droplet [144]. There are no potato cultivars significantly resistant to the main insect pests and control usually relies on pesticides [145].

Colorado potato beetle (*Leptinotarsa decemlineata*). Trichome-mediated resistance has been reported in wild potatoes such as *S. berthaultii* [146] and *S. polyadenium* [147]. Defense reported in *S. chacoense* is due to the high content of the glycoalkaloid leptin [148] synthesized in aerials tissues but not in tubers [149]. Resistance found in *S. acroglossum* is attributed to trichomes lacking high concentrations of glycoalkaloids [150]. Secondary metabolites may be an interesting approach for resistance and has been found in *S. acaule*, *S. acroscopicum*, *S. albornozii*, *S. berthaultii*, *S. brevicaule*, *S. boliviense*, *S.*

candolleanum, *S. cardiophyllum*, *S. chacoense*, *S. chomatophilum*, *S. clarum*, *S. commersonii*, *S. demissum*, *S. infundibuliforme*, *S. jamesii*, *S. kurtzianum*, *S. malmeanum*, *S. morelliforme*, *S. neocardenasii*, *S. oxycarpum*, *S. paucissectum*, *S. pinnatisectum*, *S. piurae*, *S. polyadenium*, *S. stoloniferum*, and *S. tarnii* [24,66,141,147,151–157].

Flea beetle (*Epitrix harilana rubia*). Resistance has been found in *S. berthaultii* [158] and *S. bulbocastanum* [159].

Green peach aphid (*Myzus persicae*) and potato aphid (*Macrosiphum euphorbiae*). Damages are caused by feeding on leaves with high concentration of carbohydrates, uptaking of nutrients from suction, cytotoxic effect of salivary secretion, and virus transmission [160]. Wild potatoes possessing some resistant level are *S. berthaultii*, *S. brevicaule*, *S. bulbocastanum*, *S. candolleanum*, *S. cardiophyllum*, *S. chacoense*, *S. chomatophilum*, *S. clarum*, *S. guerreroense*, *S. infundibuliforme*, *S. hjertingii*, *S. neocardenasii*, *S. lignicaule*, *S. medians*, *S. microdontum*, and *S. polyadenium* [161–163]. Protoplast fusion transferred resistant genes from *S. bulbocastanum* to *S. tuberosum* [164].

Mites (*Tetranychus urticae*). Somewhat resistant level has been found in *S. berthaultii* [158] and *S. bulbocastanum* [159]. The glandular type-B trichome which confers resistance in *S. berthaultii* has been introgressed into the cultivated potato [146,147].

Tuber worms or tuber moths. Andean tuber moth (*Symmetrischema tangolias*), Guatemalan tuber moth (*Tecia solanivora*), and potato tuber moth or tuberworm (*Phthorimaea operculella*) are important pests at subtropical and tropical regions, causing yield losses in field but mainly during storage [165]. Along growing season, *P. operculella* lays eggs preferentially on the leaves, over tubers, or on soil next to the host plant [166]. The larvae damages through feeding, digging tunnels in tubers, stems, petioles and leaves [167], and increases the risk of infections [168]. Yield losses happen during the post-harvest, caused by weight and quality loss, and unmarketable tubers [167]. Wild potatoes as potential sources of resistance are *S. chacoense*, *S. chiquidenum*, *S. commersonii*, *S. pinnatisectum*, and *S. tarijense* [169,170].

Potato leafhopper (*Empoasca fabae*). This pest is a severe defoliator and reduces the accumulation of reserves by affecting leaves [171]. Resistance is reported in *S. berthaultii*, *S. chacoense*, *S. commersonii*, *S. malmeanum*, and *S. polyadenium* [141,159,172].

Potato leaf miner (*Liriomyza huidobrensis*). This highly invasive pest causes severe damage to the leaves and favors infection by *Alternaria solanii* [173]. Some resistance is reported to *Solanum chacoense* [94].

12.2.3.4 Nematodes

These are critical pests causing severe physiological disorders, resulting in yield losses and unmarketable tubers [174]. Chemicals treatment are realized by fumigants and non-fumigants nematicides with high monetary, environmental, and health costs. Integrated pest management can adopt crop rotations but when the infestation is high, quarantine is the only decision treatment [175,176].

Potato cyst nematodes (*Globodera rostochiensis*, *G. pallida*). Potato breeding using wild relatives as resistance source against cyst nematode is an option especially since the discovery of it in *S. vernei* [177,178]. Other sources are *S. acaule*, *S. berthaultii*, *S. boliviense*, *S. brevicaule*, *S. bulbocastanum*, *S. candolleanum*, *S. cardiophyllum*, *S. chacoense*, *S. demissum*, *S. flahaultii*, *S. kurtzianum*, *S. malmeanum*, *S. megistacrobolum*, *S. microdontum*, *S. multiinterruptum*, *S. stipuloideum*, and *S. polyadenium* [24,66,94,179–181].

Root-knot nematode (*Meloidogyne chitwood*, *M. incognita*). Some level of resistance is reported to *S. acaule*, *S. boliviense*, *S. brevicaule*, *S. bulbocastanum*, *S. candolleanum*, *S. cardiophyllum*, *S. chacoense*, *S. demissum*, *S. hjertingii*, *S. hougasii*, *S. iopetalum*, *S. jamesii*, *S. kurtzianum*, *S. microdontum*, *S. pinnatisectum*, *S. polyadenium*, and *S. raphanifolium* [24,182]. Resistance in *S. bulbocastanum* is controlled by a single monogene $R_{Mc1(blb)}$, two genes $R_{Mc1(blb)}$ and $R_{Mctuber(blb)}$, while *S. hougasii* and *S. stoloniferum* also have the resistance genes $R_{Mc1(hou)}$ and $R_{Mc1(fen)}$, respectively, but only partial resistance was introgressed into elite potato germplasm by the mean of protoplast fusion [183,184].

12.2.3.5 Viruses

At least 50 viruses are reported to cause diseases in potato [185], reducing yield up to 80% [186]. Primary transmission occurs through infection in growing plants as well as in the post-harvest [187]. Seed certification is constrained by virotic tubers [188] and plants grown from previously infected seed tubers (secondary infection) commonly produce unmarketable ones. Aphids are the main vectors spreading virus in a circulative non-propagative manner such as for PLRV, PVA, PVV, and PVY, while PVX is only transmitted by contact as propagative manner, and PVM and PVS can be transmitted in both ways [187]. Breeding for resistance to viruses started in Europe by mid-1900 with *S. stoloniferum*, and since then many wild potatoes have been characterized for resistance to diverse strains. Symptoms include mosaic pattern on the leaves, leaf drop as well as stem necrosis [187].

Alfalfa mosaic virus (AMV). Resistance found in *S. mochiquense*, *S. neocardenasii*, *S. neorosssii*, and *S. paucissectum* [189].

Potato spindle tuber viroid (PSTVd). Resistance found in *S. acaule*, *S. berthaultii*, *S. candolleanum*, *S. chacoense*, *S. guerreroense*, *S. hjertingii*, *S. stoloniferum*, and *S. sucrense* [24].

Potato virus A (PVA). Resistance found in *S. chacoense*, *S. hougasii*, *S. maglia*, and *S. stoloniferum* [190].

Potato leafroll virus (PLRV). Resistance found in *S. acaule*, *S. acroscopicum*, *S. andreanum*, *S. boliviense*, S. *brevicaule*, *S. brevicaule*, *S. brevidens*, *S. candolleanum*, *S. chacoense*, *S. demissum*, *S. flahaultii*, *S. gourlayi*, *S. infundibuliforme*, *S. kurtzianum*, *S. malmeanum*, *S. nerossii*, *S. maglia*, *S. mochiquense*,

S. piurae, *S. polyadenium*, *S. raphanifolium*, *S. stenophyllidium*, *S. stipuloideum*, *S. stoloniferum*, *S. trifidum*, and *S. verrucosum* [24,66,77,96,106].

Potato virus M (PVM). Resistance found in *S. brevicaule*, *S. boliviense*, *S. brevidens*, *S. chomatophilum*, *S. infundibuliforme*, *S. maglia*, *S. raphanifolium*, and *S. stoloniferum* [106].

PVS (Potato virus S). Resistance found in *S. berthaultii*, *S. boliviense*, *S. brevicaule*, *S. laxissimum*, *S. lignicaule*, *S. maglia*, *S. michoacanum*, *S. multiinterruptum*, and *S. stoloniferum* [106].

Potato virus X (PVX). Resistance found in *S. acaule*, *S. ajanhuiri*, *S. albornozii*, *S. berthaultii*, *S. brevicaule*, *S. chacoense*, *S. commersonii*, *S. demissum*, *S. guerreroense*, *S. iopetalum*, *S. jamesii*, *S. kurtzianum*, *S. maglia*, *S. mochiquense*, *S. neorossii*, (*S. sparsipilum*), *S. sucrense*, *S. tarijense*, *S. tarnii*, and *S. vernei* [24,66,77,94,189–195].

Potato virus Y (PVY). Resistance found in *S. acaule*, *S. acroscopicum*, *S. andreanum*, *S. berthaultii*, *S. boliviense*, *S. brevicaule*, *S. brevidens*, *S. candolleanum*, *S. cardiophyllum*, *S. chacoense*, *S. chomatophilum*, *S. demissum*, *S. ehrenbergii*, *S. guerreroense*, *S. hougasii*, *S. iopetalum*, *S. jamesii*, *S. kurtzianum*, *S. multiinterruptum*, *S. neocardenasii*, *S. pinnatisectum*, *S. stenophyllidium*, *S. stoloniferum*, and *S. tarnii* [24,59,66,94,106,111,189,190,194,196,197].

Tobacco Rattle Virus (TRV). Resistance found in *S. paucissectum* [189].

12.3 Hurdles to be overcome

There are plenty of documented cases on the wild potatoes as a rich source of novel commercial traits and resistance for abiotic and biotic stresses. However, introgression of these traits into commercial cultivars is challenging. For the past decades, breeders have become focusing on the introgression of diploid wild germplasm into cultivated potato at the diploid level but in doing so there are many hurdles and barriers which obstruct the goal of achieving the tetraploid level for cultivar development [198]. Although wild potatoes are important for genetic improvement, some genetically based biological barriers prevent the introgression [199,200]. Common barriers are prezygotic barrier and cytoplasmic-genetic male sterility in which either pollen or style are incompatible [98,201,202]. Prezygotic barriers can be overcome by reciprocal crosses. Endosperm balance number is the most important postzygotic barrier. It affects the endosperm development and production of viable seeds. For normal endosperm development, it is important to keep the 2:1 maternal: paternal ratio of endosperm balance factors [203]. Wild potatoes could be 2x/1EBN, 2x/2EBN, 4x/2EBN, 4x/4EBN, and 6x/4EBN based on their ability to hybridize [2]. If EBN number does not match in interspecies crosses, abnormal development of endosperm, and inviable seed production occur. Due to different ploidy levels and EBN, it is difficult to directly cross wild potatoes with the tetraploid cultivars. This barrier could be overcome by different techniques such as bridge crosses, 2n gametes, somatic fusion, ploidy manipulation, and sexual polyploidization [204]. In ploidy manipulation technique, doubling of chromosome results in doubling of ploidy and EBN. This scheme could be done by somatic doubling of 2x and by 2n gamete formation resulting from the failure of meiosis to reduce chromosome number [205,206]. It is also possible to manipulate the ploidy level and EBN through bridge crosses [207]. This scheme has potential to introgress tertiary genepool species to the cultivated species [208,209].

Other strategy to overcome incompatibility is somatic fusion followed by in vitro plant regeneration which is useful for producing hybrids such as somatic fusions between the 2x, 1EBN species *S. bulbocastanum* and 4x, 4EBN cultivated potato [183]. Other techniques to overcome the biological barriers are mentor pollination and embryo rescue. In mentor pollination, compatible species pollen is applied 1 or 2 days after pollinating with an incompatible species to minimize the premature fruit drop [198].

12.4 Future of wild potatoes in breeding

There is probably no other crop that has such a diversity of species distributed throughout its wild gene pool such as potatoes. Potato genepool has, more than 100 species that have evolved in many environmental conditions and provide a rich source of desirable agronomic traits as resistance for diverse biotic and abiotic stresses, and tuber quality traits that can be introgressed into cultivated potato [2,3,210,211].

Food demand by 2050 are estimated to be 70% more than required by 2006, considering a population rise to 10 billion people [212]. Potatoes are the crop that mostly produces calories per area and 2008 was considered the International Year of the Potato [213] due its important role to the economy and world food security [214]. Cultivated potato is a tetraploid and largely heterozygous crop, and its tetrasomic inheritance makes breeding through traditional crossbreeding a challenge [215,216], in the other hand, 70% of wild potato species are diploid [2]. Pre-breeding is an essential step to incorporate traits from wild relatives for the improvement of new commercial cultivars, as well as to delivering wild potato diversity in a readily manageable form.

Because the genetic base of modern potato cultivars is narrow, breeders are looking for wild potatoes with more interest in Genebanks. Unfortunately, much of species are under characterized, which makes it difficult to use this germplasm efficiently. The advance of technologies such as application of omic-scale for gene discovery could accelerate introgression of interesting traits [217].

The genome of the cultivated potatoes and some of its wild relatives [218] intensified the whole-genome level studies. Genomic analysis makes it possible to find genes encoding valuable agronomic traits such as resistance for several abiotic and biotic stresses, nutritional compounds, vitamins as even yield traits [219–221]. Furthermore, genomic analysis also supports potato breeding programs for increasing genetic gains [222].

Biotechnological tools are being used to transform some traits of potato germplasm. Transcription activator-like effector nucleases (TALENs) to knockout *VInv* improving cold storage and processing traits for the commercial variety Ranger Russet [223]. CRISPR-Cas9 is an alternative tool to put desirable genes into elite germplasm, for example, increases the amounts of beta-carotene to provide a richer supply of vitamin A [224], reduce herbicide sensitivity phenotype [225], causes the complete abolition of the steroidal glycoalkaloids α-solanine and α-chaconine accumulation in potato [226], or transform diploid wild potatoes from self-incompatible to self-compatible [227,228].

Recently, the efforts increased to transform tetraploid potato at diploid hybrid potatoes that are propagated by true potato seeds, as advantage to avoid virus transmissions and reduce operational production costs. Along more than 100 species, around 70% of wild potato relatives are diploid but are self-incompatible. Tetraploid potatoes need to take more than 20 generations to have 99% of homozygosity, whereas diploid potatoes can reduce heterozygosity by 50% each generation. By this way, gene fixing becomes easily at hybrid diploid level compared to tetraploid level, and marker-assisted selection could accelerate the introgression of desirable traits [8,9]. This approach is discussed by many research groups. In *S. chacoense* a gene that allows self-compatible fertilization by response of alleles *Sli* was identified [229]. The discovery of a self-compatible gene made it possible to create an inbred line diploid potato [230,231]. Particularly, *S. commersonii* (Fig. 12.1] is the first wild potato which had the whole genome sequenced, what makes possible to exploit more genes families [17,220,221,232]. Nowadays, with the developing of several new biotechnological techniques, the potential of using wild potatoes for breeding the cultivated potato is more promising than ever.

References

[1] Ross H. The use of wild *Solanum* species in German potato breeding of the past and today. Am Potato J 1966;43(3):63–80.

[2] Spooner DM, Ghislain M, Simon R, Jansky SH, Gavrilenko T. Systematics, diversity, genetics, and evolution of wild and cultivated potatoes. Bot Rev 2014;80(4):283–383.

[3] Hawkes JG. The potato: evolution, biodiversity and genetic resources. Belhaven Press; 1990.

[4] Hijmans RJ, Spooner DM. Geographic distribution of wild potato species. Am J Bot 2001;88(11):2101–12.

[5] Spooner DM, Alvarez N, Peralta IE, Clausen AM. Taxonomy of wild potatoes and their relatives in Southern South America (*Solanum* sect. *Petota* and *Etuberosum*). Syst Bot Monogr. 2016;100:240.

[6] Carputo D, Aversano R, Frusciante L. Breeding potato for quality traits. Acta Horticult 2005;(684):55–64.

[7] Kumari M, Kumar M, Solankey SS. Breeding potato for quality improvement. Potato—From incas to all over the worldIntechOpen; 2018.

[8] Barone A. Molecular marker-assisted selection for potato breeding. Am J Potato Res 2004;81(2):111–7.

[9] Barone A, Sebastiano A, Carputo D, Della Rocca F, Frusciante L. Molecular marker-assisted introgression of the wild *Solanum commersonii* genome into the cultivated *S. tuberosum* gene pool. Theor Appl Genet 2001;102(6–7):900–7.
[10] Iovene M, Barone A, Frusciante L, Monti L, Carputo D. Selection for aneuploid potato hybrids combining a low wild genome content and resistance traits from *Solanum commersonii*. Theor Appl Genet 2004;109(6):1139–46.
[11] Osman SF, Herb SF, Fitzpatrick TJ, Schmiediche P. Glycoalkaloid composition of wild and cultivated tuber-bearing *Solanum* species of potential value in potato breeding programs. J Agric Food Chem 1978;26(5):1246–8.
[12] Akeley RV, Mills WR, Cunningham CE, Watts J. Lenape: a new potato variety high in solids and chipping quality. Am Potato J 1968;45(4):142–5.
[13] Ali A, Jansky S. Fine screening for resistance to cold-induced sweetening in potato hybrids containing *Solanum raphanifolium* germplasm. Adv Agric 2015;2015:1–4.
[14] Hanneman RE Jr. Evaluation of wild species for new sources germplasm that chip directly from cold storage. Am J Potato Res 1996;73:360.
[15] Ross H, Hunnius W. Potato breeding: problems and perspectives, 13. Fortschritte der Pflanzenzüchtung, P. Parey; 1986.
[16] Pérez W, Salas A, Raymundo R, Huaman Z, Nelson R, Bonierbale M. Evaluation of wild potato species for resistance to late blight. Scientist and Farmer, Partners in Research for the 21st Century, Program Report 1999-2000; 2000. p. 49–62.
[17] Aversano R, Contaldi F, Ercolano MR, Grosso V, Iorizzo M, Tatino F, et al. The *Solanum commersonii* genome sequence provides insights into adaptation to stress conditions and genome evolution of wild potato relatives. Plant Cell 2015;27(4):954–68.
[18] Kikuchi A, Huynh HD, Endo T, Watanabe K. Review of recent transgenic studies on abiotic stress tolerance and future molecular breeding in potato. Breed Sci 2015;65(1):85–102.
[19] Bamberg JB, Hanneman RE Jr, Palta JP, Harbage JF. Using disomic 4 x (2EBN) potato species' germplasm via bridge species *Solanum commersonii*. Genome 1994;37(5):866–70.
[20] Cardi T, D'Ambrosio E, Consoli D, Puite KJ, Ramulu KS. Production of somatic hybrids between frost-tolerant *Solanum commersonii* and *S. tuberosum:* characterization of hybrid plants. Theor Appl Genet 1993;87(1–2):193–200.
[21] Carputo D, Barone A, Cardi T, Sebastiano A, Frusciante L, Peloquin SJ. Endosperm balance number manipulation for direct in vivo germplasm introgression to potato from a sexually isolated relative (*Solanum commersonii* Dun.). Proc Natl Acad Sci 1997;94(22):12013–7.
[22] Carvallo MA, Pino MT, Jeknic Z, Zou C, Doherty CJ, Shiu SH, et al. A comparison of the low temperature transcriptomes and CBF regulons of three plant species that differ in freezing tolerance: *Solanum commersonii*, *Solanum tuberosum*, and *Arabidopsis thaliana*. J Exp Bot 2011;62(11):3807–19.
[23] Li PH. Frost killing temperatures of 60 tuber-bearing *Solanum* species. Am Potato J 1977;54(9):452–6.
[24] Machida-Hirano R. Diversity of potato genetic resources. Breed Sci 2015;65(1):26–40.
[25] Pino MT, Ávila A, Molina A, Jeknic Z, Chen THH. Enhanced in vitro drought tolerance of *Solanum tuberosum* and *Solanum commersonii* plants overexpressing the ScCBF1 gene. Cien Inv Agr 2013;40(1):171–184.
[26] Hassanpanah D. In vitro and in vivo screening of potato cultivars against water stress by polyethylene glycol and potassium humate. Biotechnology 2009;8(1):132–7.
[27] Watanabe KN, Kikuchi A, Shimazaki T, Asahina M. Salt and drought stress tolerances in transgenic potatoes and wild species. Potato Res 2011;54(4):319–24.
[28] Smith O. Potatoes: production, storing, processing. AVI Publishing Company; 1968.

[29] Borah MN. Growth of the potato as influenced by temperature. Indian J Plant Physiol 1962;5:53–72.

[30] Bodlaender KBA, Lugt C, Marinus J. The induction of second-growth in potato tubers. Eur Potato J 1964;7(1):57–71.

[31] Levy D, Veilleux RE. Adaptation of potato to high temperatures and salinity-a review. Am J Potato Res 2007;84(6):487–506.

[32] Dimenstein L, Lisker N, Kedar N, Levy D. Changes in the content of steroidal glycoalkaloids in potato tubers grown in the field and in the greenhouse under different conditions of light, temperature and daylength. Physiol Mol Plant Pathol 1997;50(6):391–402.

[33] Sinden SL, Sanford LL, Webb RE. Genetic and environmental control of potato glycoalkaloids. Am Potato J 1984;61(3):141.

[34] Veilleux RE, Paz MM, Levy D. Potato germplasm development for warm climates: genetic enhancement of tolerance to heat stress. Euphytica 1997;98(1–2):83–92.

[35] Levy D, Itzhak Y, Fogelman E, Margalit E, Veilleux RE. Ori, Idit, Zohar and Zahov: tablestock and chipstock cultivars bred for adaptation to Israel. Am J Potato Res 2001;78(3):167–73.

[36] Temperado EC. Available from: https://www.embrapa.br/clima-temperado; 2020.

[37] Maas EV, Hoffman GJ. Crop salt tolerance–current assessment. J Irrig Drain Div 1977;103(2):115–34.

[38] Amacher JK, Koenig R, Kitchen B. Salinity and plant tolerance. AG-SO 2000;3:1.

[39] Bruns S, Caesar K. Shoot development and tuber yield of several potato cultivars under high salt concentrations at different stages of development. Potato Res 1990;33(1):23–32.

[40] Bruns S, Hecht-Buchholz C. Light and electron microscope studies on the leaves of several potato cultivars after application of salt at various development stages. Potato Res 1990;33(1):33–41.

[41] Fidalgo F, Santos A, Santos I, Salema R. Effects of long-term salt stress on antioxidant defence systems, leaf water relations and chloroplast ultrastructure of potato plants. Ann Appl Biol 2004;145(2):185–92.

[42] Ayers RS, Westcot DW. Water quality for agriculture. Food and Agriculture Organization of the United Nations Rome; 1985.

[43] Bilski JJ, Nelson DC, Conlon RL. Response of six wild potato species to chloride and sulfate salinity. Am Potato J 1988;65(10):605.

[44] Sabbah S, Tal M. Salt tolerance in *Solanum kurzianum* and *S. tuberosum* cvs Alpha and Russet Burbank. Potato Res 1995;38(4):319–30.

[45] Silva JAB, Otoni WC, Martinez CA, Dias LM, Silva MAP. Microtuberization of Andean potato species (*Solanum* spp.) as affected by salinity. Sci Horticul 2001;89(2):91–101.

[46] Glendinning DR. Potato introductions and breeding up to the early 20th century. N Phytol 1983;94(3):479–505.

[47] Hawkes JG. Genetic poverty of the potato in Europe. In: Proceedings of the conference broadening the genetic base of crops. Wageningen: Centre for Agricultural Publishing and Documentation; 1979. p. 19–27.

[48] Brown CR. Modern evolution of the cultivated potato gene pool. The molecular and cellular biology of the potatoCAB International; 1990. p. 1–11.

[49] Goodwin SB, Drenth A. Origin of the A2 mating type of Phytophthora infestans outside Mexico. Phytopathology 1997;87(10):992–9.

[50] Zadoks JC. The potato murrain on the European continent and the revolutions of 1848. Potato Res 2008;51(1):5–45.

[51] Hawkes JG. Significance of wild species and primitive forms for potato breeding. Euphytica 1958;7(3):257–70.

[52] Tsror L, Erlich O, Mordechai-Lebiush S, Galilov I, Hazanovsky M, Chalupowicz L, et al. First report of *Pectobacterium parmentieri* one of the causal agents of potato blackleg and tuber soft rot diseases in Israel. Plant Dis 2020;104:2288.

[53] Jansky SH, Rouse DI. Multiple disease resistance in interspecific hybrids of potato. Plant Dis 2003;87(3):266–72.

[54] Hayward AC. Biology and epidemiology of bacterial wilt caused by *Pseudomonas solanacearum*. Ann Rev Phytopathol 1991;29:65–87.

[55] Genin S, Denny TP. Pathogenomics of the *Ralstonia solanacearum* species complex. Ann Rev Phytopathol 2012;50:67–89.

[56] Xu GW, Gross DC. Selection of fluorescent pseudomonads antagonistic to Erwinia carotovora and suppressive of potato seed piece decay. Phytopathology 1986;76(4):414–22.

[57] Fegan M, Prior P. How complex is the *Ralstonia solanacearum* species complex. Bacterial wilt disease and the *Ralstonia solanacearum* species complex, 1. APS Press; 2005. p. 449–61.

[58] Carputo D, Aversano R, Barone A, Di Matteo A, Iorizzo M, Sigillo L, et al. Resistance to *Ralstonia solanacearum* of Sexual Hybrids Between *Solanum commersonii* and S. tuberosum. Am J Potato Res 2009;86(3):196–202.

[59] Chung YS, Holmquist K, Spooner DM, Jansky SH. A test of taxonomic and biogeographic predictivity: resistance to soft rot in wild relatives of cultivated potato. Phytopathology 2011;101(2):205–12.

[60] French ER, Anguiz R, Aley P. The usefulness of potato resistance to *Ralstonia solanacearum*, for the integrated control of bacterial wilt. Bacterial wilt diseaseSpringer; 1998. p. 381–5.

[61] González M, Galván G, Siri MI, Borges A, Vilaró F. Resistencia a la marchitez bacteriana de la papa en *Solanum commersonii* Dun. Agrociencia (Uruguay) 2013;17(1):45–54.

[62] Hawkes JG. Origins of cultivated potatoes and species relationships. Potato GeneticsCAB International; 1994.

[63] Laferriere LT, Helgeson JP, Allen C. Fertile *Solanum tuberosum + S. commersonii* somatic hybrids as sources of resistance to bacterial wilt caused by *Ralstonia solanacearum*. Theor Appl Genet 1999;98(8):1272–8.

[64] Kim-Lee H, Moon JS, Hong YJ, Kim MS, Cho HM. Bacterial wilt resistance in the progenies of the fusion hybrids between haploid of potato and *Solanum commersonii*. Am J Potato Res 2005;82(2):129–37.

[65] Norman DJ, Yuen JMF, Bocsanczy AM. Threat of brown rot of potato and existing resistance. Am J Potato Res 2020;97:272–7.

[66] Srivastava A, Bhardwaj V, Singh BP, Khurana SMP. Potato diversity and its genetic enhancement. Gene pool diversity and crop improvementSpringer; 2016. p. 187–226.

[67] Franc GD. Persistence and latency of *Clavibacter michiganensis* subsp. sepedonicus in field-grown seed potatoes. Plant Dis 1999;83(3):247–50.

[68] Nelson GA. Long-term survival of *Corynebacterium sepedonicum* on contaminated surfaces and in infected potato stems. Am Potato J 1980;57(12):595–600.

[69] Li X, Tambong J, Yuan KX, Chen W, Xu H, Lévesque CA, et al. Re-classification of *Clavibacter michiganensis* subspecies on the basis of whole-genome and multi-locus sequence analyses. Int J Syst Evol Microbiol 2018;68(1):234.

[70] Whitworth JL, Selstedt RA, Westra AAG, Nolte P, Duellman K, Yellareddygari SKR, et al. Symptom expression of mainstream and specialty potato cultivars to bacterial ring rot (*Clavibacter sepedonicus*) and evaluation of in-field detection. Am J Potato Res 2019;96(4):427–44.

[71] Gudmestad NC. Recommendations of the national task force for the eradication of bacterial ring rot. Am Potato J 1987;64(12):695–7.

[72] Kriel CJ, Jansky SH, Gudmestad NC, Ronis DH. Immunity to *Clavibacter michiganensis* subsp. sepedonicus: screening of exotic *Solanum* species. Euphytica 1995;82(2):125–32.

[73] Rokka V-M, Laurila J, Tauriainen A, Laakso I, Larkka J, Metzler M, et al. Glycoalkaloid aglycone accumulations associated with infection by *Clavibacter michiganensis* ssp. sepedonicus in potato species *Solanum acaule* and *Solanum tuberosum* and their interspecific somatic hybrids. Plant Cell Rep 2005;23(10–11):683–91.

[74] Pérombelon MCM. Potato diseases caused by soft rot Erwinias: an overview of pathogenesis. Plant Pathol 2002;51(1):1–12.

[75] Bains PS, Bisht VS, Lynch DR, Kawchuk LM, Helgeson JP. Identification of stem soft rot (*Erwinia carotovora* sub-species atroseptica) resistance in potato. Am J Potato Res 1999;76(3):137–41.

[76] Carputo D, Monti LM, Speggiorin M, Garreffa P, Raio A. Screening for resistance to tuber soft rot and blackleg in diploid *Solanum* species and *S. tuberosum* haploids [1996]. J Genet Breed (Italy) 1998;50(3):221–6.

[77] Jansky S. Breeding for disease resistance in potato. Plant Breed Rev 2000;19:69–156.

[78] Braun S, Gevens A, Charkowski A, Allen C, Jansky S. Potato common scab: A review of the causal pathogens, management practices, varietal resistance screening methods, and host resistance. Am J Potato Res 2017;94(4):283–96.

[79] Loria R, Bignell DRD, Moll S, Huguet-Tapia JC, Joshi MV, Johnson EG, et al. Thaxtomin biosynthesis: the path to plant pathogenicity in the genus *Streptomyces*. Antonie Van Leeuwenhoek 2008;94(1):3–10.

[80] Leiner RH, Fry BA, Carling DE, Loria R. Probable involvement of thaxtomin A in pathogenicity of *Streptomyces* scabies on seedlings. Phytopathology (USA) 1996;86(7):709–13.

[81] Braun SR, Endelman JB, Haynes KG, Jansky SH. Quantitative trait loci for resistance to common scab and cold induced sweetening in diploid potato. Plant Genome 2017;10(3):1–9.

[82] Lopes CA, Rossato M. History and status of selected hosts of the *Ralstonia solanacearum* species complex causing bacterial wilt in Brazil. Front Microbiol 2018;9:1228.

[83] Dionne LA, Lawrence CH. Early scab resistant derivatives of *Solanum chacoense* x *Solanum phureja*. Am Potato J 1961;38(1):6–8.

[84] Hosaka K, Matsunaga H, Senda K. Evaluation of several wild tuber-bearing *Solanum* species for scab resistance. Am J Potato Res 2000;77(1):41–5.

[85] Jansky S, Douches D, Haynes K. Germplasm release: Three tetraploid potato clones with resistance to common scab. Am J Potato Res 2018;95(2):178–82.

[86] Munyaneza JE. Zebra chip disease of potato: biology, epidemiology, and management. Am J Potato Res 2012;89(5):329–50.

[87] Butler CD, Gonzalez B, Manjunath KL, Lee RF, Novy RG, Miller JC, et al. Behavioral responses of adult potato psyllid, *Bactericera cockerelli* (Hemiptera: Triozidae), to potato germplasm and transmission of *Candidatus Liberibacter psyllaurous*. Crop Prot 2011;30(9):1233–8.

[88] Cooper WR, Bamberg JB. Variation in *Bactericera cockerelli* (Hemiptera: Triozidae) oviposition, survival, and development on *Solanum bulbocastanum* germplasm. Am J Potato Res 2014;91(5):532–7.

[89] Cooper WR, Bamberg JB. Variation in susceptibility to potato psyllid, *Bactericera cockerelli* (Hemiptera: Triozidae), among *Solanum verrucosum* germplasm accessions. Am J Potato Res 2016;93(4):386–91.

[90] Levy J, Tamborindeguy C. *Solanum habrochaites*, a potential source of resistance against *Bactericera cockerelli* (Hemiptera: Triozidae) and "Candidatus Liberibacter solanacearum". J Econom Entomol 2014;107(3):1187–93.

[91] Jansky SH, Dempewolf H, Camadro EL, Simon R, Zimnoch-Guzowska E, Bisognin DA, et al. A case for crop wild relative preservation and use in potato. Crop Sci 2013;53(3):746–54.
[92] Boiteux LS, Reifschneider FJB, Fonseca MEN, Buso JA. Search for sources of early blight (*Alternaria solani*) field resistance not associated with vegetative late maturity in tetraploid potato germplasm. Euphytica 1995;83(1):63–70.
[93] Dita Rodriguez MA, Brommonschenkel SH, Matsuoka K, Mizubuti ESG. Components of resistance to early blight in four potato cultivars: effect of leaf position. J Phytopathol 2006;154(4):230–5.
[94] Ochoa CM. The potatoes of South America: Bolivia. Cambridge University Press; 1990.
[95] Prasad B, Dutt BL. Reaction of *Solanum* spp. against early blight of potatoes. JIPA J Indian Potato Asso 1980;7:68–72.
[96] USDA, NRCS. The plants database (Greensboro: National Plant Data Team). NC 27401-4901 USA. 2020.
[97] Jansky SH, Simon R, Spooner DM. A test of taxonomic predictivity: Resistance to early blight in wild relatives of cultivated potato. Phytopathology 2008;98(6):680–7.
[98] Weber BN, Hamernik AJ, Jansky SH. Hybridization barriers between diploid *Solanum tuberosum* and wild *Solanum raphanifolium*. Genet Resour Crop Evol 2012;59(7):1287–93.
[99] Weber BN. Breeding for early blight resistance in potato using wild species *Solanum raphanifolium*. In: Fourth annual plant breeding meeting. USDA; 2010.
[100] Powelson ML, Rowe RC. Biology and management of early dying of potatoes. Ann Rev Phytopathol 1993;31(1):111–26.
[101] Kerkeni A, Daami-Remadi M, Khedher MB. In vivo evaluation of compost extracts for the control of the potato *Fusarium* wilt caused by *Fusarium oxysporum* f. sp. tuberosi. Afr J Plant Sci Biotechnol 2013;7(1):36–41.
[102] Fry W. Phytophthora infestans: the plant (and R gene) destroyer. Mol Plant Pathol 2008;9(3):385–402.
[103] Aguilera-Galvez C, Champouret N, Rietman H, Lin X, Wouters D, Chu Z, et al. Two different R gene loci co-evolved with Avr2 of Phytophthora infestans and confer distinct resistance specificities in potato. Stud Mycol 2018;89:105–15.
[104] Black W, Gallegly ME. Screening of *Solanum* species for resistance to physiologic races of *Phytophthora infestans*. Am Potato J 1957;34(10):273–81.
[105] Champouret N. Functional genomics of Phytophthora infestans effectors and *Solanum* resistance genes. Thesis, Wageningen, The Netherlands: Wageningen University; 2010.
[106] De Galarreta JIR, Carrasco A, Salazar A, Barrena I, Iturritxa E, Marquinez R, et al. Wild *Solanum* species as resistance sources against different pathogens of potato. Potato Res 1998;41(1):57–68.
[107] Douches DS, Bamberg JB, Kirk W, Jastrzebski K, Niemira BA, Coombs J, et al. Evaluation of wild *Solanum* species for resistance to the US-8 genotype of Phytophthora infestans utilizing a fine-screening technique. Am J Potato Res 2001;78(2):159–65.
[108] Inglis DA, Brown CR, Gundersen BG, Porter LD, Miller JS, Johnson DA, et al. Assessment of *Solatium hougasii* in Washington and Mexico as a source of resistance to late blight. Am J Potato Res 2007;84(3):217.
[109] Micheletto S, Boland R, Huarte M. Argentinian wild diploid *Solanum* species as sources of quantitative late blight resistance. Theor Appl Genet 2000;101(5–6):902–6.
[110] Ochoa CM. The Potatoes of South America: Peru. The wild species. International Potato Center; 2004.
[111] Thieme R, Rakosy-Tican E, Gavrilenko T, Antonova O, Schubert J, Nachtigall M, et al. Novel somatic hybrids (*Solanum tuberosum* L. + Solanum tarnii) and their fertile BC 1

progenies express extreme resistance to potato virus Y and late blight. Theor Appl Genet 2008;116(5):691–700.

[112] Malcolmson JF. Races of *Phytophthora infestans* occurring in Great Britain. Trans Br Mycol Soc 1969;53(3):417–23.

[113] Malcolmson JF, Black W. New R genes in *Solanum demissum* Lindl. and their complementary races of *Phytophthora infestans* (Mont.) de Bary. Euphytica 1966;15(2):199–203.

[114] Danan S, Chauvin JE, Caromel B, Moal JD, Pelle R, Lefebvre V. Major-effect QTLs for stem and foliage resistance to late blight in the wild potato relatives *Solanum sparsipilum* and *S. spegazzinii* are mapped to chromosome X. Theor Appl Genet 2009;119(4):705–19.

[115] Rauscher GM, Smart CD, Simko I, Bonierbale M, Mayton H, Greenland A, et al. Characterization and mapping of RPi-ber, a novel potato late blight resistance gene from *Solanum berthaultii*. Theor Appl Genet 2006;112(4):674–87.

[116] Naess SK, Bradeen JM, Wielgus SM, Haberlach GT, McGrath JM, Helgeson JP. Resistance to late blight in *Solanum bulbocastanum* is mapped to chromosome 8. Theor Appl Genet 2000;101(5–6):697–704.

[117] Park T-H, Gros J, Sikkema A, Vleeshouwers VGAA, Muskens M, Allefs S, et al. The late blight resistance locus Rpi-blb3 from *Solanum bulbocastanum* belongs to a major late blight R gene cluster on chromosome 4 of potato. Mol Plant Microbe Interact 2005;18(7):722–9.

[118] van der Vossen EAG, Gros J, Sikkema A, Muskens M, Wouters D, Wolters P, et al. The Rpi-blb2 gene from *Solanum bulbocastanum* is an Mi-1 gene homolog conferring broad-spectrum late blight resistance in potato. Plant J 2005;44(2):208–22.

[119] Lindqvist-Kreuze H, Carbajulca D, Gonzalez-Escobedo G, Perez W, Bonierbale M. Comparison of transcript profiles in late blight-challenged *Solanum cajamarquense* and B3C1 potato clones. Mol Plant Pathol 2010;11(4):513–30.

[120] Ochoa C. Northern Peru, a possible new source of potatoes resistant to *Phytophthora infestans*. Phytopathology 1954;44(9).

[121] Tan MYA, Hutten RCB, Celis C, Park T-H, Niks RE, Visser RGF, et al. The RPi-mcd1 locus from *Solanum microdontum* involved in resistance to *Phytophthora infestans*, causing a delay in infection, maps on potato chromosome 4 in a cluster of NBS-LRR genes. Mol Plant Microbe Interact 2008;21(7):909–18.

[122] Smilde WD, Brigneti G, Jagger L, Perkins S, Jones JD. *Solanum mochiquense* chromosome IX carries a novel late blight resistance gene Rpi-moc1. Theor Appl Genet 2005;110(2):252–8.

[123] Vleeshouwers VG, Rietman H, Krenek P, Champouret N, Young C, Oh SK, et al. Effector genomics accelerates discovery and functional profiling of potato disease resistance and *Phytophthora infestans* avirulence genes. PLoS One 2008;3(8):e2875.

[124] Kuhl JC, Hanneman RE Jr, Havey MJ. Characterization and mapping of Rpi1, a late-blight resistance locus from diploid (1EBN) Mexican *Solanum pinnatisectum*. Mol Genet Genom 2001;265(6):977–85.

[125] Liu Z, Halterman D. Identification and characterization of RB-orthologous genes from the late blight resistant wild potato species *Solanum verrucosum*. Physiol Mol Plant Pathol 2006;69(4–6):230–9.

[126] Liu Z, Halterman D. Different genetic mechanisms control foliar and tuber resistance to *Phytophthora infestans* in wild potato *Solanum verrucosum*. Am J Potato Res 2009;86(6):476.

[127] Tan MY, Hutten RC, Visser RG, van Eck HJ. The effect of pyramiding *Phytophthora infestans* resistance genes R Pi-mcd1 and R Pi-ber in potato. Theor Appl Genet 2010;121(1):117–25.

[128] Haverkort AJ, Boonekamp PM, Hutten R, Jacobsen E, Lotz LAP, Kessel GJT, et al. Durable late blight resistance in potato through dynamic varieties obtained by *Cisgenesis*: Scientific and societal advances in the DuRPh project. Potato Res 2016;59(1):35–66.

[129] Obidiegwu JE, Flath K, Gebhardt C. Managing potato wart: a review of present research status and future perspective. Theor Appl Genet 2014;127(4):763–80.

[130] Eppo/Oepp. Diagnostic protocols for regulated pests, European and Mediterranean plant protection organisation. OEPP/EPPO Bull 2004;34:309–14.

[131] Przetakiewicz J. The viability of winter sporangia of *Synchytrium endobioticum* (Schilb.) Perc. from Poland. Am J Potato Res 2015;92(6):704–8.

[132] Rowe RC, Powelson ML. Potato early dying: management challenges in a changing production environment. Plant Dis 2002;86(11):1184–93.

[133] Willetts HJ. The survival of fungal sclerotia under adverse environmental conditions. Biol Rev 1971;46(3):387–407.

[134] Isaac I, Harrison JAC. The symptoms and causal agents of early-dying disease (*Verticillium* wilt) of potatoes. Ann Appl Biol 1968;61(2):231–44.

[135] Corsini DL, Pavek JJ, Davis JR. *Verticillium* wilt resistance in noncultivated tuber-bearing *Solanum* species. Plant Dis 1988;72(2):148–51.

[136] Jansky SH, Rouse DI. Identification of potato interspecific hybrids resistant to*Verticillium* wilt and determination of criteria for resistance assessment. Potato Res 2000;43(3):239–51.

[137] Jansky SH, Hamernik A, Bethke PC. Germplasm release: Tetraploid clones with resistance to cold-induced sweetening. Am J Potato Res 2011;88(3):218–25.

[138] Lynch DR, Kawchuk LM, Hachey J, Bains PS, Howard RJ. Identification of a gene conferring high levels of resistance to *Verticillium* wilt in *Solanum chacoense*. Plant Dis 1997;81(9):1011–4.

[139] Concibido VC, Secor GA, Jansky SH. Evaluation of resistance to *Verticillium* wilt in diploid, wild potato interspecific hybrids. Euphytica 1994;76(1–2):145–52.

[140] Alves FM, Gonring AHR, Zanuncio JC, de Sena Fernandes ME, Plata-Rueda A, Fernandes FL. Economic damage levels and treatment thresholds for leafminer insects in *Solanum tuberosum* crops. Crop Prot 2017;100:81–6.

[141] Flanders KL, Hawkes JG, Radcliffe EB, Lauer FI. Insect resistance in potatoes: sources, evolutionary relationships, morphological and chemical defenses, and ecogeographical associations. Euphytica 1992;61(2):83–111.

[142] Jansky SH, Simon R, Spooner DM. A test of taxonomic predictivity: resistance to the Colorado potato beetle in wild relatives of cultivated potato. J Econ Entomol 2009;102(1):422–31.

[143] Ryan JD, Gregory P, Tingey WM. Phenolic oxidase activities in glandular trichomes of *Solanum berthaultii*. Phytochemistry 1982;21(8):1885–7.

[144] Gregory P, Tingey WM, Ave DA, Bouthyette PY. Potato glandular trichomes: A physicochemical defense mechanism against insects. Natural resistance of plants to pests. ACS Symposium Series;1986. p. 160–167.

[145] Tingey WM, Yencho GC. Insect resistance in potato: a decade of progress. In: Zehnder GW, Powelson ML, Jansson RK, editors. Advances in potato pest biology and management. St. Paul Minn.: APS Press; 1994. p. 405–425.

[146] Gibson RW. Aphid-trapping glandular hairs on hybrids of *Solanum tuberosum* and *S. berthaultii*. Potato Res 1974;17(2):152–4.

[147] Gibson RW. Glandular hairs on *Solatium polyadenium* lessen damage by the Colorado beetle. Ann Appl Biol 1976;82(1):147–50.

[148] Sinden SL, Sanford LL, Cantelo WW, Deahl KL. Leptine glycoalkaloids and resistance to the Colorado potato beetle (Coleoptera: *Chrysomelidae*) in *Solanum chacoense*. Environ Entomol 1986;15(5):1057–62.

[149] Lorenzen JH, Balbyshev NF, Lafta AM, Casper H, Tian X, Sagredo B. Resistant potato selections contain leptine and inhibit development of the Colorado potato beetle (Coleoptera: *Chrysomelidae*). J Econ Entomol 2001;94(5):1260–7.
[150] Kuhn R, Löw I. Zur konstitution der leptine. Chem Berichte 1961;94(4):1088–95.
[151] Dimock MB, Tingey WM. Host acceptance behaviour of Colorado potato beetle larvae influenced by potato glandular trichomes. Physiol Entomol 1988;13(4):399–406.
[152] Dimock MB, Lapointe SL, Tingey WM. *Solanum neocardenasii*: a new source of potato resistance to the Colorado potato beetle (Coleoptera: *Chrysomelidae*). J Econom Entomol 1986;79(5):1269–75.
[153] Flanders KL, Radcliffe EB, Hawkes JG. Geographic distribution of insect resistance in potatoes. Euphytica 1997;93(2):201–21.
[154] Jansky SH. Identification of *Verticillium* wilt resistance in U.S. Potato Breeding Programs. Am J Potato Res 2009;86(6):504–12.
[155] Pelletier Y, Grondin G, Maltais P. Mechanism of resistance to the Colorado potato beetle in wild *Solanum* species. J Econom Entomol 1999;92(3):708–13.
[156] Pelletier Y. Level and genetic variability of resistance to the colorado potato beetle (*Leptinotarsa decemlineata* (say)) in wild solarium species. Am J Potato Res 2007;84(2):143–8.
[157] Pompon J, Quiring D, Giordanengo P, Pelletier Y. Characterization of *Solanum chomatophilum* resistance to 2 aphid potato pests, *Macrosiphum euphorbiae* (Thomas) and *Myzus persicae* (Sulzer). Crop Prot 2010;29(8):891–7.
[158] Gibson RW. Resistance in glandular-haired wild potatoes to flea beetles. Am Potato J 1978;55(11):595–9.
[159] Sleesman JP. Resistance in wild potatoes to attack by the potato leafhopper and the potato flea beetle. Am Potato J 1940;17(1):9–12.
[160] Halbert SE, Corsini DL, Wiebe MA. Potato virus Y transmission efficiency for some common aphids in Idaho. Am J Potato Res 2003;80(2):87–91.
[161] Lapointe SL, Tingey WM. Glandular trichomes of *Solanum neocardenasii* confer resistance to green peach aphid (Homoptera: *Aphididae*). J Econom Entomol 1986;79(5):1264–8.
[162] Radcliffe EB, Lauer FI, Lee M-H, Robinson DP. Evaluation of the United States potato collection for resistance to green peach aphid and potato aphidAgricultural Experiment Station, University of Minnesota; 1981.
[163] Radcliffe EB, Lauer FI. Further studies on resistance to green peach aphid and potato aphid in the wild tuber-bearing *Solanum* species. J Econom Entomol 1970;63(1):110–4.
[164] Davis JA, Radcliffe EB, Thill CA, Ragsdale DW. Resistance to aphids, late blight and viruses in somatic fusions and crosses of *Solanum tuberosum* L. and *Solanum bulbocastanum* Dun. Am J Potato Res 2012;89(6):489–500.
[165] Dangles O, Mesías V, Crespo-Perez V, Silvain JF. Crop damage increases with pest species diversity: evidence from potato tuber moths in the tropical Andes. J Appl Ecol 2009;46(5):1115–21.
[166] Fenemore PG. Oviposition of potato tuber moth, *Phthorimaea operculella* Zell. (Lepidoptera: *Gelechildae*); the physical nature of the oviposition substrate. N Zeal J Zool 1978;5(3):591–9.
[167] Rondon SI. Decoding *Phthorimaea operculella* (Lepidoptera: *Gelechiidae*) in the new age of change. J Integr Agric 2020;19(2):316–24.
[168] Westedt AL, Douches DS, Pett W, Grafius EJ. Evaluation of natural and engineered resistance mechanisms in *Solanum tuberosum* for resistance to *Phthorimaea operculella* (Lepidoptera: *Gelechiidae*). J Econom Entomol 1998;91(2):552–6.

[169] Chávez R, Schmiediche PE, Jackson MT, Raman KV. The breeding potential of wild potato species resistant to the potato tuber moth, *Phthorimaea operculella* (Zeller). Euphytica 1988;39(2):123–32.

[170] Horgan FG, Quiring DT, Lagnaoui A, Pelletier Y. Tuber production, dormancy and resistance against *Phthorimaea operculella* (Zeller) in wild potato species. J Appl Entomol 2013;137(10):739–50.

[171] Radcliffe EB, Johnson KB. Biology and management of leafhoppers on potato. Advances in potato pest biology and managementSaint Paul, MN: APS Press; 1994. p. 71–82.

[172] Tingey WM, Gibson RW. Feeding and mobility of the potato leafhopper impaired by glandular trichomes of *Solanum berthaultii* and *S. polyadenium*. J Econom Entomol 1978;71(6):856–8.

[173] Soares WS, Plata-Rueda RA, de Sena Fernandes ME, Fernandes FL, Alves FM, da Silva ÍW. First record of *Liriomyza huidobrensis* (Diptera: Agromyzidae) disseminating *Alternaria solani* (*Pleosporaceae*) in potato crops in Brazil. Florida Entomol 2019;102(1):234–5.

[174] Evans K, Trudgill DL, Raman KV, Radcliffe EB. Pest aspects of potato production. The potato cropSpringer; 1992. p. 438–506.

[175] Dandurand L-M, Zasada IA, Wang X, Mimee B, De Jong W, Novy R, et al. Current status of potato cyst nematodes in North America. Ann Rev Phytopathol 2019;57:117–33.

[176] EPPO GD. EPPO global database. European and Mediterranean Plant Protection Organization. EPPO; 2020.

[177] Ellenby C. Resistance to the potato-root eelworm. Nature 1948;162(4122):704.

[178] Ellenby C. Resistance to the potato root eelworm, *Heterodera rostochiensis* Wollenweber. Nature 1952;170(4337):1016.

[179] Castelli L, Ramsay G, Bryan G, Neilson SJ, Phillips MS. New sources of resistance to the potato cyst nematodes *Globodera pallida* and *G. rostochiensis* in the Commonwealth Potato Collection. Euphytica 2003;129(3):377–86.

[180] Ruiz de Galarreta JI, Carrasco A, Salazar A, Barrena I, Iturritxa E, Marquinez R, et al. Wild *Solanum* species as resistance sources against different pathogens of potato. Potato Res 1998;41(1):57–68.

[181] Rothacker D, Stelter H. *Solanum tarijense* Hawk—eine weitere gegen den Kartoffelnematoden *Heterodera rostochiensis* (Wollenweber) resistente species. NW 1961;48(24):742–3.

[182] Brown CR, Mojtahedi H, Santo GS. Resistance to Columbia root-knot nematode in *Solanum* ssp. and in hybrids of *S. hougasii* with tetraploid cultivated potato. Am Potato J 1991;68(7):445–52.

[183] Austin S, Pohlman JD, Brown CR, Mojtahedi H, Santo GS, Douches DS, et al. Interspecific somatic hybridization between *Solanum tuberosum* L. and *S. bulbocastanum* DUN. as a means of transferring nematode resistance. Am Potato J 1993;70(6):485–95.

[184] Brown CR, Mojtahedi H, James S, Novy RG, Love S. Development and evaluation of potato breeding lines with introgressed resistance to Columbia root-knot nematode (*Meloidogyne chitwoodi*). Am J Potato Res 2006;83(1):1–8.

[185] Kreuze JF, Souza-Dias JAC, Jeevalatha A, Figueira AR, Valkonen JPT, Jones RAC. Viral diseases in potato. The potato cropSpringer; 2020. p. 389.

[186] Kumar R, Tiwari RK, Jeevalatha A, Kaundal P, Sharma S, Chakrabarti SK. Potato viruses and their diagnostic techniques: An overview. J Pharmacog Phytochem 2019;8(6):1932–44.

[187] Gray S, De Boer S, Lorenzen J, Karasev A, Whitworth J, Nolte P, et al. Potato virus Y: an evolving concern for potato crops in the United States and Canada. Plant Dis 2010;94(12):1384–97.

[188] Zeng Y, Fulladolsa AC, Houser A, Charkowski AO. Colorado seed potato certification data analysis shows mosaic and blackleg are major diseases of seed potato and identifies tolerant potato varieties. Plant Dis 2019;103(2):192–9.

[189] Horvath J, Hoekstra R. Reaction of some new Bolivian tuber-bearing *Solanum* species to different potato pathogenic viruses. Acta Phytopathol Entomol Hungar 1989;24(3–4):351–62.

[190] Webb RE, Schultz ES. Resistance of solanum species to potato viruses A, X, and Y. Am Potato J 1961;38(5):137–42.

[191] Cai XK, Spooner DM, Jansky SH. A test of taxonomic and biogeographic predictivity: Resistance to potato virus Y in wild relatives of the cultivated potato. Phytopathology 2011;101(9):1074–80.

[192] Ross H. Wild species and primitive cultivars as ancestors of potato varieties. Broadening the genetic base of crops Wageningen,The Netherlands: Centre for Agricultural Publishing and Documentation; 1979. p. 237–45.

[193] Tozzini AC, Ceriani MF, Saladrigas MV, Hopp HE. Extreme resistance to infection by potato virus X in genotypes of wild tuber-bearing *Solanum* species. Potato Res 1991;34(3):317–24.

[194] Turuleva LM, Balmasova MA. Wild potato species-sources of resistance to virus diseases. Wild potato species-sources of resistance to virus diseases. 1990. p. 100–105.

[195] Zoteyeva N, Chrzanowska M, Flis B, Zimnoch-Guzowska E. Resistance to pathogens of the potato accessions from the collection of NI Vavilov Institute of Plant Industry (VIR). Am J Potato Res 2012;89(4):277–93.

[196] Nouri-Ellouz O, Triki MA, Jbir-Koubaa R, Louhichi A, Charfeddine S, Drira N, et al. Somatic hybrids between potato and *S. berthaultii* show partial resistance to soil-borne fungi and potato virus Y. J Phytopathol 2016;164(7–8):485–96.

[197] Takacs AP, Horvath J, Kazinczi G, Pribek D. Solanum species as new resistance sources of C strain of potato Y potyvirus (PVYc). Mededelingen-Faculteit Landbouwkundige en Toegepaste Biologische Wetenschappen Universiteit Gent (Belgium). 1999.

[198] Bethke PC, Halterman DA, Jansky S. Are we getting better at using wild potato species in light of new tools? Crop Sci 2017;57(3):1241–58.

[199] Jackson SA, Hanneman RE. Crossability between cultivated and wild tuber-and non-tuber-bearing *Solanums*. Euphytica 1999;109(1):51–67.

[200] Jansky S. Overcoming hybridization barriers in potato. Plant Breed 2006;125(1):1–12.

[201] Grun P, Aubertin M. The inheritance and expression of unilateral incompatibility in *Solanum*. Heredity 1966;21(1):131.

[202] Grun P. Cytoplasmic sterilities that separate the cultivated potato from its putative diploid ancestors. Evolution 1970;24:750–8.

[203] Johnston SA, Den Nijs TPM, Peloquin SJ, Hanneman RE. The significance of genic balance to endosperm development in interspecific crosses. Theor Appl Genet 1980;57(1):5–9.

[204] Bradeen JM, Kole C. Genetics, genomics and breeding of potato. CRC Press; 2016.

[205] Carputo D, Basile B, Cardi T, Frusciante L. Erwinia resistance in backcross progenies of *Solanum tuberosum x S. tarijense* and *S. tuberosum* (+) *S. commersonii* hybrids. Potato Res 2000;43(2):135–42.

[206] Carputo D, Aversano R, Frusciante L, editors. Breeding potato for quality traits; 2004.

[207] Hermsen JGT, Ramanna MS. Double-bridge hybrids of *Solanum bulbocastanum* and cultivars of *Solanum tuberosum*. Euphytica 1973;22(3):457–66.

[208] Jansky S, Hamernik A. The introgression of 2× 1EBN *Solanum* species into the cultivated potato using *Solanum verrucosum* as a bridge. Genet Resour Crop Evol 2009;56(8):1107–15.

[209] Yermishin AP, Polyukhovich YV, Voronkova EV, Savchuk AV. Production of hybrids between the 2EBN bridge species *Solanum verrucosum* and 1EBN diploid potato species. Am J Potato Res 2014;91(6):610–7.

[210] Hawkes JG, Hjerting JP. The potatoes of Argentina, Brazil, Paraguay and Uruguay. A biosystematic study. The potatoes of Argentina, Brazil, Paraguay and Uruguay A biosystematic study. 1969.

[211] Camadro EL. Relevance of the genetic structure of natural populations, and sampling and classification approaches for conservation and use of wild crop relatives: potato as an example. Botany 2012;90(11):1065–72.

[212] Hickey LT, Hafeez AN, Robinson H, Jackson SA, Leal-Bertioli SCM, Tester M, et al. Breeding crops to feed 10 billion. Nat Biotechnol 2019;37(7):744–54.

[213] FAO. International year of the potato 2008 new light on a hidden treasure. FAO; 2008.

[214] Birch PRJ, Bryan G, Fenton B, Gilroy EM, Hein I, Jones JT, et al. Crops that feed the world 8: Potato: are the trends of increased global production sustainable? Food Secur 2012;4(4):477–508.

[215] Bradshaw JE, Bryan GJ, Ramsay G. Genetic resources (including wild and cultivated *Solanum* species) and progress in their utilisation in potato breeding. Potato Res 2006;49(1):49–65.

[216] Muthoni J, Kabira JN. Potato production in the hot tropical areas of Africa: progress made in breeding for heat tolerance. Int J Biol 2015;7(9):220.

[217] Zhang H, Mittal N, Leamy LJ, Barazani O, Song BH. Back into the wild—Apply untapped genetic diversity of wild relatives for crop improvement. Evol Appl 2017;10(1):5–24.

[218] Potato Genome Sequencing. C Genome sequence and analysis of the tuber crop potato. Nature 2011;475(7355):189.

[219] Aversano R, Contaldi F, Adelfi MG, D'Amelia V, Diretto G, De Tommasi N, et al. Comparative metabolite and genome analysis of tuber-bearing potato species. Phytochemistry 2017;137:42–51.

[220] Esposito S, Barteri F, Casacuberta J, Mirouze M, Carputo D, Aversano R. LTR-TEs abundance, timing and mobility in Solanum commersonii and S. tuberosum genomes following cold-stress conditions. Planta 2019;250(5):1781–7.

[221] Esposito S, Aversano R, Bradeen JM, Di Matteo A, Villano C, Carputo D. Deep-sequencing of *Solanum commersonii* small RNA libraries reveals riboregulators involved in cold stress response. Plant Biol (Stuttg) 2020;22(Suppl. 1):133–42.

[222] Ortiz R. Genomic-led potato brVeeding for increasing genetic gains: achievements and outlook. Crop Breed Genet Genom 2020;2(2):e200010.

[223] Clasen BM, Stoddard TJ, Luo S, Demorest ZL, Li J, Cedrone F, et al. Improving cold storage and processing traits in potato through targeted gene knockout. Plant Biotechnol J 2016;14(1):169–76.

[224] Chitchumroonchokchai C, Diretto G, Parisi B, Giuliano G, Failla ML. Potential of golden potatoes to improve vitamin A and vitamin E status in developing countries. PLoS One 2017;12(11):e0187102.

[225] Butler NM, Baltes NJ, Voytas DF, Douches DS. Geminivirus-mediated genome editing in potato (*Solanum tuberosum* L.) using sequence-specific nucleases. Front Plant Sci 2016;7:1045.

[226] Nakayasu M, Akiyama R, Lee HJ, Osakabe K, Osakabe Y, Watanabe B, et al. Generation of α-solanine-free hairy roots of potato by CRISPR/Cas9 mediated genome editing of the St16DOX gene. Plant Physiol Biochem 2018;131:70–7.

[227] Ye M, Peng Z, Tang D, Yang Z, Li D, Xu Y, et al. Generation of self-compatible diploid potato by knockout of S-RNase. Nat Plants 2018;4(9):651–4.

[228] Enciso-Rodriguez F, Manrique-Carpintero NC, Nadakuduti SS, Buell CR, Zarka D, Douches D. Overcoming self-incompatibility in diploid potato using CRISPR-Cas9. Front Plant Sci 2019;10:376.

[229] Hosaka K, Hanneman RE. Genetics of self-compatibility in a self-incompatible wild diploid potato species Solanum chacoense. 1. Detection of an S locus inhibitor (Sli) gene. Euphytica 1998;99(3):191–7.

[230] Lindhout P, Meijer D, Schotte T, Hutten RCB, Visser RGF, van Eck HJ. Towards F 1 hybrid seed potato breeding. Potato Res 2011;54(4):301–12.

[231] Jansky SH, Charkowski AO, Douches DS, Gusmini G, Richael C, Bethke PC, et al. Reinventing potato as a diploid inbred line–based crop. Crop Sci 2016;56(4):1412–22.

[232] Esposito S, Aversano R, D'amelia V, Villano C, Alioto D, Mirouze M, et al. Dicer-like and RNA-dependent RNA polymerase gene family identification and annotation in the cultivated *Solanum tuberosum* and its wild relative *S. commersonii*. Planta 2018;248(3):729–43.

Chapter 13

Wild Relatives of Sweetpotato; Distribution and Prospectus for Crop Improvement

Muhammad Qadir Ahmad[a,b], Mohamed Hamed Arisha[a,c] and Qiang Li[a]
[a]*Xuzhou Institute of Agricultural Sciences in Jiangsu Xuhuai District/Key Laboratory of Biology and Genetic Breeding of Sweetpotato, Ministry of Agriculture and Rural Affairs/ Sweetpotato Research Institute, CAAS, Xuzhou, Jiangsu, China;* [b]*Department of Plant Breeding and Genetics, Faculty of Agricultural Sciences and Technology, Bahauddin Zakariya University, Multan, Pakistan;* [c]*Department of Horticulture, Faculty of Agriculture, Zagazig University, Sharkia, Egypt*

13.1 Introduction

13.1.1 Importance of sweetpotato

Sweetpotato is grown on approximately 8 million hectares in 114 countries of the world particularly in tropical and sub-tropical regions of South East Asia and Sub-Saharan Africa [1]. It is a small holding crop in most developing countries and is grown under wide range of climatic conditions with minimum supply of nutrients [2]. Being moderately tolerant, the crop can be cultivated from sea level to 3000 m.a.s.l. without compromising translocation of photosynthates to storage roots. Therefore, it has become the staple food of population living at highlands of Uganda, Rawanda, and Papua New Guinea [2]. In Sub-Saharan Africa, Sweetpotato is mainly cultivated on less fertile soils under limited water conditions even though it produces good yields as compared to input supply and labor cost [1]. Sweetpotato is considered as the luxury vegetable in some temperate countries such as United States, New Zealand, and Japan. Among sweetpotato producing countries, China is the largest producer with 53.2 million tons production, approximately 58% of total world's production during year 2018 [3].

Being a versatile crop, sweetpotato offers diverse uses ranging from as direct consumption of leaves and roots by human and animals to processing for making noodles, natural colorant, starch, alcohol, desserts, snacks, roasted sweetpotato candies, and cakes. In china, nearly half of the sweetpotato produce is utilized

Wild Germplasm for Genetic Improvement in Crop Plants.
http://dx.doi.org/10.1016/B978-0-12-822137-2.00013-8

as animal's feed and remaining is used for human consumption. Homemade chips are also used in some tropical countries [2]. Sweetpotato is a rich source of energy per hectare as compared to other food crops [4]. The crop is also a rich source of vitamin A, C, E, calcium, iron, and number of other amino acids [5,6]). Orange fleshed sweetpotato contains high quantity of β-carotene as high as 50-folds more than normal sweetpotato. It also has become major crop for bio-fortification [1,7,8].

13.1.2 Wild relatives of sweetpotato

Crop wild relatives (CWR) or wild cousins of cultivated plant species are being used as sources of genetic diversity since decades [9]. CWR have enormous importance for their potential to contribute valuable traits to crop plants [10]. Crop wild relatives come from their natural habitats adapted to adverse environmental conditions under various types of abiotic and biotic stresses. Therefore, these genetic sources should be incorporated into domesticated crop plants through breeding programs. Crop wild relatives have contributed number of traits related to insect pest resistance, disease resistance, abiotic stress tolerance in important food and other crops such as wheat, cotton, banana, barley, chickpea, maize, tomato, lettuce, cassava, beans, oat, sugarcane, barley, and sunflower [1,11,12]). Even though the significant importance of CWR for improving economic traits in different crop plants, the use of CWR for certain staple crops is under-utilized including sweetpotato. In future, use of CWR is expected to increase due to advancement in knowledge about species diversity and breeding tools. Cultivated sweetpotato (*Ipomea batatas* (L.) Lam) and its closely related wild species belong to family Convolvulaceae and genus *Ipomea*. The family Convolvulacea is widely distributed in range of habitats in tropical and temperate regions. It contains both annual and perennial species. The genus *Ipomoea* comprises of 600–700 species and more than half of them are concentrated to America where they have been classified into three subgenera *Eriospermum*, *Quamocclite*, and *Ipomoea* and 400 taxa [13]. These three subgenera have been further divided into nine sections, out of which six [*Calonyction*, *Exogonium*, *Microsepalae*, *Mina*, *Leptocallis*, and *Tricolor*] were originally concentrated into America before the dispersal of species as cultigens, medicinal plants, and weeds [14,15]. Sweetpotato and its close relatives have been grouped into *Batatas* series of *Eriospermum.* There are 13 wild species within *Batatas* series considered to be the wild relatives of sweetpotato which include *I. trifida*, *I. tiliaceae*, *I. tabascana*, *I. cordatotriloba*, *I. cynanchifolia*, *I. lacunosa*, *I. umbraticola*, *I. ramosissima*, *I. tenuissima*, *I. littoralis*, *I. grandifolia*, *I. triloba*, and *I.* × *leucantha* [16–18]. Earlier two more species; (1) *I. peruviana* belong to Peru and (2) *I. gracilis* from Australia were included in *Batatas* series but now they have been included in *Ipomoea* series *Setosae* [19]. Among the species belonging to *Batatas* series, two species are considered to be hybrids, *I.* × *leucantha* an intermediate between *I. cordatotriloba* ×

FIGURE 13.1 Sweetpotato and its wild relatives (A) *I. batatas*, (B) *I. trifida*, (C) *I. triloba*, (D) *I. cordatotriloba*.

I. lacunosa and *I. grandifola* has been hypothesized a derivative of *I. batatas* × *I. cordatotribola* (Fig. 13.1).

13.2 Domestication and dispersal of sweetpotato

Despite the importance of sweetpotato, the botanical origin and time and place of its domestication is still unclear. Sweetpotato has great complexity in its origin dating back from 8000 years ago to pre-human evolution [20]. Predominantly, it is considered that sweetpotato was originally domesticated in new world at least 5000 years ago in the region between Yucatan Peninsula of Mexico and the Orinoco River in Venezuela [21–23]. Maximum diversity for sweetpotato accession and some other species in *Batatas* series have been observed in this region. Radio carbons studies conducted on the oldest remain found from caves of Chilca Canyon in Peru [24] and preserved in natural history museum suggested the origin of sweetpotato between 10,000 and 8,000 BC. Starch granules analysis of these remains further confirmed these findings [25]. Interestingly, two possible routes have been suggested by researchers for the introduction of sweetpotato to rest of the world. (1) Pre-Columbian dispersal and (2) post-Columbian dispersal. According to pre-Columbian hypotheses sweetpotato was introduced to Oceania by polynesian people. Second migration was conducted by Columbus who introduced sweetpotato to Europe in last decade of 15th century. From there, sweetpotato was reached to Africa alongwith European immigrants. After that Portuguese travelers introduced it to rest of the Africa and Asia. Recently, role of Pennsylvanian people in dispersal of sweetpotato from America to Oceania have been challenged by Munoz-Rodriguez and coworkers [20]. The authors challenged the hypotheses of development of storage roots as a human activity. They postulated that on the basis of iterative nature of storage root evolution and the prevalence of naturally occurring long distance dispersal within *Ipomoea*, the origin of sweetpotato and dispersal to Polynesia is far beyond the hypothesis that it was domesticated 8000 years ago and brought to Polynesia by humans in last 700 years. Their analysis was based on 94 specimens of *I. batatas* and *I. trifida* and they documented a pre-human origin of sweetpotato.

13.3 Compatibility groups among wild relatives of sweetpotato

Ipomoea species can be grouped into three distinctive groups based on self compatibility and interspecific hybridization [26–28]. These groups can be represented as A, B, and X group. Species belong to B group are self-incompatible but hybridization among these species is successful. This group includes four species, namely *I. batata*, *I. trifida*, *I. littoralis*, and *I. tabascana*. X group contains only *I. tiliacea,* a tetraploid species. This tetraploid species is self-incompatible. Earlier another tetraploid species *I. gracilis* was also placed in X group but now it is considered as outgroup species. Interestingly, hybridization between both these species included in X group is successful. Group A comprised of maximum number of species, which include *I. tribola*, *I. cynanchifolia*, *I. cordatotriloba*, *I. umbraticola*, *I. lacunosa*, *I.* × *leucantha*, *I. ramosissima*, *I. grandifolia, and I. tenuissima.* All these species are diploid except *I. cordatotriloba* in which tetraploid plants have also been observed [13,27,29]. Species included in A group are self-incompatible but hybridization among these species is successful. Hybridization between A and X group is successful when latter group species are used as pollen parent.

13.4 Wild relatives and origin of sweetpotato

Cultivated sweetpotato is a hexaploid species 2n = 6x = 90. Two possible origins of sweetpotato involving wild relatives have been proposed: (1) an autopolyploidy origin involving autopolyploidization within the population of *I. trifida* a diploid progenitor of sweetpotato. The wild species *I. trifida*

(Kunth.) G. Don (2n = 2x = 30) has long been considered as the progenitor of sweetpotato [23,30]. Shiotani [31] also supported the autoploidization for the development of sweetpotato. In sweetpotato, occurrence of unreduced 2n gametes might have been involved in the polyploidy series. Production of 2n gametes in *I. trifida* and tetraploid *I. batatas* have been widely reported [32–34] which supports the possibility of autopolyploidy origin of sweetpotato. Occurrence of 2n gametes have also played major role in autoploidization origin of other crop plants such as wheat. (2) In contrary, Austin [16] proposed an allopolyploidy origin of sweetpotato by involving the events of hybridization between *I. tifida* and *I. tribola* (2n = 2x = 30). Furthermore, hybridization between *I. trifida* and any other unknown diploid species translocated from Caribbean to mainland cannot be excluded. Another polyploidization hypotheses suggest that sweetpotato was evolved after a number of polyploidy and autoploidy events in which initially, hybridization between diploid *Ipomoea* species *I. leucantha Jacq.*, and tetraploid species *I. littoralis Blume* produced triploid *I. trifida* followed by chromosomal doubling [13,35]. As presence of polysomic inheritance supports that hexaploid sweetpotato may have been resulted in by autopolyploidy [23,36]). Some molecular studies revealed that *I. trifida* might have been involved in the evolution of sweetpotato being a close relative but not as a direct ancestor because there are no sound evidences which predict the contribution of *I. trifida* genome to the hexaploid genome of sweetpotato [37,38]. Therefore, Roullier and colleagues [30] suggested that sweetpotato may have been developed by domestication of two independent tuber bearing populations in South America and Caribbean/Central America, following close contact between these two independent populations through human movements lead toward the development of hexaploid cultivated sweetpotato. Even though, in the presence of contrary findings in the origin of sweetpotato, the role of *I. trifida* in the origin of sweetpotato is widely accepted [39,40].

13.5 Distribution of sweetpotato across the globe

Sweetpotato was originally domesticated in the new world and then moved to other parts of the world. Currently, sweetpotato is being cultivated across the globe. Hijmans and colleagues [41] provided a very comprehensive report about the distribution of sweetpotato. According to this report, China has striking aspect of the crops distribution with high cropping area. In China, three areas have shown highest distribution of the crop including (1) Sichuan Basin (Sichuan and Chongqing provinces), (2) east central provinces (Anhui, Henan, Shandong), and (3) south eastern coast. In addition to China's significant distribution of crop is observed in the Caribbean region, In java, Islands of New Guinea, Africa particularly lake Victoria, Uganda, Ghana, Nigeria, Malawi, and Madagascar.

13.5.1 Distribution of new world species

It is predominantly believed that originally sweetpotato was domesticated in America and then it was spread to other parts of the world. Then it reached from a latitude between 42 N° and 35 S° from sea level to up to an altitude of 3000 m [13]. Species in *Batatas* series are perennial although some barely sustain for 2 years except *I. lacunosa* which grow annual in southeastern United States. Austin [21] explained distribution of wild relatives of sweetpotato. Two species (*I. lacunosa and I. cordatotriloba*) belong to North America; three were from Caribbean (*I. tenuisissima, I. tiliacea, I. triloba*) whereas other species were distributed in central and southern America (*I. cynanchifolia, I. grandifolia, I. Peruviana, I. ramosissima, and I. trifida*). Among Caribbean species, *I. triloba* is pantropical.

Previously, several researchers have represented the Mexican and Central American regions of distribution as the posited centers of origin and primary diversity of cultivated sweetpotato [18,30,42]. Recently, a study on the distribution of CWR of sweetpotato in the new world was conducted by Khoury and coworkers [1]. The study included all the 14 species in *Batatas* series. Occurrence records of these species were acquired from various online data sources, herbaria and germplasm databases by personal communications and/or by visiting the site and collecting information and samples. The distribution pattern showed maximum species richness in central Mexico to the northern Andean region through Central America, and in the southeastern United States. Nine species showed overlapping in Mexico from the states of Veracruz through the Yucatan peninsula. In this study, Northwestern South America, which is the home of oldest archeological remains of cultivated sweetpotato displayed a considerably lesser but still notable degree of CWR species richness. Earlier, Austin [21] observed abundance of *I. trifida* in northwestern South America and parts of Central America. Two other species *I. triloba* and *I. tiliacea* were also found abundant in Caribbean Islands particularly Circum Caribbean region. On the basis of eco-geographic distribution some researchers have selected core collections of sweetpotato and its related species[43,44]. Khoury and coworkers [1] also identified a pattern of eco-geographic distribution into temperate North America (*I. lacunosa, I. tenuissima*), Mesoamerica (e.g., *I. Splendor-sylvae, I. tabascana*), wide tropical (e.g., *I. triloba and I. trifida*) and South Americas (e.g., *I. grandifolia, I. cynanchifolia*) areas. Furthermore, Principal component analysis divided CWR into four major PCs explaining 75.6% of observed variation. Among these PCs, first PCA being most diverse accounted 32% variation and showed relationship with temperature fluctuation and extremes. Second and third PCs accounted for 21.6% and 13.9% and were related to variation in the occurrence of precipitation and altitudes where these species were located, respectively. Least variation (8.1%) was accounted by final component and showed relationship with soil properties (texture). They also observed a pattern of adaptation in species distribution. Species which showed notable adaptation to high temperature include *I. tabascana, I. trifida, I. × leucantha, I. tiliacea, I. tenuissima, I. tribola, and wild I. batatas*. The species which showed adaptation to low temperatures include *I. lacunosa, I. grandifolia, I. cordatotriloba, I. tenuissima,* and *I. ramossissima. I. tenuissima* showed adaptation both under low and high temperature areas. The species *I. trifida* and *I. batatas* appeared to be adapted under wide range of conditions.

13.5.2 Distribution of old world species

I. littoralis is the only member of *Batatas* section of *Ipomoea* which is found in old world outside the regions of new world. Austin and coworkers [45] reported wide scale distribution of *I. littoralis* around the pacific and Indian oceans. The species have been reported from southern Asia (Sri Lanka, India, Indo-china, Thailand), Madagascar, Mauritius, the Seychelles, Taiwan, Malaysia (Philippines, New Guinea, Malay Peninsula), around the pacific (Micronesia, including the Marianas, the Marshalls, the Carolinas, the Gilberts, Hawaii, Guam, the Ryukyu Island, the Australia Island, Palau Island, the Society Islands, the Tuamotu Archipelago, Fiji, New Caledonia), and in Australia (N Queensland on Cape York Peninsula). Previously, many researchers have reported the distribution of *I. littoralis* across the pacific and Indian oceans [16,46–49]. Some other authors also reported the distribution of *I. littoralis* in Pacific and Indian Ocean, but they erroneously called it *I. gracillis* [50–52].

Answering to the question that how *I. littoralis* belongs to old world instead new world where all other species of *Batatas* series are confined, Austin [45] discussed two possibilities: (1) ancestral stock might be transported by ocean current places where it evolved into distinct species as *I. littoralis* and (2) dispersal to old world by human or some other agents may resulted in evolution of this distinct species. Austin [45] was in favor of the first hypotheses on the basis of growth and development of vines. Previously, Nishiyama [35] stated the involvement of *I. littoralis Blume* in the evolution of sweetpotato. Although, recent studies consider *I. trifida* and *I. tribola* as most closely related to *I. batatas* and expected parents of *I. batatas*. Although more evidences needed for the divergence of sweetpotato and its wild relatives in prehistoric time. But we cannot roughly exclude prehistoric distribution and dispersal of *Ipomoea* species in old world as suggested by Munoz-Rodriguez and coworkers [20].

Fosberg and Sachet [47] observed *Ipomoea* plants at higher elevations, at the edges of taro pits, interior opening, and thin places in woods and coconut grooves. They concluded that these plants may be accidentally introduced by man at these places (Table 13.1).

TABLE 13.1 List of wild relatives of sweetpotato, ploidy level, self compatibility and geographic range.

Ipomoea species	Characteristics	Ploidy level X = 15	Self compatibility	Geographic range	Genome type
I. batatas	A vine and perennial herb with deep purple flowers treated as an annual. Produces storage roots.	6x = 90	I	American tropics, China, Peru, Japan, etc.	B
I. tiliaceae	Spreading type vines having deep purple color. Triangular leaves without lobes. Flowering is very scarce.	4x-60	I	Caribbean	X
I. tabascana	Vines with three deep lobed leaves. Flowers are pentagonal and purple colored.	4x = 60	C	Mexico.	B
I. trifida	Spreading type with dark purple leaves and green tips. Cordite leaves with sparse tip pubescence. Flowering is scarce.	2x,4x,6x	I	Mexico, Central and South America.	B
I. cordatotriloba	A vine with heart shape and three lobed leaves. Flower color varies from pink to deep purple. Flowers are large and numerous.	2x, 4x	C	North and South America.	A
I. cynanchifolia	Thinly pubescence and cordate leaves. Pink funnel shaped flower with dark center.	2x = 30	C	Central and South America, Brazil, Guyana.	A
I. lacunosa	It is called white star due to small white flowers. Leaves are cordate, thinly pubescence, and margins of leaf are typically purple and smooth.	2x = 30	C	North America. Native to USA.	A
I. umbraticola	It is perennial plant with ovate leaf blades. It has three lobes usually, cordate type.	2x = 30	C	Costa Rica/Mexico.	A
I. ramosissima	A slender twining annual herb. Leaves are ovate or shallowly three-lobed. Funnel shaped pink color flower with dark center.	2x = 30	C	Central and South America.	A

I. tenuissima	It has hairy stems and leaves. Leaves vary in shape but often arrowhead shape. Solitary flowers with purple color.	2x = 30	C	Caribbean Islands/West Indies.	A
I. × leucantha	Spreading type wines, mostly twining. Triangular leaves slightly lobed and no pubescence. Flowering is scarce.	2x = 30	C	Tropics.	A
I. grandifolia	It is twining herbaceous plant with tri lobed leaves. Flower color is pink to purple but red center. Flowers are five lobed fused together but each lobe has a pointed tip in center.	2x-30	C	Tropics of Central and South America.	A
I. triloba	Spreading type with dark purple spots. Leaves are without lobes and have sparse hairing. Flowering is scarce.	2x-30	C	Central America and Caribbean.	A
I. littoralis	These are perennial vines with scattered trichomes. Flowers color is from lavender to pinkish purple. Pubescence on the lower half.	2x-30	I	Old World/Australia/ Indian Ocean.	B

Self incompatible, C-self compatible.
Source: Adapted and modified from Refs. [2,13].

13.6 Prospectus for crop improvement

13.6.1 Constraints to sweetpotato production

A range of abiotic and biotic stresses effects the production of raw sweetpotato in most of sweetpotato producing regions. Among abiotic stresses, drought has become severe challenge along with combined effect of heat stress in the areas facing consequences of climate change. Susceptibility to drought stress has become constrain in the adoption of orange fleshed high β-carotene sweetpotato cultivar [1,53,54]. Similarly, among biotic stresses sweetpotato weevils, soil borne pathogens and sweetpotato virus are main constraints to yield [55–57]. Worldwide sweetpotato is susceptible to more than 40 diseases, 40 types of insects and pests, and more than 20 viruses [58–60]. Among viruses, Sweetpotato Chlorotic Stunt Virus (SPCSV) and the aphid born Sweetpotato Feathery Mottle Virus (SPFMV) are severe constraint to sweetpotato production in sub-Saharan Africa [61]. Yield losses due to virus disease ranges from 50% to 98% in the United States and sub-Saharan Africa [62–64]. In China, 20% yield losses due to SPVD have been reported [65]. The disease caused by sweetpotato virus are difficult to manage due to complexity of disease complex resulting from synergistic dual infection caused by aphid (*Aphis gossypii* (Glovers)) transmitted SPFMV and silver leaf white fly (*Bemissia taba*ci (Gennadius) [61]. Development of weevils larvae in the storage roots, concealed feeding behavior and oviposition make control of these insects very difficult, therefore, improved management practices and identification of genetic sources of resistance is need of the hour.

13.6.2 Wild relatives of sweetpotato and crop improvement

Sweetpotao is mainly breed for storage roots: storage ability, earliness, yield, nutritive value, eating quality, insect resistance, low temperature tolerance and disease resistance [61,66,67]. Using wild species for the improvement of sweetpotato started in 1956 with the introduction of K-123 from Mexico to Japan [35,68]. The accession K123 has been considered as *I. trifida* by some investigators [69,70]. The lines derived from K-123 received high attention due to improved starch content and resistance against two important Nematodes. Afterward genotype K-123 was used as a parent in the development of number of cultivars. In Japan, the cultivar Minamiyutaka was registered in 1975, which was developed using K-123 and it was found that this cultivar's one-eighth genetic background was from K-123. Furthermore, the cultivar "Minamiyutaka" surpassed the yield over the local cultivar at several locations in south Japan. High yield and starch content might be due to heterosis resulted in by the introgression of alien chromosomes [71,72]. Minamiyutaka also showed strong resistance to root knot nematode and root lesion nematode, tuberous roots of this variety have large storage tolerance than other cultivars. Same cultivar was used in China for improving yield and starch contents. This showed importance of wild germplasm in nematode resistance and starch improvement. Along with

K-123 many other wild species of *Ipomoea* were also introduced from Mexico, Colombia, Ecuador, Guatemala, and the United States. Researchers found many important disease and insect resistance traits in these wild relatives which further can be used for the improvement of sweetpotato [73–76]. The results influenced many researchers to obtain interspecific hybrids by wide crosses through controlled pollination or somatic cell culture. This significant achievement demonstrates the importance of exotic germplasm for the introduction of specific traits.

13.6.3 Production of interspecific hybrids by controlled pollination

Differences in ploidy levels and presence of interspecific compatibility among sweetpotato and its wild relatives make it difficult for researchers to obtain interspecific hybrids. Some successful attempts have been made by researchers to improve sweetpotato quality and disease resistance in B-genome group species [66,71,77]. Use of A-genome species have been restricted in sweetpotato improvement due to cross incompatibility although *I. triloba* which is predominantly found at both dry and wet conditions may be a rich source of crop improvement against drought tolerance, resistance to root rots, and some foliar fungal diseases [78,79]. Ting and coworkers [80] performed a series of crosses between sweetpotato and other 24 related wild species of genus *Ipomoea* and could not succeed to obtain even a single F_1 hybrid. Similarly, Wedderburn [78] also made number of crosses between *I. batatas* and *I. trichocrpa* or *I. gracilis* and hardly obtained developed embryo only when *I. batatas* was used as a male parent. The seeds produced were unviable. Iwanaga and coworkers [71] performed number of crosses among diploid and tetraploid species of *I. trifida*. The triploid obtained were further treated with colchicine to obtain hexaploid synthetics. It was the first report on production of 2n gametes in *Ipomoea* species. Earlier, Kobayashi [81] reported very low seed set in crosses between sweetpotato and diploid *I. trifida*. Although the ploidy level differences acted as a reproductive barrier, but it was not strong enough to completely restrict the gene flow between these related species.

Introgression of genes from *I. trifida* to sweetpotato was reported by many researchers. Orjeda and coworkers [82] developed 4X interspecific hybrids by crossing hexaploid *I. batatas* and diploid *I. trifida* for the improvement of sweetpotato. Significant yield improvement showed by these hybrids revealed the importance of development and utilization of these hybrids for commercial purpose. These hybrids have two major advantages over 6X hybrids. First, lower ploidy level (tetrasomic) inheritance is simpler than 6X inheritance. Second, gene flow between sweetpotato and related wild relatives become easier. Studies have also been conducted to improve agronomic traits and fertilities of these 4X populations to improve the chances of developing such cultivars of sweetpotato [32]. Komaki [75] obtained very low crossing rate between *I. trifida* and sweetpotato.

China being the biggest producer of sweetpotato contributes more than half in total world's production. However, quality and yield of storage yield of sweetpotato in China and some other sweetpotato producing countries is adversely affected by sweetpotato viruses and nematodes. Cao and coworkers [83] produced some novel interspecific hybrids between sweetpotato and two other diploid wild species *I. grandifolia* and *I. purpurea*. Their findings revealed that *I. grandifolia* is resistant to sweetpotato stem nematode (SSN) caused by *Ditylencus destructor* and moderately resistant to sweetpotato virus (SPV) while *I. purpurea* is highly resistant to both SSN and SPV. However, *I. purpurea* is included in *Ipomoea* section Pharbitis which is distantly related to sweetpotato [17]. Hybrids obtained by crossing these wild species yielded storage roots of intermediate size which were much larger than their wild parents. These results demonstrated that it is quite possible to integrate important characters in sweetpotato from these wild relatives.

13.6.4 Tetraploid sweetpotato and gene transfer

To overcome the barriers of gene transfer between sweetpotato and other wild species, Bohac and coworkers [84] discovered some wild tetraploid sweetpotatoes which is significant to researchers interested in sweetpotato improvement. These tetraploids observed as the same species as sweetpotatoes and enhancing the probability of successful gene introgression. These tetraploids also seem to overcome the obstacles resulted in by ploidy gap from hexaploid to tetraploid.

13.6.5 Interspecific hybrids by somatic cell hybridization

Somatic cell hybridization was attempted to produce hybrids between sweetpotato and its relatives. Liu and coworkers [85]) made first successful attempt and produced interspecific somatic hybrid between sweetpotato and *I. triloba* by fusing petiole protoplast of two species using polyethylene glycol (PEG) method. Wang and coworkers [86]) also used petiole protoplast fusion method to produce somatic hybrids between sweetpotato and *I. triloba*. Somatic hybridization between sweetpotato and its relatives was greatly advanced by utilization of embryogenic suspension culture for protoplast isolation. Using embryogenic suspension culture and petiole protoplast fusing method approximately, 5000 somatic hybrids between sweetpotato and its wild relatives (*I. lacunosa*) [87] and *I. cairica* [88]) have been produced.

Yang and coworkers [89] developed interspecific hybrids between sweetpotato and *I. triloba*. *I. triloba* is proved to have drought tolerance traits. Somatic hybrids obtained between *I. triloba* and sweetpotato cultivar Xushu 18, and between *I. triloba* and sweetpotato cultivar kokei No.14 produced storage roots and showed improved drought tolerance through analyzing water potential, proline concentration, SOD activity, and other important correlative physiological characters under the drought stress condition. Similarly, a somatic hybrid produced between *I. triloba* and sweetpotato cultivar Xushu18 produced storage roots. *I. cairica,* a distantly related wild relative of sweetpotato have shown the virus resistance characteristics, but this species cannot be used directly in the conventional virus resistance breeding programs due to its cross incompatibility with sweetpotato. Therefore, protoplasts obtained from embryogenic suspension cultures of sweetpotato cvs. Xushu 18 and Xu 55-2 were fused with leaf protoplasts of *I. cairica*. These somatic hybrids need further exploration for virus resistance.

13.6.6 Genome sequences of wild species and sources of genetic improvement

Poor understanding of molecular and physiological basis of important agronomic traits is a major constraint to sweetpotato improvement across the globe. As agronomic improvement of sweetpotato is lagging due to lack of genetic and genomic resources therefore, exploration of wild relatives of sweetpotato to discover the targets for genetic improvement is a way to move forward. Hirakawa and coworkeras [90] surveyed the genome of *I. trifida* and discovered important genes associated with starch biosynthesis in storage roots. Recently, Wu and coworkers [23] obtained genome sequences of diploid species *I. trifida* and *I. triloba* and found interesting insights in the genome of these wild relatives. They identified important gene families in *I. triloba* and *I. trifida* genome involved in stress resistance and adaptation in roots including genes which could be involved in storage root development in the hexaploid sweetpotato. In this study, Mwanga diversity panel which consisted of 16 sweetpotato cultivars, landraces, breeding lines, selected from various germplasm sources in some countries including Uganda was also used. This panel has been extensively used in breeding programs in different African countries. The panel was resequenced and aligned to reference assemblies derived from *I. trifida* and *I. triloba* to reveal alleles and genes associated with important agronomic traits.

Orange flesh type of sweetpotato is a rich source of β-carotene content and also highly promoted to combat vitamin A deficiency in sub-Saharan Africa. This panel is consisted of both orange fleshed and white accessions therefore it enables tagging genomic regions associated with high β-carotene contents along with other agronomic traits. In this study, various SNPs associated with flesh color were identified. These SNPs were present in the homologs of genes involved in key carotenoid biosynthetic enzymes including phytoene desaturase, phytoene synthase, lycopene β-cyclase, and ζ-carotene isomerase. These enzymes except lycopene β-cyclase showed increased expression in orange fleshed storage roots as compared to other root types which reveal their importance in development of orange fleshed high beta carotene sweetpotato.

References

[1] Khoury CK, Heider B, Castaneda-Alvarez NP, Achicanoy HA, Sosa CC, Miller RE, et al. Distributions, ex situ conservation priorities, and genetic resource potential of crop wild relatives of sweetpotato [*Ipomoea batatas* (L.) Lam., I. series *Batatas*. Front Plant Sci 2015;6:251. doi: 10.3389/fpls.2015.00251.

[2] Lebot V. Sweetpotato. In: Bradshaw JE, editor. Root and tuber crops, handbook of plant breeding, 7. New York: Springer Science Business Media, LLC; 2010. p. 97–125.

[3] FAOSTAT data. Available from: http://faostat.fao.org/; 2020.

[4] De Vries CA, Ferwerda JD, Flach M. Choice of food crops in relation to actual and potential production in the tropics. Neth J Agr Sci 1967;15:241–8.

[5] Kays SJ, Kays SE. Sweetpotato chemistry in relation to health. In: Proceedings of the sweetpotato production system towards the twenty-first century. Miyakonojo: Kyushu National Agricultural Experimental Station; 1998. p 231–272.

[6] Tumwegamire S, Kapinga R, Rubaihayo PR, LaBonte DR, Gruneberg WJ, Burgos G, et al. Evaluation of dry matter: protein, starch, β-carotene, iron, zinc, calcium, and magnesium in East African sweetpotato [*Ipomoea batatas* (L.) Lam] germplasm. Hort Sci 2011;46:348–57.

[7] Low JW, Arimond M, Osman N, Cunguara B, Zano F, Tschirley D. A food-based approach introducing orange-fleshed sweetpotatoes increased vitamin a intake and serum retinol concentrations in young children in rural Mozambique. J Nutr 2007;137:1320–7.

[8] Hotz C, Loechl C, Lubowa A, Tumwine JK, Ndeezi G, Nandutu Masawi A, et al. Introduction of β-carotene-rich orange sweetpotato in rural Uganda resulted in increased vitamin a intakes among children and women and improved vitamin a status among children. J Nutr 2012;142:1871–80. doi: 10.3945/jn.111.151829.

[9] Castañeda-Álvarez P, Khoury CK, Achicanoy HA, Bernau V, Dempewolf H, Eastwood RJ, et al. Global conservation priorities for crop wild relatives. Nat Plants 2016;2:16022.

[10] Dempewolf H, Eastwood RJ, Guarino L, Khoury CK, Müller JV, Toll J. Adapting agriculture to climate change: a global initiative to collect, conserve, and use crop wild relatives. Agroecol Sust Food 2014;38:369–77. doi: 10.1080/21683565.2013.870629.

[11] Hajjar R, Hodgkin T. The use of wild relatives for crop improvement: a survey of developments over the past 20 years. Euphytica 2007;156:1–13.

[12] Khoury C, Laliberte B, Guarino L. Trends in ex situ conservation of plant genetic resources: a review of global crop and regional conservation strategies. Genet Resour Crop Evol 2010;57:625–39. doi: 10.1007/s10722-010-9534-z.

[13] Nimmakayala P, Vajja G, Reddy UK. Ipomoea. In: Kole C, editor. Wild crop relatives: Genomic and breeding resourcesBerlin Heidelberg: Industrial Crops, Springer-Verlag; 2011. 123–113.

[14] Austin DF, Bianchini RS. Additions and corrections in American *Ipomoea* (Convolvulaceae). Taxon 1998;47:833–8.

[15] Austin DF, Staples GW, Simao-Bianchini R. A synopsis of *Ipomoea* (Convolvulaceae) in the Americas: Further corrections, changes, and additions. Taxon 2015;64:625–33.

[16] Austin DF. The *Ipomoea* batatas complex-I. Taxonomy. Bull Torrey Bot Club 1978;105:114–29.

[17] Austin DF. An infrageneric classification for *Ipomoea* (Convolvulaceae). Taxon 1979;28:359–61.

[18] Austin DF, Huáman Z. A synopsis of *Ipomoea* (Convolvulaceae) in the Americas. Taxon 1996;45:3–38.

[19] McDonald JA, Austin DF. Changes and additions in Ipomoea section *Batatas* (Convolvulaceae). Brittonia 1990;42:116–20.

[20] Munoz-Rodriguez P, Carruthers T, Wood JRI, Williams BRM, Weitemier K, Kronmiller B, et al. Reconciling conflicting phylogenies in the origin of sweetpotato and dispersal to Polynesia. Curr Biol 2018;28:1246–56.

[21] Austin DF. The taxonomy, evolution and genetic diversity of sweetpotatoes and related wild species. In: Gregory P, editor. Exploration, maintenance and utilization of sweetpotato genetic resourcesLima, Peru: International Potato Center; 1988. p. 27–60.

[22] Zhang DP, Rossel G, Kriegner A, Hijmans R. AFLP assessment of diversity from Latin America and the Pacific region: Its implications on the dispersal of the crop. Genet Resour Crop Evol 2004;51:115–20.

[23] Wu S, Lau KH, Cao Q, Hamilton JP, Sun H, Zhou C, et al. Genome sequences of two diploid wild relatives of cultivated sweetpotato reveal targets for genetic improvement. Nat Com 2018;9:4580. doi: 10.1038/s41467-018-06983-8.

[24] Woolfe JA. Sweetpotato, an untapped food resourceCambridge, UK: Cambridge University Press; 1992. p. 643.

[25] Perry L. Starch granule size and the domestication of manioc (*Manihot esculenta*) and sweetpotato (*Ipomoea batatas*). Econ Bot 2002;56:335–49.

[26] Jones A, Deonier MT. Interspecific crosses among *Ipomoea lacunosa, I. ramoni, I. trichocarpa* and *I. triloba*. Bot Gaz 1965;126:226–32.

[27] Nishiyama I, Miyazaki T, Sakamoto S. Evolutionary autoploidy in the sweetpotato (*Ipomoea batatas*) and its progenitors. Euphytica 1975;24:197–208.

[28] Oracion MZ, Niwa K, Shiotani I. Cytological analysis of tetraploid hybrids between sweetpotato and diploid *Ipomoea trifida* (H.B.K.) Don. Theor Appl Genet 1990;80:617–724.

[29] Jarret RL, Gawel N, Whittemore A. Phylogenetic relationship of the sweetpotato (*Ipomoea batatas* (L.) Lam.). J Am Soc Hortic Sci 1992;117:633–47.

[30] Roullier C, et al. Disentangling the origins of cultivated sweetpotato (*Ipomoea batatas* (L.) Lam. PLoS ONE 2013;8:e62707.

[31] Shiotani I, Kawase T. Synthetic hexaploids derived from wild species related to sweetpotato. Jpn J Breed 1987;37:367–76.

[32] Orejda J, Freyre R, Iwanaga M. Production of 2n pollen in diploid *I. trifida*, a putative wild ancestor of sweetpotato. J Hered 1990;81:462–7.

[33] Freyre R, Iwanga M, Orjeda G. Use of Ipomoea trifida (H.B.K.) G. Don. germplasm for sweetpotato improvement. 2. Fertility of synthetic hexaploids and triploids with 2n gametes of I. trifida, and their interspecific crossability with sweetpotato. Genome 1991;34:209–14.

[34] Lopez-Lavalle LAB, Orjeda G. Occurrence and cytological mechanism of 2n pollen formation in a tetraploid accession of *Ipomoea batatas* (sweetpotato). Heredity 2002;93:185–92.

[35] Nishiyama I. Evolution and domestication of the sweetpotato. Bot Mag Tokyo 1971;84:377–87.

[36] Kriegner A, Cervantes JC, Burg K, Mwanga ROM, Zhang DA. Genetic linkage map of sweetpotato [*Ipomoea batatas* (L.) Lam.] based on AFLP markers. Mol Breed 2003;11: 169–85.

[37] Roullier C, Duputie A, Wennekes P, Benoit L, Fernandez Bringas VM, Rossel G, et al. Disentangling the origins of cultivated sweetpotato (*Ipomoea batatas* (L.) Lam.). PloS ONE 2013;8(5):e62707. https://doi.org/10.1371/journal.pone.0062707.

[38] Srisuwan S, Sihachakr D, Siljak-Yakovlev S. The origin and evolution of sweetpotato (*Ipomoea batatas* Lam.) and its wild relatives through the cytogenetic approaches. Plant Sci 2006;171:424–33.

[39] Huang JC, Sun M. Genetic diversity and relationships of sweetpotato and its wild relatives in *Ipomoea* series *Batatas* (Convolvulaceae) as revealed by inter-simple sequence repeat (ISSR) and restriction analysis of chloroplast DNA. Theo Appl Genet 2000;100:1050–60.

[40] Buteler M, Jarret R, LaBonte D. Sequence characterization of microsatellites in diploid and polyploid *Ipomoea*. Theor Appl Genet 1999;99:123–32.

[41] Hijmans, Huaccho RJL, Zhang DP. Global distribution of sweetpotato. CIP Program Report. 1999–2000.

[42] Gichuki ST, Berenyi M, Zhang D, Hermann M, Schmidt J, Glössl J, et al. Genetic diversity in sweetpotato (*Ipomoea batatas* (L.) Lam.) in relationship to geographic sources as assessed with RAPD markers. Genet Resour Crop Evol 2003;50:429–37.

[43] Huaman Z, De la Puente F. Development of a sweetpotato genebank at CIP. CIP Circ 1988;16:1–7.

[44] Huamán Z, Aguilar C, Ortiz R. Selecting a Peruvian sweetpotato core collection on the basis of morphological, eco-geographical, and disease and pest reaction data. Theor Appl Genet 1999;98:840–5.

[45] Austin DF. *Ipomoea littoralis* (Convolvulaceae)—taxonomy, distribution and ethnobotany. Econ Bot 1991;45:251–6.

[46] Abel WE, Austin DF. Introgressive hybridization between *Ipomoea trichocarpa* and *I. lacunosa* (*Convolvulaceae*). Bul Torrey Bot Club 1980;108:231–9.

[47] Fosberg FR, Sachet MH. Flora of Micronesia. 3. Convolvulaceae. Contrib Bot 1977;36:1–34.

[48] Heine H. Florede la Nouvelle-Caledonie et dependances, 13. ConvolvulaceaeMuseum National D' Histoire Naturelle Paris; 1984. p. 1–91.

[49] Stone B. Flora of Guam. Micronesica 1970;6:1–659.

[50] Chang C. Flora of Taiwan. Convolvulaceae. Epoch Publ 1978;4:347–89.

[51] Cralb WG, Kerr AFGSiam Society Bangkok. . Florae siamensis enumeratio, 3. Bangkok Times Press; 1954. p. 13.

[52] Smitinand T. Thai plant names. Privately published, Funny Publ Ltd Bangkok; 1980.

[53] Low J, Lynam J, Lemaga B, Crissman C, Barker I, Thiele G, et al. Sweetpotato in Sub-Saharan Africa. In: Loebenstein G, Thottappilly G, editors. The sweetpotatoAmsterdam: Springer; 2009. p. 359–90.

[54] Mwanga R, Ssemakula G. Orange-fleshed sweetpotatoes for food, health and wealth in Uganda. IJSA 2011;9:42–9. doi: 10.3763/ijas.2010.0546.

[55] Sutherland JA. A review of the biology and control of the sweetpotato weevil, cylas formicarius elegaluntus (F). Trop Pest Manage 1986;32:304–15. doi: 10.1080/09670878609371084.

[56] Ngailo S, Shimelis H, Sibiya J, Mtunda K. Sweetpotato breeding for resistance to sweetpotato virus disease and improved yield: progress and challenges. Afr J Agric Res 2013;8:3202–15.

[57] Okada Y, Saito A, Nishiguchi M, Kimura T, Mori M, Hanada K, et al. Virus resistance in transgenic sweetpotato [*Ipomoea batatas* L. (Lam)] expressing the coat protein gene of sweetpotato feathery mottle virus. Theor Appl Genet 2001;103:743–51.

[58] Clark CA, Holmes GJ, Ferrin DM. Major fungal and bacterial diseases. In: Loebenstein G, Thottappilly G, editors. The sweetpotatoNetherlands: Springer; 2009. p. 81–103.

[59] Loebenstein G, Thottappilly G, Fuentes S, Cohen J. Virus and phytoplasma diseases. In: Loebenstein G, Thottappilly G, editors. The sweetpotatoSpringer, Sci Bussiness Media; 2009. p. 105–34.

[60] Sorensen KA. Sweetpotato insects: Identification, biology and management. In: Loebenstein G, Thottappilly G, editors. The sweetpotatoSpringer, Sci Bussiness Media; 2009. p. 161–88.

[61] Yada B, Alajo A, Ssemakula GN, Mwanga ROM, Brown-Guedira G, Yencho GC. Selection of simple sequence repeat markers associated with inheritance of sweetpotato virus disease resistance in sweetpotato. Crop Sci 2017;57:1421–30.

[62] Gibson RW, Mpembe I, Alicai T, Carey EE, Mwanga ROM, Seal SE, et al. Symptoms, aetiology and serological analysis of sweetpotato virus disease in Uganda. Plant Pathol 1998;47:95–102.

[63] Karyeija R, Gibson R, Valkonen J. The significance of sweetpotato feathery mottle virus in subsistence sweetpotato production in Africa. Plant Dis 1998;82:4–15.
[64] Tairo F, Kullaya A, Valkonen JPT. Incidence of viruses infecting sweetpotato in Tanzania. Plant Dis 2004;88:916–20. doi: 10.1094/PDIS.2004.88.9.916.
[65] Feng G, Yifu G, Pinbo Z. Production and deployment of virus free sweetpotato in China. Crop Prot 2000;19:105–11.
[66] Iwanaga M. Use of wild germplasm for sweetpotato breeding. In: Gregory P, editor. Exploration, maintenance, and utilization of sweetpotato genetic resourcesInternational Potato Center Lima; 1988. p. 199–210.
[67] Kuranouchi T, Nishinaka M, Takada A, Kumagai T, Nakatani M, Katayama K. Recent progress in breeding low-temperature-tolerant sweetpotato Lines in Japan. Japan Agric Res Quart 2020;54:1–6.
[68] Nishiyama I, Teramura T. Mexican wild forms of sweetpotato. Econ Bot 1962;16:305–14.
[69] Shiotani I, Kawase T. Synthetic hexaploids derived from wild species related in *Ipomoea trifida*. Jpn J Breed 1987;39:57–66.
[70] Shiotani I. Genomic structure and the gene flow in sweetpotato and related species. Exploration and maintenance and utilization of sweetpotato genetic resources. First planning conferenceLima, Peru: International Potato Centre (CIP) Lima; 1988. p. 61–73.
[71] Iwanaga M, Freyre R, Orjeda G. Use of *Ipomoea trifida* (H.B.K.) G. Don. germplasm for sweetpotato improvement. 1. Development of synthetic hexaploids of I. trifida by ploidy-level manipulations. Genome 1991;34:201–8.
[72] Sakamoto S. Utilization of related species on breeding of sweetpotato in Japan. Japan Agri Res Quart 1970;5:1–4.
[73] Komaki K, Katayama K. Root thickness of diploid *Ipomoea trifida* (H. B. K.) G. Don and performance of progeny derived from the cross with sweetpotato. Breed Sci 1999;49: 123–9.
[74] Sakamoto S. Breeding of a new sweetpotato variety. Minarniyutaka, by the use of wild relatives. Japan Agric Res Quart 1976;10:183–6.
[75] Komaki K. Breeding value of wild species closely related to sweetpotatoKorea: Proceedings of international workshop on production, utilization and development of sweetpotato; 2004. p. 164–72.
[76] Zhang SS, Liu LF. Utilization of the wild relatives in sweetpotato breeding. In: Liu QC, editor. Sweetpotato breeding and industrialization in China, Beijing: China Agricultural University Press; 2005. p. 29–32.
[77] Hozyo Y, Kato S. The plant production of wild type plants in *lpomoea trifida* (H.B.K.) Don. Tokyo Natl Inst Agric Sci Bull Ser D 1973;24:35–60.
[78] Wedderburn MM. A study of hybridization involving the sweetpotato and related species. Euphytica 1967;16:69–75.
[79] Martin FW, Jones A. The species of *Ipomoea* closely related to the sweetpotato. Econ Bot 1973;26:201–15.
[80] Ting YC, Kehr AE, Miller JC. A cytological study of the sweetpotato plant (*Ipomoea batatas* (L.) Lam.) and its related species. Am Nat 1957;91:197–203.
[81] Kobayashi RS, Bouwkamp JC, Sinden SL. Interspecific hybrids from cross incompatible relatives of sweetpotato. Euphytica 1994;80:159–64.
[82] Orjeda G, Freyre R, Iwanaga M. Use of *Ipomoea trifida* germplasm for sweetpotato improvement. 3. Development of 4x interspecific hybrids between *Ipomoea batatas* (L.) Lam. ($2n = 6x = 90$) and *I. trifida* (H.B.K) G. Don. ($2n = 2x = 30$) as storage-root initiators for wild species. Theor Appl Genet 1991;83:159–63.

[83] Cao Q, Zhang A, Ma D, Li H, Li Q, Li P, et al. Novel interspecific hybridization between sweetpotato (*Ipomoea batatas* (L.) Lam.) and its two diploid wild relatives. Euphytica 2009;169:345–52.

[84] Bohac JR, Austin DF, Jones A. Discovery of wild tetraploid sweetpotatoes. Econ Bot 1993;47:193–201.

[85] Liu QC, Wang JX, Li WJ, Zhou HY. Protoplast fusion and regeneration of interspecific somatic hybrid plants between sweetpotato (*Ipomoea batatas* (L.) Lam.) and its related species. J Agr Biotechnol 1994;2:85–90.

[86] Wang JS, Liu QC, Meng XX, Wang WH, Xu LJ, Zhai H. Regeneration of interspecific somatic hybrids between sweetpotato and its wild relative *Ipomoea triloba*. J Agr Biotechnol 2003;11:40–3.

[87] Zhang BY, Liu QC, Zhai H, Zhou HY, Zhang DP, Wang Y. Production of fertile interspecific somatic hybrid plants between sweetpotato and its wild relative, *Ipomoea lacunosa*. Acta Hort 2002;583:81–5.

[88] Guo JM, Liu QC, Zhai H, Wang YP. Regeneration of plants from Ipomoea cairica L. protoplasts and production of somatic hybrids between *I. cairica* L. and sweetpotato, *I. batatas* (L.) Lam. Plant Cell Tiss Organ Cult 2006;87:321–7.

[89] Yang Y, Guo J, Guan S, Zhai H, Liu Q. Production and characteristics evaluation of interspecific somatic hybrid plants between sweetpotato and its wild relatives. In: Liu Q, editor. Proceedings of third China-Japan-Korea workshop on sweetpotatoChina Agriculture University Press; 2008. p. 55–6.

[90] Hirakawa H, Okada Y, Tabuchi H, Shirasawa K, Watanabe A, Tsuruoka H. Survey of genome sequences in a wild sweetpotato, *Ipomoea trifida* (H. B. K.) G. Don. DNA Res 2015;22(2):171–9.

Chapter 14

Generation of New Landraces of Forage Species: Red Fescue and Clover

Zeeshan Shamim[a], Humera Razzaq[b], Muhammad Naveed Shahid[c] and Muhammad Tariq Awan[a]
[a]Department of Biotechnology, Mirpur University of Science and Technology (MUST), Mirpur, Pakistan; [b]Department of Plant Breeding and Genetics, University of Agriculture, Faisalabad, Pakistan; [c]Department of Botany, Division of Science and Technology, University of Education, Lahore, Pakistan

14.1 Introduction

14.1.1 Forage

Crops have been an integral and essential part of human society since ages and extensively been used by human and animals for their need and survival. The crops, being used by animals for feeding, are referred as forage, and can be traced back to 1840s when human had started livestock domestication for domestic and beneficial uses [1]. Leaves and stem parts of the plant material make the forage for livestock in most of the cases. In past, forage only included pastures, immature cereal crops, or crop residues but now, the paradigm has been shifted by inclusion of whole crop systems like various legumes and grasses under the forage umbrella. These crops could be conserved/used in different ways for livestock, as hay or silage depending upon the state of crop and the demand of farmers, as it tends to release pressure on farmers during lean periods. It is often used as a synonym for fodder in field, but a striking difference between the two would help us understand in technical terms, fodders are certain crops those are used in stall feeding whereas forage is the vegetative material that could be preserved or used as fresh for animal feeding. Forage crops include legumes, crucifers, grasses, and other crops that are cultivated and used in various forms such as fodder hay, silage, and pasture. Some important forages are as follows:

Grasses

- Grass forages include:
- *Agrostis* spp.—bentgrasses
  - *Agrostis capillaris*—common bentgrass

Wild Germplasm for Genetic Improvement in Crop Plants. http://dx.doi.org/10.1016/B978-0-12-822137-2.00014-X

    - *Agrostis stolonifera*—creeping bentgrass
- *Andropogon hallii*—sand bluestem
- *Arrhenatherum elatius*—false oat-grass
- *Bothriochloa bladhii*—Australian bluestem
- *Bothriochloa pertusa*—hurricane grass
- *Brachiaria decumbens*—Surinam grass
- *Brachiaria humidicola*—koronivia grass
- *Bromus* spp.—bromegrasses
- *Cenchrus ciliaris*—buffelgrass
- *Chloris gayana*—Rhodes grass
- *Cynodon dactylon*—bermudagrass
- *Dactylis glomerata*—orchard grass
- *Echinochloa pyramidalis*—antelope grass
- *Entolasia imbricata*—bungoma grass
- *Festuca* spp.—fescues
    - *Festuca arundinacea*—tall fescue
    - *Festuca pratensis*—meadow fescue
    - *Festuca rubra*—red fescue
- *Heteropogon contortus*—black spear grass
- *Hymenachne amplexicaulis*—West Indian marsh grass
- *Hyparrhenia rufa*—jaragua
- *Leersia hexandra*—southern cutgrass
- *Lolium spp.*—ryegrasses
    - *Lolium multiflorum*—Italian ryegrass
    - *Lolium perenne*—perennial ryegrass
- *Megathyrsus maximus*—Guinea grass
- *Melinis minutiflora*—molasses grass
- *Paspalum conjugatum*—carabao grass
- *Paspalum dilatatum*—dallisgrass
- *Phalaris arundinacea*—reed canarygrass
- *Phleum pratense*—timothy
- *Poa* spp.—bluegrasses, meadow-grasses
    - *Poa arachnifera*—Texas bluegrass
    - *Poa pratensis*—Kentucky bluegrass
    - *Poa trivialis*—rough bluegrass
- *Setaria sphacelata*—African bristlegrass
- *Themeda triandra*—kangaroo grass
- *Thinopyrum intermedium*—intermediate wheatgrass

Legumes

- *Arachis pintoi*—pinto peanut
- *Chamaecrista rotundifolia*—roundleaf sensitive pea
- *Clitoria ternatea*—butterfly-pea
- *Lotus corniculatus*—bird's-foot trefoil
- *Macroptilium atropurpureum*—purple bush-bean

- *Macroptilium bracteatum*—burgundy bean
- *Medicago* spp.—medics
- *Medicago sativa*—alfalfa, lucerne
- *Medicago truncatula*—barrel medic
- *Melilotus* spp.—sweetclovers
- *Neonotonia wightii*—perennial soybean
- *Onobrychis viciifolia*—common sainfoin
- *Stylosanthes* spp.—stylo
- *Stylosanthes humilis*—Townsville stylo
- *Stylosanthes scabra*—shrubby stylo
- *Trifolium* spp.—clovers
- *Trifolium hybridum*—alsike clover
- *Trifolium incarnatum*—crimson clover
- *Trifolium pratense*—red clover
- *Trifolium repens*—white clover
- *Vicia* spp.—vetches
- *Vicia articulata*– oneflower vetch
- *Vicia ervilia*—bitter vetch
- *Vicia narbonensis*—narbon vetch
- *Vicia sativa*—common vetch, tare
- *Vicia villosa*—hairy vetch
- *Vigna parkeri*—creeping vigna

Others

- Acacia aneura—mulga
- Albizia spp.—silk trees
- Albizia canescens—Belmont siris
- Albizia lebbeck—lebbeck
- Enterolobium cyclocarpum—earpodtree
- Leucaena leucocephala—leadtree
- Alfalfa
- Maize (corn)
- Sorghums
- Oats
- Sorghum
- Corn or soybean

14.1.2 Landraces

Although it is quite precarious to characterize a landrace, it was implicitly defined landraces exploring its historic background. Many scientists have contributed toward defining it with their perspectives [2–16]. The most agreed definition proposed on the annual meeting of on farm conservation and management (see www.ecpgr.cgiar.org/Networks/Insitu_onfarm/OnfarmTF_intro.htm) is as following:

"*A landrace of a seed-propagated crop is a variable population, which is identifiable and usually has a local name. It lacks 'formal' crop improvement, is characterized by a specific adaptation to the environmental conditions of the area of cultivation (tolerant to the biotic and abiotic stresses of that area) and is closely associated with the uses, knowledge, habits, dialects, and celebrations of the people who developed and continue to grow it."*

Landraces develop as a result of non-anthropogenic activities and were presented naturally to some scientist while other argues that land races were created intentionally by human selection in past which got socio-cultural adherence and acceptance to a particular region. A generally perceived fact is that farmers grow normally landraces who have limited access to modern techniques and varieties whereas breeders grow cultivars to improve the agricultural output. Landraces are the deposits of gene pool that enhances the biodiversity and keep up the balance of our environment in a certain manner to make it more useful and beneficial. These are highly diverse, heterogeneous populations, and are genetically not uniformed but are well adapted to the local environments they live in. Cultivation of landraces sustain biodiversity in agriculture across the globe, in addition to benefit other needs of our environment, like, carbon sequestration, nutrient cycling, and control of soil erosion. Landraces are great sources of valuable genes responsible for efficient nutrients utilization and their uptake, adaptation to various environment stresses, hence provide important genetic resources for future breeding program to be utilized in other crops.

Moreover, landraces have been used extensively for crop improvement programs to broaden the genetic basis by creation of new cultivars [17,18]. These are heterogeneous [19], genetically not uniform and containing high levels of diversity [20], "local diverse crop varieties" [21], "heterogeneous crop populations" [4], "materials with variable levels of heterogeneity" (Ambrose, personal communication). It was indicated that the genetic diversity of landraces has two dimensions: between sites/populations, and within sites/populations [22].

14.2 Changing climate

The foreseen effects of environmental change on grassland and related ecosystem and related management practices have been assessed extensively [23,24]. Environmental changes at global level are probably going to move the adaptive areas for long term to the most of forage species. Thus, there is a need to recognize and fuse the significant adaptive qualities into existing and new forage species to maintain the grasslands and forage crops. Climate has always been one of the great abiotic threats to the crop biodiversity worldwide. Since the population is increasing exponentially and is determined to hit 9.2 billion mark in 2050, serious and concerted efforts are needed to counter the demand by improving existing germplasm, and forages can be exploited to meet the needs of animals for its part. Climate change is believed to affect the forages in various ways for which following considerations would help us achieve sustainability and stability:

- Improvement in drought tolerance and water use efficiency
- Salinity tolerance water logging tolerance

- Disease tolerance
- Efficient nutrient uptake
- Forage quality
- Introduction of new species or ecotypes

14.2.1 Impact of climate change

14.2.1.1 Increased CO_2 uptake

Increased plant canopy and soil breathing under raised CO_2 is associated with the total gross osmosis [25]. Purple moor grass (*Molinia caerulea* (L.) Moench) plants presented to increase CO_2 demonstrated a decrease in total leaf area, an expanded number of senescent leaves, and an increased rootstock development [26]. It is believed that increase CO_2 will lead to increase the net primary and will decrease canopy evapotranspiration because of reduced stomatal conductance. However, increase in yield of certain grasses has also recorded. So, overall this is ought to have both positive and negative impact on forages.

14.2.1.2 Increase temperature

Increase in temperature will have profound effects with increased production in the northern temperate zone to decrease production in dry land. Whereas winters are more likely to be affected as survival of the perennial crops will be at stake, although it's difficult to predict. Moreover, increase in temperature is associated with changes in duration of winter hardening periods. To survive the low temperature in winters, various crops have modified themselves and this includes the specialized sensitive tissues that can escape freezing or chilling and resumes their growth after the winter has passed. Thus, changing climate could adversely affect this property [27].

14.2.1.3 Unpredictable rainfall

Rainfall is a very important factor for crops in a land where water scarcity ruins the agriculture especially dry land. Since crops activity is highly impacted by rainfall, their production is also affected. Intensity, duration, and time of rainfall are extremely crucial for any crop development.

14.2.1.4 Variability due to environment

Increase in environmental variability, for example, certain crops will manage to cope with the environment and become resistant or tolerant by modification in their internal systems creating thus the variability as compare to others.

14.2.2 Potential of improvement in forages

Forage species are highly adapted to particular climatic regimes and at the middle of their life cycle, they show more endurance by surviving both extremes of temperature [28]. Adaptation is highly influenced by phenotypic and genotypic plasticity; the genotypic depends upon the survival of genotypes that makes up the population and the latter results from the interaction between the genotype and the environment. Dry season is a significant ecological factor constraining the efficiency of production of forages around the world. Environmental variation predicts more prominent variability in precipitation during expanded periods of summer dry spell that would influence numerous regions including fields. Ever increasing population will likewise requires a greater amount of food and the accessible water for household etc. [29]. The increasing demand for crops having drought tolerance has emphasized the forage plant breeders to select candidate genes which would help to cope with the unprecedented changes in future and improve crops.

With the recent advancement in breeding techniques and molecular approaches a lot of work is underway but to find the best potential reliable source has always been in question. Therefore, crop wild relatives and landraces are always the best option to look for certain genes as these have wider adaptability and variability having resistant genes.

14.3 New generation landraces of forages

This incorporates adjustment of synthetic landraces created by breeding of different local landraces to farming system either by intentionally or unintentionally using latest technologies. In this regard, a mixed framework may be set up in which farmers and breeders create new landraces from the shared and available materials. This is the new revisited concept that proposes the use of landraces with a modern twist, that is, breeding; biotechnological tools, molecular genetics, and many more for instances, back cross breeding, which leads to the production of new cultivars changing the landraces and hybrids as well. In forages, yield and climatic impact is great and with this revolutionary approach, we can undermine the problems that persist in the forages.

14.3.1 Red fescue

Festuca rubra (red fescue) belongs to graminoid species of plants of the family of true grasses, with an astonishing self supporting growth habit. Belonging to freshwater habitat, it exhibits plain, broad leaves and yellow inflorescence. Moreover, it is also a photoautotroph (Fig. 14.1).

Red fescue is tolerant to cutting or grazing and reacts by producing fine thick grass so is broadly used to make yards. Fescue turf is resistant to wear and tear and holds promise for valuable incorporation in our farming system. If the field is left without cutting this crop, it would cover densely and protect against soil deterioration. These can last for long and withstand nutrient deficiency and drought. It is highly diversified and has the capability to withstand the environmental stresses. Interestingly, these are utilized for heavy metals accumulation [30].

14.3.2 Clovers

As a crop of interest, clovers have potential of nitrogen fixation and have symbiotic relationship with bacteria to fix nitrogen. Clover forages are appropriate

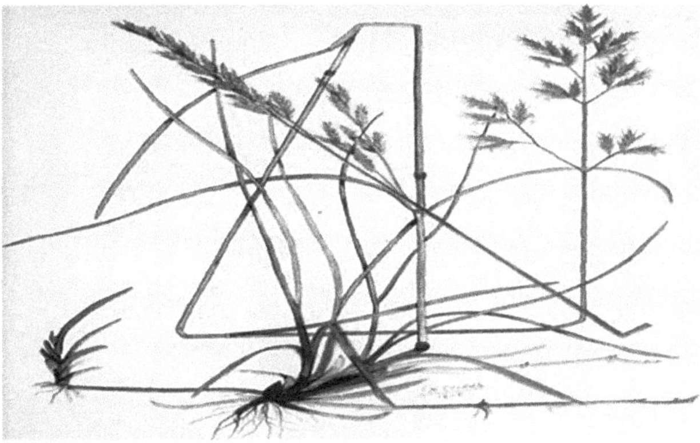

FIGURE 14.1 Tall red fescue high resolution stalk and leaves.

for reasonable low-info or natural dairying, since they require less nitrogen treatment, while giving higher protein content in feed, than pure feed grasses (Table 14.1). They are highly resistant to varying degree of stresses.

TABLE 14.1 Various forms of clovers being used to feed animals.

Picture	Title	Caption	Copyright
	Flower	*Trifolium hybridum* (Alsike clover); flowering stem. Vääna Landscape Reserve, nr Tallinn, Estonia. June, 2013.	Ivar Leidus-2013/ via Wikipedia—CC BY-SA 3.0
	Habit	*Trifolium angustifolium* (narrow-leaf clover); habit, showing flowers and leaves.	Dick Culbert-2008, Gibsons, British Columbia, Canada/ via Wikipedia—CC BY 2.0
	Habit	*Trifolium hybridum* (Alsike clover); flowering habit. Hohenlohe, Germany. July, 2011.	BerndH-2011/via Wikipedia—CC BY-SA 3.0

(Continued)

TABLE 14.1 Various forms of clovers being used to feed animals. (*Cont.*)

Picture	Title	Caption	Copyright
	Flowers	*Trifolium angustifolium* (narrow-leaf clover); flowers and leaves. Keila, Estonia. August 2013.	Ivar Leidus-2013, Estonia/via Wikipedia—CC BY-SA 3.0
	Flower.	*Trifolium hybridum* (Alsike clover); close view of flower. Kirchwerder, Hamburg, Germany. August, 2008.	Aiwok-2008/via Wikipedia—CC BY-SA 3.0
	Habit	*Trifolium hybridum* (Alsike clover); habit. Bozeman, Montana, USA. June, 2007.	Prof Matt Lavin-2012/Bozeman, Montana, USA—CC BY-SA 2.0
	Leaves and stem	*Trifolium hybridum* (Alsike clover); leaves and stem. Netherlands. October, 2007.	Jan Bakker-2007/ via Wikipedia (Rasbak)—CC BY-SA 3.0
	Leaf	*Trifolium hybridum* (Alsike clover); close-up of leaf. USA.	Bruce Ackley/The Ohio State University/Bugwood.org—CC BY 3.0 US
	Seeds	*Trifolium hybridum* (Alsike clover); seeds, en masse. Czech Republic. March, 2011.	Public Doman/ released by Pavel Šinkyr̆ík

Source: CABI.

14.4 Conclusion

Red fescue and clovers have great potential and can be utilized for improving genetic base of the landraces with the incorporation of essential genes into them so that they can help to improve the status of forages in our region. All this would help toward minimizing our dependence on import and would surely improve livestock health. In the coming era of global warming valuable genes could be worked on to enhance drought endurance and increasing yield. Such measures would be of great interest to farmers and human being.

References

[1] Horrocks RD, Valentine JF. Harvested foragesAcademic Press; 1999. p. 293–314.

[2] Anderson E, Cutler HC. Races of *Zea mays*: I. Their recognition and classification. Ann Missouri Bot Garden 1942;29(2):69–88.

[3] Asfaw Z. The barleys of Ethiopia. In: Brush SB, editor. Genes in the fieldIPGRI; 2000. p. 77–107.

[4] In: Brown AH, Frankel OH, Marshall DR, Williams JT, editors. The use of plant genetic resourcesCambridge University Press; 1989.

[5] Brush SB. Ethnoecology, biodiversity and modernization. J Ethnobiol 1992;12(2):161–85.

[6] Brush SB. *In situ* conservation of landraces in centers of crop diversity. Crop Sci 1995;35(2):346–54. https://books.google.co.uk/books?hl=en&lr=&id=c1f09mKWxNMC&oi=fnd&pg=PA3&dq=6.%09Brush+SB.+The+issues+of+in+situ+conservation+of+crop+genetic+resources.+Genes+in+the+field:+On-farm+conservation+of+crop+diversity.+2000:3-26&ots=K1KgV-ch7y&sig=wgnXJSwW_ilOGGzm0TJZR_jp4-0#v=onepage&q&f=false.

[7] Friis-Hansen E, Sthapit BR, editors. Participatory approaches to the conservation and use of plant genetic resourcesBiodiversity International; 2000.

[8] Harlan JR. Our vanishing genetic resources. Science 1975;188(4188):618–21.

[9] Harlan JR. Crops and man. American Society of Agronomy. Crop Science Society of America, Madison, Wisconsin 1992;16(2):63–262.

[10] Negri V. Landraces in central Italy: where and why they are conserved and perspectives for their on-farm conservation. Genet Res Crop Evol 2003;50(8):871–85.

[11] Negri V. Agro-biodiversity conservation in Europe: ethical issues. J Agric Environ Ethics 2005;18(1):3–25.

[12] Papa C. The 'farre de Montelione': landrace and representation. In: Hulled wheats: Promoting the conservation and use of underutilized and neglected crops. Proceedings of the first international woskshop on hulled wheats. Castelvecchio Pascoli, Italy: IPGRI; 1995. p. 154–171.

[13] Papa C. Il farro a Monteleone di Spoleto: pratiche agrarie, consuetudini giuridiche e ritualità. Il Farro. Saperi, usi e conservazione delle varietalocali. Quaderni del CEDRAV 1999;1:9–26.

[14] Saxena S, Singh AK. Revisit to definitions and need for inventorization or registration of landrace, folk, farmers' and traditional varieties. Curr Sci 2006;91(11):1451–4.

[15] Villa TC, Maxted N, Scholten M, Ford-Lloyd B. Defining and identifying crop landraces. Plant Genet Resour 2005;3(3):373–84.

[16] Zeven AC. Landraces: a review of definitions and classifications. Euphytica 1998;104(2):127–39.

[17] Frankel OH. Natural variation and its conservation. Genetic diversity in plants, Boston, MA: Springer; 1977. p. 21–44.

[18] Frankel OH, Brown AH, Burdon JJ. The conservation of plant biodiversity. Cambridge University Press; 1995.

[19] National Research Council. Managing global genetic resources. U.S. Natl. Plant Germplasm System, Natl. Acad. Sci., Washington, D.C. 1991.

[20] FAO. Technical meeting on the methodology of the world information and early warning system on plant genetic resources. FAO Report. Available from: www.apps3.fao.org/views/prague/technical; 1999.

[21] Brush SB. The issues of in situ conservation of crop genetic resources. Genes in the field: On-farm conservation of crop diversity; 2000. p. 3–26.

[22] Frankel O, Soulé ME. Conservation and evolution. Cambridge University Press; 1981.

[23] Del Prado A, Merino P, Estavillo JM, Pinto M, González-Murua C. N 2 O and NO emissions from different N sources and under a range of soil water contents. Nutr Cycl Agroecosys 2006;74(3):229–43.

[24] Nosberger J, Blum H, Fuhrer J. Crop ecosystem responses to climatic change: productive grasslands. In: Hodges HF, editor. Climate change and global crop productivityWallingford, UK: CAB International; 2000. p. 271–91.

[25] Casella E, Soussana JF. Long-term effects of CO_2 enrichment and temperature increase on the carbon balance of a temperate grass sward. J Exp Bot 1997;48(6):1309–21.

[26] Franzaring J, Högy P, Fangmeier A. Effects of free-air CO_2 enrichment on the growth of summer oilseed rape (*Brassica napus* cv. Campino). Agric Ecosys Environ 2008;128(1–2): 127–34.

[27] Guy CL. Cold acclimation and freezing stress tolerance: role of protein metabolism. Ann Rev Plant Biol 1990;41(1):187–223.

[28] Nelson CJ, Moser LE. Plant factors affecting forage quality. Forage quality, evaluation, and utilizationAmerican Society of Agronomy, Inc.; 1994. p. 115–54.

[29] Condon AG, Richards RA, Rebetzke GJ, Farquhar GD. Breeding for high water-use efficiency. J Exp Bot 2004;55(407):2447–60.

[30] Dradrach A, Karczewska A, Szopka K. Arsenic accumulation by red fescue (*Festuca rubra*) growing in mine affected soils-Findings from the field and greenhouse studies. Chemosphere 2020;248:126045.

Chapter 15

Chickpea Wild Relatives: Potential Hidden Source for the Development of Climate Resilient Chickpea Varieties

Melike Bakir[a], Duygu Sari[b], Hatice Sari[b], Muhammad Waqas[c] and Rana Muhammad Atif[c,d]

[a]*Department of Agricultural Biotechnology, Faculty of Agriculture, Erciyes University, Kayseri, Turkey;* [b]*Department of Field Crops, Faculty of Agriculture, Akdeniz University, Antalya, Turkey;* [c]*Centre for Advanced Studies in Agriculture and Food Security, University of Agriculture, Faisalabad, Pakistan;* [d]*Department of Plant Breeding and Genetics, University of Agriculture, Faisalabad, Pakistan*

15.1 Introduction

Chickpea (*Cicer arietinum* L.) is a self-pollinating diploid (2n = 2X = 16) species with a genome size of 738 Mb [1]. According to Food and Agriculture Organization's statistical database (FAOSTAT) in 2018, chickpea has the widest sowing area among cool-season food legumes, while it is the fourth rank after soybean, groundnut, and common bean among food legumes [2]. It is an important source of protein in the human diet and animal feed [3], and it is also used as a cover crop, intercrop, and rotation crop due to the ability to fix atmospheric nitrogen into the soil [4]. Well-known two distinctive chickpeas namely (1) kabuli chickpea with larger seed size, non-pigmented plants, and flowers and (2) desi chickpea with smaller seed size, pigmented plants, and flowers are sown from temperate to subtropics and tropics, and grown from sea level to 110–2400 m [5] in 50 countries in the world [2]. Although chickpea is adapted to marginal areas where drought and heat stresses are prevailing, it is suffered from several environmental stresses including drought, salinity, high and low temperatures [6–8]. The major barrier affecting the productivity of chickpea is forecasted to be unpredictable stressors due to climate changes [9,10]. Like the other domesticated plant species, domesticated chickpeas have already been absent from special genes that tolerate a/biotic stresses during domestication and post-domestication [11,12]. In contrast to domesticated chickpea, wild *Cicer*

Wild Germplasm for Genetic Improvement in Crop Plants. http://dx.doi.org/10.1016/B978-0-12-822137-2.00015-1

species does not only possess special suppressed genes to create transgressive segregations when crossed with domesticated chickpea [13–16], but they are also resistant to a/biotic stressors [16–29]. In the present chapter, we therefore reviewed environmental stresses hampered on the yield of chickpea and updated potential hidden resources of resistance to these stressors to improve climate-resilient chickpea varieties.

15.2 The genus *Cicer* L.

The genus *Cicer* L. is composed of 49 taxa of which 9 are annual and 40 are perennial and separated into 4 sections as of Monocicer, Chamaecicer, Polycicer, and Acanthocicer [14,30,31]. Among nine annual species of section Monocicer, *Cicer arietinum* L. is the only domesticated one. The remaining eight annual species consisting of *C. reticulatum* Ladiz., *C. echinospermum* P. H. Davis., *C. pinnatifidum* Jaub. & Sp., *C. judaicum* Boiss., *C. bijugum* K.H. Rech., *C. cuneatum* Hochst. ex Rich., *C. chorassanicum* (Bunge) M. Pop., and *C. yamashitae* Kitamura are wild relatives of domesticated chickpea. Annual species namely *C. chorassanicum* and perennial species namely *C. incisum* K. Mally were placed in section Chamaecicer, 23 perennial species were placed in section Polycicer and 7 woody perennial species were placed in section Acanthocicer [32].The numbers of taxa in the genus were updated to be 45 taxa by van der Maesen and coworkers [33] and then 49 taxa have been reported with new subspecies and varieties [34–36].

Although 49 taxa in the genus have been detailed in the present chapter, only 2 species namely *C. echinospermum* and *C. reticulatum* can be crossed with domesticated chickpea [14,15,37,38]. According to the introgression of desirable alien gene(s) from wild species to domesticated plants, each crop and related plant species was classified in gene pools by Harlan and de Vet [39]. Crossing in the primary gene pool (GP 1) is easy, the progeny is usually fertile due to good chromosome pairing, and gene transfer is possible from wild to domesticated with traditional methods. In this sense, GP 1 consists of *C. reticulatum* and *C. echinospermum* but the integration of desirable gene(s) from *C. reticulatum* to domesticated chickpeas including kabuli and desi chickpeas is easy in general, while it depends on accessions of *C. echinospermum* [40]. Rate of successful hybrids between domesticated chickpea and accessions of *C. echinospermum* may be dependent on genetically affinity to *C. reticulatum* because Talip and coworkers [41] outlined two different plant morphologies in accessions of *C. echinospermum* as "echinate spiny" and "warty" seed coat. The secondary gene pool (GP 2) contains *C. bijugum, C. judaicum,*and *C. pinnatifidum* (Table 15.1). Species in the GP 2 can be crossed with domesticated chickpeas and produced at least some fertile progeny, while hybrids are weak, partly sterile, and recovery of progeny in the subsequent generation is difficult due to post-fertilization problem [42]. Although embryo rescue is therefore needed [43], achievement deserves attention since most of the resistance resources of wild *Cicer* species

TABLE 15.1 Gene pool concept in the genus *Cicer* L.

Primary gene pool	Secondary gene pool	Tertiary gene pool
C. arientinum	*C. bijugum*	*C. chorassanicum*
C. reticulatum	*C. pinnatifidum*	*C. yamashitae*
C. echinospermum	*C. judaicum*	*C. cuneatum*
		C. canariense A. Santos & G.P. Lewis
		C. anatolicum Alef.

belong to GP 2 (Table 15.2). The tertiary gene pool (GP 3) consists of the other perennial wild species, and gene transfer from these species to domesticated chickpeas in GP 3 is impossible with nowadays knowledge. Species in the GP1, GP 2, and GP 3 can be inter-mated and successful bridge crosses and interspecific crosses in GP 2 and GP 3 was reported by Abbo and coworkers [44].The use of rather extreme radical methods can be an aid to rescue embryo. Mutagen treatments to pollen and polyploidy in F_1 progeny and bridge crosses between species may be solved post-zygotic problems. Most of the perennial species have not been attempted to cross with domesticated chickpea so far and species having potential traits in perennials should be considered in short term breeding programs.

The progenitor of domesticated chickpea is considered to be *C. reticulatum* [11,37]. The oldest carbonized seed samples of *C. reticulatum* were dated to 7260 BC from Tell-el-Kerkh, Syria [45]. According to archeological findings, distribution of progenitor, and genetic diversity in the wild species, chickpea was reported to be originated from Northern Syria and Southeastern Turkey [11,37,44,46,47]. *C. reticulatum* and *C. echinospermum*that is the closest species after progenitor to domesticated chickpea are endemic species to Turkey [36,48,49]. A mission to collect for expanding the number of accessions of *C. reticulatum* and *C. echinospermum*was initiated in 2013 in South-East Turkey [36], and 763 new accessions were collected [41]. All these evidences indicated that chickpea may probably be domesticated before 10,000 years ago. The annual wild *Cicer* species are mostly distributed from east to west (from Afghanistan to Turkey) and from north to south (from Turkey to the Middle East and Ethiopia), while perennial wild species were found in the Himalayan region, Central Asia (Turan), Caucasian and Balkan regions to Atlas mountains, and Canary island [30,47].

15.3 Effect of climate changes on chickpea

Abiotic stress term contains many stresses brought about by multiple environmental cases, such as strong light, UV, high and low temperature, drought, salinity, nutritional imbalance. According to reports from the Intergovernmental

TABLE 15.2 Resources for resistance to abiotic stresses in *Cicer* species.

Species	Stresses	References
C. bijugum K.H. Rech.	Cold	[16]
	Cold	[150]
	Cold	[22]
	Cold	[125]
	Cold	[24]
	Drought	[124]
	Drought	[123]
C. echinospermum P.H. Davis	Cold	[125]
	Cold	[16]
	Cold	[150]
	Cold	[24]
	Cold	[117]
C. pinnatifidum Jaub. & Sp.	Cold	[16]
	Cold	[150]
	Cold	[125]
	Cold	[24]
	Drought and heat	[26]
C. reticulatum Ladiz.	Cold	[150]
	Cold	[125]
	Cold	[24]
	Cold	[117]
	Drought and heat	[26]
	Drought	[124]
	Salinity	[195]
C. anatolicum Alef.	Drought and heat	[25]
C. microphyllum Benht.	Cold	[67]
	Drought and heat	[25,59]
C. montbretii Jaub et. Sp.	Drought and heat	[25,59]
C. oxydon Boiss. et Hoh.	Drought and heat	[25,59]
C. songaricum Steph ex DC.	Drought and heat	[25,59]

Panel of Climate Change [50], the effects of stresses will be increased in near future due to global climate change. Therefore, the knowledge of abiotic stress responses is presently considered to be one of the most essential issues in crop production.

Among abiotic stresses, terminal drought, high temperature, cold sensitivity, and salinity are important abiotic stresses that caused yield decreases in chickpea [51,52]. Abiotic stresses enhanced by climate change will affect chickpea production and yield and result in seed yield decrease of as much as 19% [8]. Drought and heat stresses are major limitations for agricultural production in the context of climate change [53] and lead to 50% of the yield losses in crop plants [6]. In chickpea, drought stress causes 40%–45% yield losses in the world [54] and heat stress at the reproductive stages is also an important factor in chickpea yield losses due to abortion or retardation of flowering and prevention pod development [6,55]. It has been noticed that every 1°C temperature increase above the optimum temperature leads to 10%–17% yield loss in chickpea [56,57]. This demonstrates that extreme heat and drought stress are the main factors that affect chickpea production. Low temperatures, which are classified chilling and freezing, are the other important temperature factor limiting chickpea production. Chilling temperatures during the reproductive stage of chickpea are detrimental in reproductive stages and cause up to 30%–40% yield loss [55]. On the other hand, winter-sown chickpea during the vegetative stage of chickpea exposes to freezing stress [58,59]. Freezing temperatures result in significant yield losses in chickpea at the late vegetative stage [22]. Soil salinity is the other important problem of arid and semi-arid areas since it is related to temperature and precipitation [60]. Under greater temperature and less humidity, evaporation may enormously increase. Therefore, the final movement of water remains upward during a great part of the year [61]. Thus, it is expected that soil salinity problem will increase the future under changing climate conditions determined by global warming. Chickpea is very sensitive to salinity and salinity stress affects germination, plant establishment, N_2 fixation, growth, flowering and inhibits pod formation, and seed filling in chickpea [59,62,63]. Global yield losses of chickpea attributed to salinity stress are between 8% and 10% [63].

For developing the resistant crops to changing environmental conditions, agro-morphological adaptive traits are fairly essential in plant breeding [41,64,65]. Adaptive traits display their adjusting resilience in different climate change conditions and resistance to biotic and abiotic stress conditions [66]. Wild relatives are major sources when looking for valuable adaptive traits or genes. It is very important to reveal the desirable traits in *C. reticulatum* and *C. echinospermum* for chickpea breeding, because these species are easily crossable with the cultivated species. So, agro-morphological traits were investigated of new germplasm sources of *C. reticulatum* and *C. echinospermum* and showed resistance to biotic and abiotic stresses by Talip and coworkers [41]. They suggested that the most remarkable agro-morphological traits were canopy width, number of branches, pods per plant, and biological yield especially seed yield that decreased under climate change conditions.

To summarize, drought, high and low temperatures, and salinity stresses enhanced by climate change have a destructive effect on chickpea production.

On the other hand, tolerance of wild species of *Cicer* against a/biotic stressors has demonstrated in previous studies [17–25,27,28,67–73]. Therefore, wild relatives of chickpea have vital importance to enrich the domesticated varieties and can be employed as a source to be gained tolerance to the domesticated varieties.

15.3.1 Drought stress

Drought (water stress) is one of the most significant environmental stresses and occurs for different causes, including scanty rainfall, high temperatures, and high intensity of light. One of the main abiotic stresses causing globally yield loss is drought in chickpea. Chickpea is largely grown in arid and semi-arid areas in the world [74]. Chickpea is exposed to decreasing soil moisture and temperature increase during flowering and pod filling. Chickpea faces two types of drought such as terminal and intermittent. Moreover, terminal and intermittent drought are the most responsible for chickpea yield loss [75]. Terminal drought occurs due to the temperature increase and the gradual decrease of precipitation as the season progress after the rainy season, and the effect of drought increases toward flowering and maturity stages. Intermittent or unpredictable drought occurs due to insufficient rainfall in the season, but the occurrence of intermittent drought is not predictable [76,77].

The most sensitive developmental periods of chickpea during the drought period are flowering and pod filling stages [78]. Drought can weaken pollination and decreases pollen germination, pollen viability, and fertility [79,80]. Due to the lack of water, drying occurs in the soil and this causes some changes in the plant root and soil. Soil dryness restricts the regular growth of the node root system, for this reason, can prevent the plant from getting enough water and nutrients. Drought also disrupts symbiotic nitrogen (N_2) fixation and causes a serious loss of yield in chickpea [81]. Although it varies according to the region and season grown, drought reduces chickpea yield up to 50% [25,59,82].

15.3.1.1 Drought stress tolerance mechanisms

Chickpea has different mechanisms to overcome drought stresses including escape, avoidance, and tolerance [9,83]. The mechanism of drought escape is based on completing the life cycle by accelerating phenological events before the plant is exposed to drought stress [83,84]. Chickpea can escape drought stresses thanks to early flowering, early podding, and early maturity [26,85,86]. Flowering time and vegetation periods are the most considerable traits for escaping drought in chickpea [87,88].

Drought avoidance mechanism is based on minimizing water loss and using water most effectively [83]. While dehydration avoidance is related with the size, depth, length, and density of plant roots [68,89–96], dehydration tolerance is substantially related with canopy form [93], leaf movements, leaf surface such as tiny and thick [9,93], stoma [97], and cuticular features. A substantial amount of genetic diversity was found in terms of root length density in the wild chickpea mini-core collection [89]. While evaluating the drought stress in chickpea, chlorophyll content, polyamine, and proline accumulation are also important features. Polyamines are synthesized in cells when plants are exposed to high-temperature and drought stresses. Thermo-polyamines are among the most commonly synthesized polyamines [98,99]. The proline accumulation has been accepted as a biochemical factor to evaluate stress responses in chickpea [100,101]. It has been reported that the amount of proline increases faster than other amino acids under drought and high-temperature stress conditions [102].

15.3.2 Extreme temperatures

Above 30°C and below 15°C, temperatures are called high and low extreme temperatures, respectively; since chickpea prefers to grow between 15°C–30°C [103].

15.3.2.1 High temperature

Two types of high-temperature stresses have been reported, detrimental for chickpea especially during flowering and pod formation stages [104,105]. Of these, the heat shock is explained by the fact that the plant is exposed to deadly temperatures from a few minutes to several hours, while the moderate heat is explained by the plant's exposure to higher temperatures than the optimum conditions during the growing season [59,81,106].

Upadhyaya and coworkers [107] reported that a 31% decrease in seasonal precipitation and every 0.1°C temperature increase reduced chickpea yield by 38.5%. It has been noticed that every 1°C temperature increase above the optimum temperature in cultivation leads to 10%–17% and 53 kg/ha yield loss [56,57]. High temperatures negatively affect seed germination. Although chickpea has a wide temperature range for germination, it has been reported that it cannot germinate above 45°C [108]. There are many different ranges such as 10°C–15°C [109] and 28°C–33°C [110] at optimum temperatures for seed germination. In addition, it was found that chickpea could germinate between about 32°C and 34°C [111]. High-temperature stress causes many physiological and morphological changes such as burning in the leaves and stems of chickpeas, drying in the whole plant, loss of flower and pods, slowing of shoot and root growth, stunting and pod damage [112–116], as well as reduces pollen viability and causing pod loss [52,117]. High temperatures also significantly reduce biomass yield. High temperatures lead to disruption of symbiotic N_2 fixation in chickpea [59,112]. The most sensitive period for heat stress is the reproductive phase in chickpea [104]. High temperatures in reproductive stages cause more yield loss in the pod filling period compared to early periods. While 59% yield loss has been reported in the pod filling period under heat stress, it has been reported that there is a yield loss between 22% and 34% in the early flowering period [118]. Although chickpea has a significant genetic variation for high-temperature stress tolerance, many accessions cannot form pods at temperatures above 35°C [119].

However, in a recent study, it was reported that in the F_2 population obtained from the crosses of *C. arietinum* × *C. reticulatum*, the greenhouse temperature was over 40°C–45°C during the flowering period and over 50°C in the pod filling period, and consequently offspring was formed pods [16].

15.3.2.1.1 High-temperature stress tolerance mechanisms

Heat stress resistance is defined as the ability to perform better when the plant is exposed to high temperatures [81]. Plants have developed various mechanisms to show their metabolic functions to maintain plant water status, and survive when exposed to intense evaporation due to excessive heat [25,59]. Mechanisms of resistance to high-temperature are listed as heat escape, heat avoidance, and heat tolerance [81].

1. Heat escape occurs by completing the life cycle quickly when chickpea is exposed to heat shock. Phenological traits such as earlier flowering and early maturity can be a powerful resistance mechanism against heat shock [59].
2. Major physiological components of heat avoidance are leaf reflectance, reduction of non-photosynthetic energy intercepted by the canopy, and transpiration. Leaves assume a significant function in heat avoidance by altering their orientations, transpiration rate, and reflectance [81].
3. These are associated with heat tolerance membrane stability, change of membrane lipid composition, accumulation of heat shock proteins, and specific solutes such as proline and glycine particularly in pollen [106].

Chickpea has shown different responses to heat stress according to their growth stages [120]. It is difficult to distinguish heat stress from drought stress in the field. Since both drought and heat stresses are generally observed in chickpea-grown regions, both stresses should be screened and evaluated together [59].

15.3.2.1.2 Drought and high-temperature tolerance resources in *Cicer* species

Drought stress is generally accompanied by high-temperature stresses in most of chickpea growing areas since chickpea is grown in rainfed conditions. Thus, about 90% of chickpea in the world is subjected to drought and high-temperature stresses [121]. Wild species and landraces supply a broader genetic base and also, they have more resistant sources than the cultivated chickpea to several a/biotic stresses [9,20,23,24,26,59,67,68,89,122–124]. AWC 605, AWC 616, AWC 620, and AWC 625 belonging to *C. reticulatum*, one of the wild relatives of chickpeas, were determined to be resistant to drought and heat stresses [26]. Moreover, AWC 500 belonging to the *C. pinnatifidum* species has also been found to be tolerant to heat and drought stresses [26]. According to the

International Center for Agricultural Research in the Dry Areas (ICARDA) report, *C. reticulatum*, *C. bijugum*, and *C. judaicum* were discovered to be more tolerant to drought than the other species (Table 15.2) [123].

C. reticulatum genotypes ILWC 36, ILWC 116 [124], and *C. bijugum* genotypes ILWC 34 and ILWC 65 were determined to be drought resistant [123,124]. As *C. reticulatum* is easily crossable with the domesticated varieties, the high-yielding genotypes could be evaluated in hybridization to develop cultivated chickpea to drought and heat stresses [124]. Although there are many sources of drought and heat resistant in the cultivated chickpea, the resistance of many accessions of perennial wild *Cicer* species including *C. montbretii* Jaub et Sp., *C. anatolicum* Alef., *C. oxydon* Boiss. et Hoh., *C. songaricum* Steph ex DC. and *C. microphyllum* Benht. were found higher than the cultivated chickpea (Table 15.2) [59]. Moreover, several accessions of wild *Cicer* relatives including *C. anatolicum* (PI 561078, PI 383626, WG-14183), *C. microphyllum* (PI 532928), *C. montbretii* (WG-14189), *C. oxydon* (PI 561103), and *C. songaricum* (WG-4574) were identified as sources of resistance to drought and heat (Table 15.2) [25]. Drought and high-temperature stress-tolerant wild *Cicer* species are shown in Figs. 15.1 and 15.2 in their natural habitats and under field conditions. The development of early flowering and early maturing varieties can help the cultivars use the present soil moisture more efficiently and obtain higher yields [3]. Robertson and coworkers [125] reported a wide variation in *C. judaicum* for early flowering. This variation in *C. judaicum* can be considered for drought and heat escape/avoidance [3]. Landraces with root length density in the Mediterranean, West Asia, and Central Asia regions with high genetic diversity have been identified which promise effective water use under the drought and heat stress [89]. It has been reported that the amount of proline increases faster than the other amino acids under drought and high-temperature stress conditions. High

(A) (B) (C)

FIGURE 15.1 Drought and high temperature tolerant wild *Cicer* species (from left to right, (A) *C. echinospermum*, (B) *C. anatolicum* and (C) *C. bijugum*, respectively) in their natural habitats.

FIGURE 15.2 Drought and high temperature tolerant wild *Cicer* species (A) *C. chorassainicum*, (B) *C. montbretti* in their natural habitats and (C) *C. reticulatum* in field conditions.

proline content was determined in *C. pinnatifidum*, the wild relative of chickpea known to be drought resistant [102].

Since the last decade, ICRISAT has been focusing on the improvement of molecular markers, obtaining the wide genetic maps, identification of Quantitative Trait Loci (QTLs), and marker-trait associations for drought and heat tolerance in chickpea [126]. High temperature and drought are quantitative characters where gene tagging has been difficult and hence they significantly restrict the chickpea productivity globally [3]. Nine QTL clusters have been identified

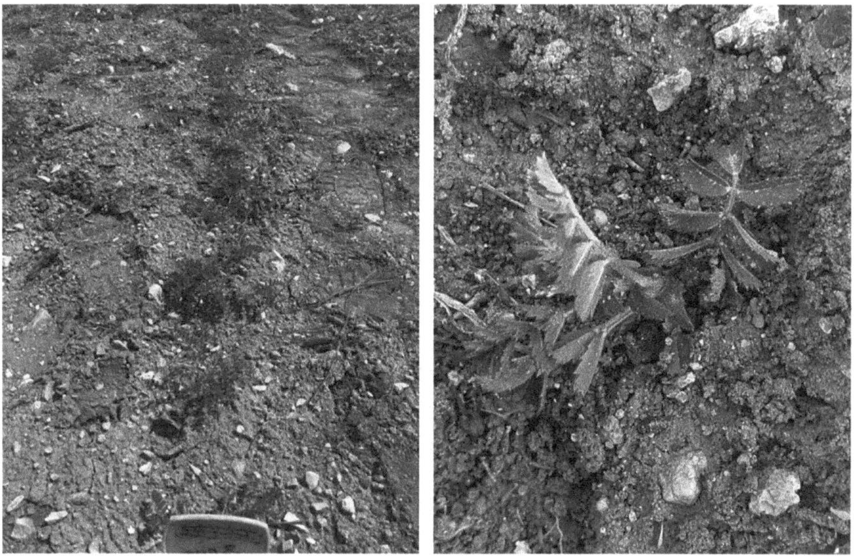

FIGURE 15.3 Cold tolerant RILs (*C. arietinum*´ *C. reticulatum*) in field conditions.

associated with drought tolerance [127] or high temperature stress-tolerance in the $F_{8\text{-}9}$ population obtained from a cross between ICC 4567 (sensitive) and ICC 15614 (tolerant) in cultivated chickpea [128], however, there has been no study in wild *Cicer* species yet.

While, there is a considerable variation for drought and high-temperature tolerance in cultivated chickpea, there is a need to broaden the search for novel genetic diversity to ensure extra allelic variation for heat and drought tolerance [129]. The wild *Cicer* relatives are a possible source of high variation and this novel allelic variation would allow chickpea breeders to overcome the current reproductive high temperature and drought borders on chickpea.

15.3.2.2 Low-temperature stress

Below temperatures of 15°C is known as low temperature or cold stress, one of the most important abiotic stress factors restricting productivity in chickpea. Considering to severity of cold, cold stress is classified as chilling stress, which occurs when temperature between 0°C and 15°C, and freezing stress, which takes place when the temperature is below 0°C [130]. Although the effect of cold stress on plants varies to types of plant, severity, and duration of cold stress, many plants show physical and physiological symptoms of cold damages when exposed to cold stress during developmental stages [131]. Reduction of growth and development rate, retardation of germination, wilting and chlorosis on leaf tissue are known symptoms of cold stress in plants [132]. Prolonged exposure to this stress may lead to necrosis and cause the death of the whole plant [124].

15.3.2.2.1 Effect of chilling temperatures

Plants vary in their ability to survive in cold temperatures. The majority of plant species grown in tropical and sub-tropical regions in the world and most of these plants are sensitive to chilling stress [133]. Temperate climate plants and some economically important tropical and subtropical plants including cotton, maize, soybean, rice, and tomato also sensitive to chilling stress although they cultivated in temperate regions [134,135]. Chickpea is grown in tropical, subtropical, and temperate regions, and has developed sensitivity to low temperature during the cultivation process involved in agricultural selection [44,136]. The optimum temperature demand for chickpea in early vegetative stages ranges from 21°C to 32°C, and the optimum temperature demand in flowering stages ranges from 18°C to 29°C [109,124,137].

The temperature below 15°C causes infertility, termination and retardation of flowering, podding, and seed filling leading to decrease in chickpea yield [138–141]. When the temperature drops below 10°C, the development of seedling can vigorously limit the vegetative stages [138] (Fig. 15.4). Chilling stress below 8°C leads to reduce or retardate germination during the germination stage in chickpea.

15.3.2.2.2 Effect of freezing temperatures

Freezing temperatures are an important defining factor in geographic distribution of plants. Freezing temperatures limit productivity and cause significant yield losses in crop plants [142]. Freezing temperatures damage in summer-sown

FIGURE 15.4 Cold sensitive (left) and cold tolerant (right) RILs (*C. arietinum´ C. reticulatum*) in controlled conditions.

crops cause considerably great product losses in Europe and Northern America during late spring and early autumn. Although snow cover helps to protect from frost damage to winter crops, repetitive freezing and thawing with cold windy weather cause freezing injury, plant death, and yield losses during the winter season [143].

When the temperature drops below the freezing point, extracellular water turns into ice form primarily [144]. Because the chemical potential of iced water is less than its liquid form, intracellular water tends to move outside of the cell to protect osmotic balance [145]. This movement leads to vigorous cellular dehydration result in frost injury including protein denaturation and cell death (cellular damage) [146]. Many crop plants are not able to tolerate freezing temperature. Freezing temperatures decrease plant development in many stages from the seedling stages to reproductive stages in chickpea [147]. The elongation zones in the chickpea are generally the first affected regions by freezing temperatures. This elongation zones affected by freezing temperature take form a sigmoidal curve known as "hockey stick" symptom [22]. Prolonged exposure to freezing temperature results in prevention of germination, reduction of vegetative growth, abortion of flowering, and podding in chickpea [148–150].

15.3.2.2.3 Low-temperature tolerance mechanisms

Cold escape, tolerance, and avoidance are three different mechanisms to tolerate low-temperature stress in plants [81]. Cold escape mechanism is generally been employed by breeders in autumn-sown crops to provide the vernalization demands of the seeds [81]. Chilling tolerance is the ability to tolerate low nonfreezing temperature without any damage [151]. On the other hand, freezing temperatures tolerance is the ability to survive in sub-zero temperatures without an important cell or membrane damage despite freezing temperatures causes forming intercellular ice [106]. Avoidance mechanisms involve impeding tissues from freezing [152]. Many plants stay in the dormant form (seed/tissue) to avoid from freezing effect [153]. Others gain freezing tolerance during low nonfreezing temperatures and this phenomenon is known as cold acclimation [142]. Cold acclimation (<10°C) entails some biochemical, physiological, and molecular changes such as modification in lipid composition, accumulation of cryoprotective solutes (sugars, proline, polyamines, and glycine betaine) and activation of cold-regulated genes encoding molecular chaperones, kinases, and polypeptides [154–156].

Membrane systems of the plant cells are first being affected by freezing injury [157]. Freezing-induced cellular dehydration is a primary reason for membrane damage; however, freezing-induced production of reactive oxygen species (ROS), which also contribute to membrane injury [158]. In this point, accumulation of sugar and polypeptides (such as late embryogenesis abundant (LEA)), stimulation of antioxidant mechanism, and freezing tolerance genes during cold acclimation help to stabilize of the plant membrane structures in freezing temperatures [159–161].

15.3.2.2.4 Low-temperature tolerance resources in *Cicer* species

Chickpea is sensitive to cold stress and the effects of cold stress show differences according to sowing season and region. While winter or autumn sown chickpea in northern south Asia and Australia suffer chilling damage at reproductive period, those in Europe, West Asia, and North Africa (WANA) suffer chilling and freezing temperature at the seedling and early vegetative period [162–164]. Therefore, chickpea is sown as a spring-sown crop to avoid the effect of cold stress and terminal drought [165]. However, conducted experiments showed that cold-tolerant winter-sown chickpea has about 50%–100% greater yields than spring-sown chickpea [105,166–169]. On the other hand, the number of genotypes available for cold-tolerant winter sowing is quite limited, and such a case then significantly hampers high yields through winter sowings.

Wild relatives of chickpea show considerably high tolerance to cold stress [17–25,27,28]. Several studies have been conducted with cultivated and wild *Cicer* species based on field-scale physiological observations, considerably higher cold tolerance was reported for wild species than for cultivated chickpea (Figs. 15.3 and 15.4). The very first study to determine cold tolerance of wild chickpea species was conducted by Van der Maesen and Pundir [67], and *C. microphyllum* was identified as cold-tolerant (Table 15.2). Singh and coworkers [170] studied the reactions of 137 accessions of 8 annual species of *Cicer* genus against freezing in a vegetative period. Researchers used a scale of 1–9 (1 = no visible damage, 9 = all plants killed), and ILC533 was sown in every 10 rows as a sensitive genotype. As a result, *C. bijigum* accessions were assessed at 1–3 scale, *C. echinospermum* accessions at 2–3 scale, *C. judaicum* and *C. pinnatifidum* accessions at scale 4, and finally *C. chorassanicum*, *C. cuneatum*, and *C. yamashiate* accessions were assessed at scale 7. In another study conducted by Singh and coworkers [150], cold tolerance of 59 lines belonging to 4284 kabuli, 2137 desi genotypes, 857 breeding lines, and 7 annual wild *Cicer* species were investigated. In this study, a germplasm line (ILC8262), a mutant (ILC8617), and a breeding line (FLIP 87-82C) of *C. arietinum* were selected as tolerant lines, and the sensitive genotypes (ILC533) was sown in every 10 rows. Among the wild genotypes, 26 lines of *C. reticulatum*, 10 lines of *C. bijugum*, 4 lines of *C. echinospermum*, 2 lines of *C. pinnatifidum*, and 1 line of *C. judaicum* were identified as cold-tolerant (Table 15.2). Of the cold-tolerant accessions, cold-tolerance of 5 accessions of *C. bijugum*, and 4 accessions of *C. echinospermum* was assessed at scale 2. In a study conducted by ICARDA, freeze-tolerance of ILWC 66 line of *C. bijugum* was assessed at scale 1 [22]. Another study reported that cold tolerance of 43 accessions of 8 annual wild *Cicer* species was assessed. The lines ILC8262 and FLIP 87-82C were sown as tolerant genotypes, and the line ILC533 was used as the sensitive genotype. This study demonstrated that all accessions of *C. bijugum* and 11 accessions of *C. reticulatum* were identified as the most tolerant accessions and cold tolerance of 4 accessions of *C. echinospermum* was assessed at scale 2–3. It was observed that *C. chorassanicum* and *C. cuneatum* accessions died almost at equal times with the sensitive genotype ILC533 [24]. Furthermore, freezing tolerance of 5 annual wild *Cicer* (*C. echinospermum* and *C. reticulatum*) and 225 cultivated chickpea accessions were evaluated grown both in the field and all accessions of wild *Cicer* was assessed as scale 1–2 in both growing conditions [171]. Lastly, reproductive phase chilling tolerance studies in wild chickpeas showed that wild *Cicer* species (*C. bijugum, C. judaicum,* and to a lesser extent *C. reticulatum, C. pinnatifidum,* and *C. echinospermum*) are more chilling tolerant than cultivated chickpea [164].

In cultivated chickpea, transcriptome analysis using different techniques such as cDNA-AFLP, or cDNA-microarray have been done to identify possible cold tolerance related genes [172–174]. However, there is no available information for global transcriptome analysis in wild *Cicer* genotypes to reveal genes having a role in cold tolerance. On the other hand, developments of genomic markers have provided considerably significant attributes to mapping studies to identify cold tolerance linked QTLs. A RILs population derived from *C. arietinum* and *C. reticulatum* was employed to identify QTLs related to tolerance of cold stress [175]. The study revealed that freezing tolerance in the cold-tolerant wild parent (*C. reticulatum*) was controlled by different QTLs placed on linkage groups (LGs) 1B, 3, and 8 [175]. Vernalization QTLs have been identified in a RIL population using the same cold-tolerant wild parent (*C. reticulatum*) and have been identified as the same cold-tolerant linked QTLs (Ct Ca-3.1) [176].

15.3.3 Salinity stress

Soil salinity is characterized by a high concentration of the chlorides and sulfates of sodium (Na^{+}), calcium (Ca^{+2}), and magnesium (Mg^{+2}) in soils. Sometimes, nitrates can be present in considerable amounts. Soils may rarely include noticeable amounts of gypsum ($CaSO_4$, $2H_2O$) in the profile. Soluble carbonates are not present in saline soils. The pH value of the soil is less than 8.2 and more [177]. Based on electrical conductivity (ECe) of the saturation extract, the soil can be identified as saline (ECe is more than 4 dS/m), and sodic (ECe is more than 15 dS/m) [178]. The value of ECe in the saline soils is equivalent to 40 mM NaCl generating an osmotic potential of approximately 0.2 MPa [179]. This definition of soil salinity based on the ECe that considerably decreases the yield in many crops [179,180]. Saline soils might be also classified as nonsaline ($<$2 dS/m), slightly saline (2–4 dS/m), moderately saline (4–8 dS/m), strongly saline (8–16 dS/m), and very strongly saline ($>$16 dS/m) [181].

There are two major causes of salinity formation in nature: Natural geologic process such as replacement and redistribution of salts causes primary salinity [61]. Secondary salinity results from human activities such as dry areas (nonirrigated landscapes, uncultivated regions), irrigation (because of ascending groundwater or the usage of low-quality water), poor drainage, and high salt concentration in wastewater effluent from intensive agriculture.

In the early stages, salinity is one of the important limiting factors that affect crop productivity [182,183]. It globally restricts crop production and almost 80 million ha of the agricultural area is exposed to salinity stress [63]. About 45 million ha of irrigated (20%) and 32 million ha of dry area (2%) are influenced by salinity [180]. Salinity extremely disturbs photosynthesis, carbon and nitrogen metabolism, and causes accumulation of Na^+ ions in plants [184–186]. These variations in physiological processes cause an appreciable reduction in plant growth resulting decrease in crop yield. Chickpea is very sensitive to salinity, and global yield loss of chickpea is between 8% and 10% attributed to salinity [63]. Salinity affects germination, plant establishment, N_2 fixation, growth, flowering and inhibits pod formation, and seed filling in chickpea [62,63]. There is an inhibitory effect of salinity on both the rate and time of germination in chickpea seeds, although there is high variation among the genotypes. Esechie and coworkers [187] identified that the germination highly decreased in 120 mM NaCl. Soltani and coworkers [188] asserted that the effects of salinity on germination were the same in kabuli and desi types. Within three kabuli genotypes, the germination time of small seeds at high salt concentration was lesser than of large seeds [189]. Photosynthetic pigments declined in chickpea grown in 100 mM NaCl [190,191], and Murumkar and Chavan [192] reported that photosynthesis was reduced by 60% in same salt concentration.

Salinity also causes an increase of senescence in chickpea [193] and activates the production of ethylene and its precursor 1-aminocycloprane-1-carboxylic acid (ACC) in roots and nodules [194]. Maliro and coworkers [195] reported the destruction of chlorophyll induced leaf necrosis, and this resulted from ion toxicity in tissues. According to these authors, visual necrosis parameters might be beneficial in the tolerance index.

15.3.3.1 Salinity stress tolerance mechanisms

Salinity tolerance appears with a combination of physiological, biochemical, and molecular mechanisms in plants [180,196–198]. There are generally three main mechanisms to salinity tolerance: (1) the tolerance to osmotic stress; increasing osmotic adjustment, which decreases inhibition of cell expansion and lateral bud development, (2) Na^+ exclusion; sequestrating of Na^+ into root vacuoles to avoid ion toxicity in chloroplasts, and (3) tissue tolerance; regulating the accumulation of compatible solutes in the cytoplasm [73,180].

Plant response to salt stress might be controlled by several genes and genotype interaction with the environmental factors [69,199,200]. Advanced genomics strategies have the potential to define transcripts or candidate gene(s) related to tolerance [7,201–203]. In chickpea, there are limited reports on the mapping of QTLs related to salinity tolerance [204]. Recently, a mapping population (ICCV 2 × JG 62) has been studied to identify a major QTL on linkage group LG 3 associated with19% phenotypic variation for high seed yield under salinity stress conditions [205]. Additionally, QTLs for seed number, pod number, and 100-seed weight under were identified on LG 6 under salinity conditions. Currently, two major QTLs associated with salinity tolerance were determined on LG 5 and LG 7 from the mapping population derived from ICCV 2 × JG 11 in chickpea [206]. The same authors proposed a group of 48 candidate genes that encode proteins responsible for ABA biosynthesis, ion transport, and transcription factors (TFs) in the QTL-containing region. The understanding of the functional mechanisms of many genes induced by stress is still not clear. In chickpea, many studies at transcriptional level have been reported that salt stress activates complex regulatory systems in the cell [7,207]. For understanding, this phenomenon, several candidate genes have been determined associated with salinity stress through evolving new generation techniques such as microarrays [172], serial analysis of gene expression (SAGE) [208] and expressed sequenced tags (ESTs) [209].

15.3.3.2 Salinity tolerance resources in Cicer species

In recent years, several studies have been made to examine genetic variation for salinity tolerance in cultivated chickpea and related species [62]. For salinity tolerance, limited genetic variations were shown in chickpea which previously studied by Johansen and coworkers [210] and Saxena [211]. Nevertheless, recent studies in chickpea germplasm have indicated a large range of genetic variation for salinity tolerance [69–71,73,212]. Serraj and coworkers [71] have identified broad variations for vegetative biomass in 252 chickpea genotypes. Furthermore, Maliro and coworkers [70] screened 200 genotypes of chickpea, including the wild relatives of *C. reticulatum* (eight genotypes), *C. echinospermum* (three genotypes), *C. bijugum* (six genotypes), *C. judaicum* (one genotype) and *C. pinnatifidum* (one genotype) for salinity tolerance at the early vegetative stage phase. Five of these genotypes (CPI 060546, ILC 01302, ICC 6474, ICC 8294, and UC 5) were scored as most tolerant. There were no wild relatives scored as tolerant to salinity. Moreover, Maliro and coworkers [195] observed rarely tolerant types in wild relatives. According to the researchers, this can be due to the narrow span of the geographic range from which these wild relative accessions originated (Syria and Turkey). Then the wild relative accessions adapted to the saline environment were small. Indeed, limited research for salinity tolerance has been conducted in wild chickpea relatives, compared with the cultivated chickpea accessions. In a chickpea dataset from East and Southeastern Turkey, a total of 221 accessions, 174 of which were *C. reticulatum* and 47 of which were C. *echinospermum* were evaluated with respect to yield and agro-morphological traits by Talip and coworkers [41]. The study was conducted to determine the allelic diversity and genetic variation in the dataset. Because allelic richness within the wild relatives [9] may reveals the desirable genes related with salt stress. Presently, it is extremely required that breeding programs included in wild relatives should be purposed of developing salt-tolerant varieties.

Acknowledgment

The authors are grateful to Prof. Dr. Cengiz Toker (Department of Field Crops, Faculty of Agriculture, Akdeniz University, Antalya, Turkey) for his critical review of the manuscript. The inspiration and support from Punjab Agricultural Research Board (PARB), and Precision Agriculture and Analytics Lab (PAAL), affiliate of National Center in Big Data and Cloud Computing (NCBC) is also acknowledged.

References

[1] Varshney RK, Song C, Saxena RK, Azam S, Yu S, Sharpe AG, et al. Draft genome sequence of chickpea (*Cicer arietinum*) provides a resource for trait improvement. Nat Biotechnol 2013;31(3):240–6.

[2] FAOSTAT. Available from: https://www.fao.org/faostat/en/#data/QC (accessed on 29 April 2020).

[3] Singh R, Sharma P, Varshney RK, Sharma SK, Singh NK. Chickpea improvement: role of wild species and genetic markers. Biotechnol Genet Eng 2008;25(1):267–314.

[4] Dalal RC, Strong WM, Weston EJ, Cooper JE, Wildermuth GB, Lehane KJ, et al. Sustaining productivity of a Vertisol at Warra, Queensland, with fertilisers, no-tillage, or legumes. 5. Wheat yields, nitrogen benefits and water-use efficiency of chickpea-wheat rotation. Aust J Exp Agric 1998;38(5):489–501.

[5] Redden RJ, Berger JD. History and origin of chickpea. In: Yadav SS, Redden R, Chen W, Sharma B, editors. Chickpea breeding and managementWallingford: CAB International; 2007. p. 1–13.

[6] Gaur PM, Jukanti AK, Samineni S, Chaturvedi SK, Basu PS, Babbar A, et al. Climate change and heat stress tolerance in chickpea. Wiley-VCH Verlag Gmbh & Co.; 2013.

[7] Garg R, Shankar R, Thakkar B, Kudapa H, Krishnamurthy L, Mantri N, et al. Transcriptome analyses reveal genotype-and developmental stage-specific molecular responses to drought and salinity stresses in chickpea. Sci Rep 2016;6:19228.

[8] Kadiyala MDM, Kumara Charyulu D, Nedumaran SD, Shyam M, Gumma MK, Bantilan MCS. Agronomic management options for sustaining chickpea yield under climate change scenario. J Agrometeorol 2016;18(01):41–7.

[9] Toker C, Yadav SS. Legume cultivars for stress environments. In: Yadav SS, McNeil DL, Redden R, Patil SA, editors. Climate change and management of cool season grain legume cropsDordrecht: Springer; 2010. p. 351–76.

[10] Toker C, Mutlu N. Breeding for abiotic stresses. Biology and breeding of food legumesCABI; 2011. p. 241–61.

[11] Toker C. A note on the evolution of kabuli chickpeas as shown by induced mutations in *Cicer reticulatum* Ladizinsky. Genet Resour Crop Evol 2009;56(1):7–12.

[12] Smýkal P, Nelson MN, Berger JD, Von Wettberg EJ. The impact of genetic changes during crop domestication. Agronomy 2018;8(7):119.

[13] Singh KB, Ocampo B. Interspecific hybridization in annual *Cicer* species. J Genet Breed 1993;47(3):199–204.

[14] Adak A, Sari D, Sari H, Toker C. Gene effects of *Cicer reticulatum* on qualitative and quantitative traits in the cultivated chickpea. Plant Breed 2017;136(6):939–47.

[15] Koseoglu K, Adak A, Sari D, Sari H, Ceylan FO, Toker C. Transgressive segregations for yield criteria in reciprocal interspecific crosses between *Cicer arietinum* L. and *C. reticulatum* Ladiz. Euphytica 2017;213(6):116.

[16] Ceylan FO, Adak A, Sari D, Sari H, Toker C. Unveiling of suppressed genes in interspecific and backcross populations derived from mutants of *Cicer* species. Crop Pasture Sci 2019;70(3):254–62.
[17] Singh KB, Malhotra RS, Saxena MC. Chickpea evaluation for cold tolerance under field conditions. Crop Sci 1989;29:282–5.
[18] Singh KB, Reddy MV. Sources of resistance to ascochyta blight in wild *Cicer* species. Neth J Plant Pathol 1993;99(3):163–7.
[19] Kaiser WJ, Alacala-Jimenez AR, Hervas-Vargas A, Trapero-Casas JL, Jimenez-Diaz RM. Screening of wild *Cicer* species for resistance to races 0 and 5 of *Fusarium oxysporum* f. sp. *ciceris*. Plant Dis 1994;78:962–7.
[20] Singh KB, Weigand S. Identification of resistant sources in *Cicer* species to *Liriomyza cicerina*. Genetic Resour Crop Evol 1994;41:75–9.
[21] Di Vito M, Singh KB, Greco N, Saxena MC. Sources of resistance to cyst nematode in cultivated and wild *Cicer* species. Genet Resour Crop Evol 1996;43:103–7.
[22] Croser JS, Clarke HJ, Siddique KHM, Khan TN. Low temperature stress: implications for chickpea (*Cicer arietinum* L.) improvement. Crit Rev Plant Sci 2003;22:185–219.
[23] Rao NK, Reddy LJ, Bramel PJ. Potential of wild species for genetic enhancement of some semi-arid food crops. Genet Resour Crop Evol 2003;50:707–21.
[24] Toker C. Preliminary screening and selection for cold tolerance in annual wild *Cicer* species. Genet Resour Crop Evol 2005;52:1–5.
[25] Toker C, Canci H, Yildirim T. Evaluation of perennial wild *Cicer* species for drought resistance. Genet Resour Crop Evol 2007;54:1781–6.
[26] Canci H, Toker C. Evaluation of annual wild *Cicer* species for drought and heat resistance under field conditions. Genet Resour Crop Evol 2009;56(1):1.
[27] Kaur L, Sirari A, Kumar D, Sandhu JS, Singh S, Kapoor K, et al. Combining Ascochyta blight and Botrytis grey mould resistance in chickpea through interspecific hybridization. Phytopathol Mediterr 2013;52:157–65.
[28] Tekin M, Sari D, Catal M, Ikten C, Smykal P, Penmetsa RV, et al. Eco-geographic distribution of *Cicer isauricum* PH Davis and threats to the species. Genet Resour Crop Evol 2018;65(1):67–77.
[29] Eker T, Erler F, Adak A, Imrek B, Guven H, Tosun HS, et al. Screening of chickpea accessions for resistance against the pulse beetle, *Callosobruchus chinensis* L. (Coleoptera: *Bruchidae*). J Stored Prod Res 2018;76:51–7.
[30] Van der Maesen LJG. *Cicer* L., a monograph of the genus, with special reference to the chickpea (*Cicer arietinum* L.), its ecology and cultivationWageningen, The Netherlands: H. Veenman & Zonen; 1972. p. 342.
[31] Donmez AA. *Cicer uludereensis* Dönmez: a new species of *Cicer* (Chickpea) (*Fabaceae*) from around the Fertile Crescent, SE Turkey. Turk J Bot 2011;35(1):71–6.
[32] Van der Maesen LJG. Origin, history and taxonomy of chickpea. In: Saxena MC, Singh RB, editors. The chickpeaWallingford: CABI; 1987. p. 139–56.
[33] Van der Maesen LJG, Maxted N, Javadi F, Coles S, Davies AMR. Taxonomy of the genus *Cicer* revisited. In: Yadav SS, Redden RJ, Chen W, Sharma B, editors. Chickpea breeding and managementWallingford: CAB International; 2007. p. 14–46.
[34] Ozturk M. Revision of the genus *Cicer* L. Turkey via morphological, palynological, cytotaxonomical, molecular phylogenetic methods and analyses of seed protein and element contents. PhD. Konya: The Graduate School of Natural and Applied Science of Selçuk University; 2011.
[35] Ozturk M, Duran A, Hakki EE. Cladistic and phylogenetic analyses of the genus *Cicer* in Turkey. Plant Syst Evol 2013;299(10):1955–66.

[36] Toker C, Berger J, Kahraman A, Aydogan A, Can C, Bukun B, et al. *Cicer reticulatum* Ladizinsky, progenitor of the cultivated chickpea (*C. arietinum* L.). Legume Perspect. 2014;5:26–7.

[37] Ladizinsky G, Adler A. Genetic relationships among the annual species of *Cicer* L. Theor Appl Genet 1976;48(4):197–203.

[38] Kazan KMFJ, Muehlbauer FJ, Weeden NE, Ladizinsky G. Inheritance and linkage relationships of morphological and isozyme loci in chickpea (*Cicer arietinum* L.). Theor Appl Genet 1993;86(4):417–26.

[39] Harlan JR, de Wet JM. Toward a rational classification of cultivated plants. Taxon 1971;20(4):509–17.

[40] Kahraman A, Pandey A, Khan MK, Lindsay D, Moenga S, Vance L, et al. Distinct subgroups of *Cicer echinospermum* are associated with hybrid sterility and breakdown in interspecific crosses with cultivated chickpea. Crop Sci 2017;57(6):3101–11.

[41] Talip M, Adak A, Kahraman A, Berger J, Sari D, Sari H, et al. Agro morphological traits of *Cicer reticulatum* Ladizinsky in comparison to *C. echinospermum* PH Davis in terms of potential to improve cultivated chickpea (*C. arietinum* L.). Genet Resour Crop Evol 2018;65:951–62.

[42] Mallikarjuna N. Ovule and embryo culture to obtain hybrids from interspecific incompatible pollinations in chickpea. Euphytica 1999;110(1):1–6.

[43] Clarke HJ, Wilson JG, Kuo I, Lülsdorf MM, Mallikarjuna N, Kuo J, et al. Embryo rescue and plant regeneration in vitro of selfed chickpea (*Cicer arietinum* L.) and its wild annual relatives. Plant Cell Tissue Organ Cult 2016;85(2):197–204.

[44] Abbo S, Berger J, Turner NC. Evolution of cultivated chickpea: Four bottlenecks limit diversity and constrain adaptation. Funct Plant Biol 2003;30:1081–7.

[45] Tanno KI, Willcox G. The origins of cultivation of *Cicer arietinum* L. and *Vicia faba* L.: early finds from Tell el-Kerkh, north-west Syria, late 10th millennium BP. Veg Hist Archaeobot 2006;15(3):197–204.

[46] Ladizinsky G, Adler A. The origin of chickpea *Cicer arietinum* L. Euphytica 1976;25(1):211–217.

[47] Berger J, Abbo S, Turner NC. Ecogeography of annual wild *Cicer* species. Crop Sci 2003;43(3):1076–90.

[48] Ladizinsky G. The origin of chickpea as indicated by seed protein electrophoresis. Israel J Bot 1975;24:183–9.

[49] Zohary D, Hopf M. 3rd ed. Domestication of plants in the Old World: The origin and spread of cultivated plants in West Asia, Europe and the Nile ValleyOxford: Oxford University Press; 2000.

[50] IPCC. Climate Change 2013: The physical science basis. Contribution of working group I to the fifth assessment report of the intergovernmental panel on climate change. In: TF., Stocker, D., Qin, GK., Plattner, M., Tignor, SK., Allen, J., Boschung, et al., editors. Cambridge, United Kingdom and New York, NY, USA: Cambridge University Press; 2018. p. 1535.

[51] Kashiwagi J, Krishnamurthy L, Crouch JH, Serraj R. Variability of root length density and its contributions to seed yield in chickpea (*Cicer arietinum* L.) under terminal drought stress. Field Crops Res 2006;95(2–3):171–81.

[52] Leport L, Turner NC, Davies SL, Siddique KHM. Variation in pod production and abortion among chickpea cultivars under terminal drought. Eur J Agron 2006;24(3):236–46.

[53] Farooq M, Wahid A, Kobayashi N, Fujita DBSMA, Basra SMA. Plant drought stress: effects, mechanisms and management. In: Sustainable agriculture. Dordrecht: Springer; 2009. p. 153–188.

[54] Ahmad F, Gaur P, Croser J. Chickpea (*Cicer arietinum* L.). In: Singh R, Jauhar P, editors. Genetic resources, chromosome engineering and crop improvement-grain legumesFlorida: CRC Press; 2005. p. 185–214.

[55] Rani A, Devi P, Jha UC, Sharma KD, Siddique KH, Nayyar H. Developing climate-resilient chickpea involving physiological and molecular approaches with a focus on temperature and drought stresses. Front Plant Sci 2020;10:1759.

[56] Kalra N, Chakraborty D, Sharma A, Rai HK, Jolly M, Chander S, et al. Effect of increasing temperature on yield of some winter crops in northwest India. Curr Sci 2008;82–8.

[57] Dubey SK, Sah U, Singh SK. Impact of climate change on pulse productivity and adaptation strategies as practiced by the pulse growers of Bundelkhand region of Uttar Pradesh. J Food Legumes 2011;24(3):230–4.

[58] Berger JD, Buck R, Henzell JM, Turner NC. Evolution in the genus *Cicer* vernalisation response and low temperature pod set in chickpea (*C. arietinum* L.) and its annual wild relatives. Aust J Agric Res 2005;56(11):1191–200.

[59] Toker C, Lluch C, Tejera NA, Serraj R, Siddique KHM. Abiotic stress. In: Yadav SS, Redden R, Chen W, Sharma B, editors. Chickpea breeding and managementWallingford: CAB International; 2007. p. 474–96.

[60] Okur B, Orcen N. Soil salinization and climate change. In: Prasad MNV, Pietrzykowski M, editors. Climate change and soil interactionsAmsterdam: Elsevier; 2020. p. 331–50.

[61] Hussain N, Sarwar G, Schmeisky H, Al-Rawahy S, Ahmad M. Salinity and drought management in legume crops. In: Yadav SS, McNeil DL, Redden R, Patil SA, editors. Climate change and management of cool season grain legume cropsNetherlands: Springer; 2010. p. 171–91.

[62] Atieno J, Li Y, Langridge P, Dowling K, Brien C, Berger B, et al. Exploring genetic variation for salinity tolerance in chickpea using image-based phenotyping. Sci Rep 2017;7:1–11.

[63] Flowers TJ, Gaur PM, Gowda CL, Krishnamurthy L, Samineni S, Siddique KH, *et al.* Salt sensitivity in chickpea. Plant Cell Environ 2010 33:490–509.

[64] Huang S, Gali KK, Tar'an B, Warkentin TD, Bueckert RA. Pea phenology: crop potential in a warming environment. Crop Sci 2017;57(3):1540–51.

[65] Kumar J, Solanki RK. Evaluation of germplasm accessions for agro-morphological traits in lentil. J Food Leg 2014;27:275.

[66] Cullis C, Kunert KJ. Unlocking the potential of orphan legumes. J Exp Bot 2017;68(8):1895–903.

[67] Van der Maesen LJG, Pundir RPS. Availability and use of wild *Cicer* germplasm. Plant Genet Resour Newslett 1984;57:282–5.

[68] Krishnamurthy L, Kashiwagi J, Upadhyaya HD, Serraj R. Genetic diversity of drought avoidance root traits in the mini-core germplasm collection of chickpea. Int Chickpea Pigeonpea Newslett 2003;10:21–4.

[69] Krishnamurthy L, Turner NC, Gaur PM, Upadhyaya HD, Varshney RK, Siddique KHM, et al. Consistent variation across soil types in salinity resistance of a diverse range of chickpea (*Cicer arietinum* L.) genotypes. J Agron Crop Sci 2011;197:214–27.

[70] Maliro MF, McNeil D, Kollmorgen J, Pittock C, Redden B. Screening chickpea (*Cicer arietinum* L.) and wild relatives germplasm from diverse country sources for salt tolerance. In: The fourth international crop science congress. Queensland, Australia; 2004.

[71] Serraj R, Krishnamurthy L, Upadhyaya HD. Screening chickpea mini-core germplasm for tolerance to soil salinity. Int Chickpea Pigeonpea Newslett 2004;11:29–32.

[72] Li Y, Ruperao P, Batley J, Edwards D, Khan T, Colmer TD, et al. Investigating drought tolerance in chickpea using genome-wide association mapping and genomic selection based on whole-genome resequencing data. Front Plant Sci 2018;9:190.

[73] Turner NC, Colmer TD, Quealy J, Pushpavalli R, Krishnamurthy L, Kaur J, et al. Salinity tolerance and ion accumulation in chickpea (*Cicer arietinum* L.) subjected to salt stress. Plant Soil 2013;365:347–61.

[74] Siddique KHM, Brinsmead RB, Knight R, Knights EJ, Paull JG, Rose IA. Adaptation of chickpea (*Cicer arietinum* L.) and faba bean (*Vicia faba* L.) to Australia. In: Knight R, editor. Linking research and marketing opportunities for pulses in the 21st centurySpringer: Dordrecht; 2000. p. 289–303.
[75] Singh S, Singh I, Kapoor K, Gaur PM, Chaturvedi SK, Singh NP, et al. Chickpea. In: Singh M, Bisht IS, Dutta M, editors. Broadening the genetic base of grain legumesNew Delhi: Springer; 2014. p. 51–73.
[76] Siddique KHM, Sedgley RH. Chickpea (*Cicer arietinum* L.), a potential grain legume for south-western Australia: seasonal growth and yield. Aust J Agric Res 1986;37(3):245–61.
[77] Siddique KMH, Sedgley RH, Davies CL, Punyavirocha S. Genotypic differences in chickpea for pod set at low temperatures during flowering. In: Proceedings of international symposium on pulses research. New Delhi; 1993.
[78] Khanna-Chopra R, Sinha SK. Chickpea: physiological aspects of growth and yield. In: Saxena MC, Singh KB, editors. The chickpea. Wallingford: C.A.B. International; 1987. p. 163–190.
[79] Leport L, Turner NC, French RJ, Barr MD, Duda R, Davies SL, et al. Physiological responses of chickpea genotypes to terminal drought in a Mediterranean-type environment. Eur J Agron 1999;11(3–4):279–91.
[80] Pang J, Turner NC, Du YL, Colmer TD, Siddique KH. Pattern of water use and seed yield under terminal drought in chickpea genotypes. Front Plant Sci 2017;8:1375.
[81] Wery J, Turc O, Lecoeur J. Mechanisms of resistance to cold, heat and drought in cool-season legumes, with special reference to chickpea and pea. In: Singh KB, Saxena MC, editors. Breeding for stress tolerance in cool-season food legumesChichester: J Wiley & Sons; 1993. p. 271–91.
[82] Sabaghpour SH, Mahmodi AA, Saeed A, Kamel M, Malhotra RS. Study on chickpea drought tolerance lines under dryland condition of Iran. Indian J Crop Sci 2006;1(1–2):70–3.
[83] Levitt J. Responses of plants to environmental stresses. New York: Academic Press; 1972. p. 697.
[84] Berger J, Palta J, Vadez V. An integrated framework for crop adaptation to dry environments: responses to transient and terminal drought. Plant Sci 2016;253:58–67.
[85] Gaur PM, Jukanti AK, Samineni S, Gowda CLL. Chickpea (*Cicer arietinum* L.). In: Bhardawaj DN, editor. Breeding of field crops. Jodhpur, India: Agrobios; 2012.
[86] Hamwieh A, Imtiaz M. Identifying water-responsive and drought-tolerant chickpea genotypes. Crop Pasture Sci 2015;66(10):1003–11.
[87] Berger JD, Ali M, Basu PS, Chaudhary BD, Chaturvedi SK, Deshmukh PS, et al. Genotype by environment studies demonstrate the critical role of phenology in adaptation of chickpea (*Cicer arietinum* L.) to high and low yielding environments of India. Field Crops Res 2006;98:230–44.
[88] Gaur PM, Samineni S, Tripathi S, Varshney RK, Gowda CL. Allelic relationships of flowering time genes in chickpea. Euphytica 2015;203(2):295–308.
[89] Kashiwagi J, Krishnamurthy L, Upadhyaya HD, Krishna H, Chandra S, Vadez V, et al. Genetic variability of drought-avoidance root traits in the mini-core germplasm collection of chickpea (*Cicer arietinum* L.). Euphytica 2005;146:213–22.
[90] Gaur PM, Krishnamurthy L, Kashiwagi J. Improving drought-avoidance root traits in chickpea (*Cicer arietinum* L.)-current status of research at ICRISAT. Plant Prod Sci 2008;11:3–11.
[91] Kashiwagi J, Krishnamurty L, Gaur PM, Chandra S, Upadhyaya HD. Estimation of gene effects of the drought avoidance root characteristics in chickpea (*C. arietinum* L.). Field Crops Res 2008;105:64–9.

[92] Kashiwagi J, Krishnamurthy L, Purushothaman R, Upadhyaya HD, Gaur PM, Gowda CLL, et al. Scope for improvement of yield under drought through the root traits in chickpea (*Cicer arietinum* L.). Field Crops Res 2015;170:47–54.
[93] Purushothaman R, Krishnamurthy L, Upadhyaya HD, Vadez V, Varshney RK. Shoot traits and their relevance in terminal drought tolerance of chickpea (*Cicer arietinum* L.). Field Crop Res 2016;197:10–27.
[94] Chen Y, Ghanem ME, Siddique KH. Characterising root trait variability in chickpea (*Cicer arietinum* L.) germplasm. J Exp Bot 2017;68:1987–99.
[95] Wasaya A, Zhang X, Fang Q, Yan Z. Root phenotyping for drought tolerance: a review. Agronomy 2018;8(11):241.
[96] Ye H, Roorkiwal M, Valliyodan B, Zhou L, Chen P, Varshney RK, et al. Genetic diversity of root system architecture in response to drought stress in grain legumes. J Exp Bot 2018;69:3267–77.
[97] Liu F, Jensen CR, Andersen MN. Hydraulic and chemical signals in the control of leaf expansion and stomatal conductance in soybean exposed to drought stress. Funct Plant Biol 2003;30:65–73.
[98] Davies J. Plant hormones: Physiology. Biochemistry and molecular biologyDordrecht, The Netherlands: Kluwer Academic; 1995.
[99] Nayyar H, Chander S. Protective effects of polyamines against oxidative stress induced by water and cold stress in chickpea. J Agron Crop Sci 2004;190(5):355–65.
[100] Hare PD, Cress WA. Metabolic implications of stress-induced proline accumulation in plants. Plant Growth Regul 1997;21(2):79–102.
[101] Mafakheri A, Siosemardeh AF, Bahramnejad B, Struik PC, Sohrabi Y. Effect of drought stress on yield, proline and chlorophyll contents in three chickpea cultivars. Aust J Crop Sci 2010;4(8):580.
[102] Kaur H, Gupta AK, Kaur N, Sandhu JS. *Cicer pinnatifidum*: A biochemically important wild species for improvement of chickpea cultivars. J Food Legumes 2010;23(1):86–8.
[103] Devasirvatham V. The basis of chickpea heat tolerance under semi-arid environments. Thesis: The University of Sydney; 2013.
[104] Malhotra RS, Saxena MC. Screening for cold and heat tolerance in cool-season food legumes. John Wiley & Sons; 1993.
[105] Singh KB, Malhotra RS, Saxena MC. Relationship between cold severity and yield loss in chickpea (*Cicer arietinum* L.). J Agron Crop Sci 1993;170:121–7.
[106] Blum A. Plant breeding for stress environmentsBoca Raton, Florida, USA: CRC Press; 1988. p. 1–223.
[107] Upadhyaya HD, Dronavalli N, Gowda CLL, Singh S. Identification and evaluation of chickpea germplasm for tolerance to heat stress. Crop Sci 2011;51(5):2079–94.
[108] Singh NT, Dhaliwal GS. Effect of soil temperature on seedling emergence in different crops. Plant Soil 1972;37(2):441–4.
[109] Ellis RH, Covell S, Roberts EH, Summerfield RJ. The influence of temperature on seed germination rate in grain legumes. XI. Intraspecific variation in chickpea (*Cicer arietinum* L.) at constant temperatures. J Exp Bot 1986;37:1503–15.
[110] Singh F, Diwakar B. Chickpea botany and production practices. Manual. India: International Crops Research Institute for the Semi-Arid Tropics (ICRISAT); 1995. Available from: http://www.icrisat.org/what-we-do/learning-opportunities/lsu-pdfs/sds.16.pdf.
[111] Covell S, Ellis RH, Roberts EH, Summerfield RJ. The influence of temperature on seed germination rate in grain legumes: I. A comparison of chickpea, lentil, soyabean and cowpea at constant temperatures. J Exp Bot 1986;37(5):705–15.

[112] Saxena MC, Saxena NP, Mohamed AK. High temperature stress. In: Summerfield RJ, editor. World crops: Cool season food legumesDordrecht: Springer; 1988. p. 845–56.
[113] Paulsen GM. High temperature responses of crop plants. In: Boote KJ, Bennett JM, Sinclair TR, Paulsen GM, editors. Physiology and determination of crop yieldMadison; 1994. p. 365–89.
[114] Guilioni L, Wery J, Tardieu F. Heat stress-induced abortion of buds and flowers in pea: is sensitivity linked to organ age or to relations between reproductive organs. Ann Bot 1997;80(2):159–68.
[115] Ismail AM, Hall AE. Reproductive-stage heat tolerance, leaf membrane thermostability and plant morphology in cowpea. Crop Sci 1999;39(6):1762–8.
[116] Vollenweider P, Günthardt-Goerg MS. Diagnosis of abiotic and biotic stress factors using the visible symptoms in foliage. Environ Pollut 2005;137(3):455–65.
[117] Boote KJ, Allen LH, Prasad PV, Baker JT, Gesch RW, Snyder AM, et al. Elevated temperature and CO_2 impacts on pollination, reproductive growth, and yield of several globally important crops. J Agric Meteorol 2005;60(5):469–74.
[118] Wang J, Gan YT, Clarke F, McDonald CL. Response of chickpea yield to high temperature stress during reproductive development. Crop Sci 2006;46(5):2171–8.
[119] Basu PS, Ali M, Chaturvedi SK. Terminal heat stress adversely affects chickpea productivity in Northern India-strategies to improve thermotolerance in the crop under climate change. In: W3 Workshop Proceedings: Impact of climate change on agriculture; 2009. p. 189–193.
[120] Laurie S, Stewart GR. Effects of nitrogen supply and high temperature on the growth and physiology of the chickpea. Plant Cell Environ 1993;16(6):609–21.
[121] Kumar J, Abbo S. Genetics of flowering time in chickpea and its bearing on productivity in semiarid environments. Advances in agronomyElsevier; 2001.
[122] Hawkes JG. The diversity of crop plants. Cambridge: Harvard University Press; 1977.
[123] International Center for Agricultural Research in the Dry Areas, ICARDA. Legume Program Annual Report, Aleppo, Syria; 1996.
[124] Imtiaz M, Malhotra RS, Yadav SS. Genetic adjustment to changing climates: Chickpea. Crop adaptation to climate changeWiley Blackwell; 2011. p. 251–68.
[125] Robertson LD, Ocampo B, Singh KB. Morphological variation in wild annual *Cicer* species in comparison to the cultigen. Euphytica 1997;95(3):309–19.
[126] Sharma S, Upadhyaya HD, Roorkiwal M, Varshney RK, Gowda CL. Chickpea. In: Singh M, Upadhyaya HD, Bisht IS, editors. Genetic and genomic resources of grain legume improvementAmsterdam: Elsevier; 2013. p. 81–111.
[127] Varshney RK, Thudi M, Nayak SN, Gaur PM, Kashiwagi J, Krishnamurthy L, et al. Genetic dissection of drought tolerance in chickpea (*Cicer arietinum* L.). Theor Appl Genet 2014;127(2):445–62.
[128] Paul PJ, Samineni S, Thudi M, Sajja SB, Rathore A, Das RR, et al. Molecular mapping of QTLs for heat tolerance in chickpea. Int J Mol Sci 2018;19(8):2166.
[129] Devasirvatham V, Tan DK, Trethowan RM. Breeding strategies for enhanced plant tolerance to heat stress. Advances in plant breeding strategies: Agronomic, abiotic and biotic stress traitsSpringer, Cham; 2016. p. 447–69.
[130] Bhandari K, Nayyar H. Low temperature stress in plants: An overview of roles of cryoprotectants in defense. Physiological mechanisms and adaptation strategies in plants under changing environmentSpringer; 2013. p. 193–265.
[131] Chen LJ, Xiang HZ, Miao Y, Zhang L, Guo ZF, Zhao XH, et al. An overview of cold resistance in plants. J Agron Crop Sci 2014;200(4):237–45.

[132] Rasool S, Singh S, Hasanuzzaman M, Rehman MU, Azooz MM, Lone HA, et al. Plant resistance under cold stress. In: Ahmad P, Rasool S, editors. Emerging technologies and management of crop stress toleranceAmsterdam: Elsevier; 2014. p. 79–98.

[133] Junttila O. Plant adaptation to temperature and photoperiod. Agr Food Sci 1996;5(3):251–60.

[134] Lukatkin AS, Brazaityte A, Bobinas C, Duchovskis P. Chilling injury in chilling-sensitive plants: a review. Agriculture 2012;99(2):111–24.

[135] Wilson JM. The economic importance of chilling injury. Outlook Agric 1985;14(4):197–203.

[136] Abbo S, Shtienberg D, Lichtenzveig J, Lev-Yadun S, Gopher A. The Chickpea, summer cropping, and a new model for pulse domestication in the ancient near east. Q Rev Bio 2003;78:435–48.

[137] Muehlbauer FJ, Singh KB. Genetics of chickpea. In: Saxena MC, Singh KB, editors. The chickpeaWallingford: CAB International; 1987. p. 99–125.

[138] Srinivasan A, Saxena NP, Johansen C. Cold tolerance during early reproductive growth of chickpea (*Cicer arietinum* L.): genetic variation in gamete development and function. Field Crops Res 1999;60:209–22.

[139] Bakht J, Bano A, Dominy P. The role of abscisic acid and low temperature in chickpea (*Cicer arietinum* L.) cold tolerance. II. Effects on plasma membrane structure and function. J Exp Bot 2006;57:3707–15.

[140] Kaur G, Kumar S, Thakur P, Malik JA, Bhandhari K, Sharma KD, et al. Involvement of proline in response of chickpea (*Cicer arietinum* L.) to chilling stress at reproductive stage. Sci Hortic 2011;128:174–81.

[141] Kumar S, Malik J, Thakur P, Kaistha S, Sharma KD, Upadhyaya HD, et al. Growth and metabolic responses of contrasting chickpea (*Cicer arietinum* L.) genotypes to chilling stress at reproductive phase. Acta Physiol Plant 2011;33:779–87.

[142] Thomashow MF. Plant cold acclimation: freezing tolerance genes and regulatory mechanisms. Annu Rev Plant Biol 1999;50(1):571–99.

[143] Hohl RM. Agricultural risk transfer: From insurance to reinsurance to capital markets. John Wiley & Sons; 2019.

[144] Olien CR, Smith MN. Ice adhesions in relation to freeze stress. Plant Physiol 1977;60(4):499–503.

[145] Steponkus PL. Role of the plasma membrane in freezing injury and cold acclimation. Annu Rev Plant Physiol 1984;35(1):543–84.

[146] Thomashow MF, Browse J. Plant cold tolerance. Molecular responses to cold, drought, heat, and salt stress in higher plantsAustin, Texas: RG Landes Company; 1999. p. 61–80.

[147] John R, Anjum NA, Sopory SK, Akram NA, Ashraf M. Some key physiological and molecular processes of cold acclimation. Biol Plant 2016;60(4):603–18.

[148] Arslan Ö, Eyidoğan F, Ekmekçi Y. Freezing tolerance of chickpea: biochemical and molecular changes at vegetative stage. Biol Plant 2018;62(1):140–8.

[149] Wery J. Adaptation to frost and drought stress in chickpea and implication in plant breeding. In: Saxena MC, Cubero JI, Wery J, editors. Present status and future prospect of chickpea crop production and improvement in the Mediterranean countriesZaragoza: CIHEAM; 1990. p. 77–85.

[150] Singh KB, Malhotra RS, Saxena MC. Additional sources of tolerance to cold in cultivated and wild *Cicer* species. Crop Sci 1995;35:1491–7.

[151] Somerville C. Direct tests of the role of membrane lipid composition in low-temperature-induced photo inhibition and chilling sensitivity in plants and cyanobacteria. Proc Natl Acad Sci U S A 1995;92(14):6215.

[152] Janská A, Maršík P, Zelenková S, Ovesná J. Cold stress and acclimation–what is important for metabolic adjustment. Plant Biol 2010;12(3):395–405.

[153] Nilsen ET, Orcutt DM. Physiology of plants under stress. Abiotic factorsChichester: John Wiley & Sons Ltd; 1996.

[154] Hannah MA, Wiese D, Freund S, Fiehn O, Heyer AG, Hincha DK. Natural genetic variation of freezing tolerance in Arabidopsis. Plant Physiol 2006;142(1):98–112.

[155] Chinnusamy V, Zhu J, Zhu JK. Gene regulation during cold acclimation in plants. Physiol Plant 2006;126(1):52–61.

[156] Zhu J, Jeong JC, Zhu Y, Sokolchik I, Miyazaki S, Zhu JK, et al. Involvement of *Arabidopsis* HOS15 in histone deacetylation and cold tolerance. Proc Natl Acad Sci 2008;105(12):4945–50.

[157] Levitt J. 2nd ed. Responses of plants to environmental stresses. Volume II. Water, radiation, salt, and other stressesAcademic Press; 1980.

[158] McKersie BD, Bowley SR. Active oxygen and freezing tolerance in transgenic plants. In: Li PH, Chen THH, editors. Plant cold hardinessBoston: Springer; 1997. p. 203–14.

[159] Anchordoguy TJ, Rudolph AS, Carpenter JF, Crowe JH. Modes of interaction of cryoprotectants with membrane phospholipids during freezing. Cryobiology 1987;24(4):324–31.

[160] Strauss G, Hauser H. Stabilization of lipid bilayer vesicles by sucrose during freezing. Proc Natl Acad Sci 1986;83(8):2422–6.

[161] Livingston DP, Henson CA. Apoplastic sugars, fructans, fructan exohydrolase, and invertase in winter oat: responses to second-phase cold hardening. Plant Physiol 1998;116(1):403–8.

[162] Singh KB, Malhotra RS, Halila MH, Knights EJ, Verma MM. Current status and future strategy in breeding chickpea for resistance to biotic and abiotic stresses. Euphytica 1994;73:137–49.

[163] Berger JD. Ecogeographic and evolutionary approaches to improving adaptation of autumn-sown chickpea (*Cicer arietinum* L.) to terminal drought: the search for reproductive chilling tolerance. Field Crop Res 2007;104:112–22.

[164] Berger JD, Kumar S, Nayyar H, Street KA, Sandhu JS, Henzell JM, et al. Temperature-stratified screening of chickpea (*Cicer arietinum* L.) genetic resource collections reveals very limited reproductive chilling tolerance compared to its annual wild relatives. Field Crop Res 2012;126:119–29.

[165] Singh KB, Omar M, Saxena MC, Johansen C. Screening for drought resistance in spring chickpea in the Mediterranean region. J Agron Crop Sci 1997;178(4):227–35.

[166] Hawtin GC, Singh KB. Prospects and potential of winter sowing of chickpea in the Mediterranean region. In: Saxena MC, Singh KB, editors. Ascochyta blight and winter sowing of chickpeaThe Hague, The Netherlands: MartinusNijhoff/Dr. W. Junk Publishers; 1984. p. 7–16.

[167] Toker C, Uzun B, Çağırgan Mİ. Utilization of winter-sowing of chickpea (*Cicer arietinum* L.) in dry and semi-drylands; 1997 Haz 19-22. İstanbul: Su Kongresi ve Sergisi; 1997. p. 10–107.

[168] Toker C, Cagirgan MI. Assessment of response to drought stress of chickpea (*Cicer arietinum* L.) lines under rainfed conditions. Tr J Agric For 1998;22:615–21.

[169] Clarke HJ, Siddique KHM. Response of chickpea genotypes to low temperature stress during reproductive development. Field Crop Res 2004;90:323–34.

[170] Singh KB, Malhotra RS, Saxena MC. Sources for tolerance to cold in *Cicer* species. Crop Sci 1990;30:1136–8.

[171] Saeed A, Darvishzadeh R, Hovsepyan H, Asatryan A. Tolerance to freezing stress in *Cicer* accessions under controlled and field conditions. Afr J Biotechnol 2010;9:2618–26.

[172] Mantri NL, Ford R, Coram TE, Pang EC. Transcriptional profiling of chickpea genes differentially regulated in response to high-salinity, cold and drought. BMC Genom 2007;8:1–14.

[173] Dinari A, Niazi A, Afsharifar AR, Ramezani A. Identification of upregulated genes under cold stress in cold-tolerant chickpea using the cDNA AFLP approach. PloS ONE 2013;8:1–7.

[174] Sharma KD, Nayyar H. Cold stress alters transcription in meiotic anthers of cold tolerant chickpea (*Cicer arietinum* L.). BMC Res Notes 2014;7:1–13.

[175] Mugabe D, Coyne CJ, Piaskowski J, Zheng P, Ma Y, Landry E, et al. Quantitative trait loci for cold tolerance in chickpea. Crop Sci 2019;59:573–82.

[176] Samineni S, Kamatam S, Thudi M, Varshney R, Gaur PM. Vernalization response in chickpea is controlled by a major QTL. Euphytica 2016;207:453–61.

[177] Abrol IP, Chhabra R, Gupta RK. A fresh look at the diagnostic criteria for sodic soils. In: International symposium on salt affected soils. Karnal, India: Central Soil Salinity Research Institute; 1980.

[178] Munns R. Genes and salt tolerance: bringing them together. New Phytol 2005;167(3):645–63.

[179] Rengasamy P. Transient salinity and subsoil constraints to dryland farming in Australian sodic soils: an overview. Aust J Exp Agric 2002;42(3):351–61.

[180] Munns R, Tester M. Mechanisms of salinity tolerance. Annu Rev Plant Biol 2008;59:651–81.

[181] Agri-Facts. Salt tolerance of plants. Agri-Facts: Practical information for Alberta's Agriculture Indurstry. Government of Alberta. Canada: Alberta Agriculture and Forestry 2001. Available from: http://www.agric.gov.ab.ca/; 2001.

[182] Devasirvatham V, Tan DKY, Gaur PM, Raju TN, Trethowan RM. High temperature tolerance in chickpea and its implications for plant improvement. Crop Pasture Sci 2012;63:419–28.

[183] Turner NC, Abbo S, Berger JD, Chaturvedi SK, French RJ, Ludwig C, et al. Osmotic adjustment in chickpea (*Cicer arietinum* L.) results in no yield benefit under terminal drought. J Exp Bot 2007;58:187–94.

[184] Balibrea ME, Cuartero J, Bolarín MC, Pérez-Alfocea F. Sucrolytic activities during fruit development of *Lycopersicon* genotypes differing in tolerance to salinity. Physiol Plant 2003;118(1):38–46.

[185] Mengel K, Kirkby EK. In: Mengel K, Kirkby EA, Kosegarten H, Appel T, editors. Principles of plant nutritionDordrecht: Kluwer Academic Publishers; 2001.

[186] Santos CV, Falcao IP, Pinto GC, Oliveira H, Loureiro J. Nutrient responses and glutamate and praline metabolism in sunflower plants and calli under Na_2SO4 stress. J Plant Nutr Soil Sci 2002;165:366–72.

[187] Esechie HA, Al-Saidi A, Al-Khanjari S. Effect of sodium chloride salinity on seedling emergence in chickpea. J Agron Crop Sci 2002;188:155–60.

[188] Soltani A, Galeshi S, Zeinali E, Latifi N. Germination, seed reserve utilization and seedling growth of chickpea as affected by salinity and seed size. Seed Sci Technol 2002;30:51–60.

[189] Kaya M, Kaya G, Kaya MD, Atak M, Saglam S, Khawar KM. Interaction between seed size and NaCl on germination and early seedling growth of some Turkish cultivars of chickpea (*Cicer arietinum* L.). J Zhejiang Univ Sci 2008;9:371–7.

[190] Beltagi MS. Exogenous ascorbic acid (vitamin C) induced anabolic changes for salt tolerance in chickpea (*Cicer arietinum* L.) plants. Afr J Plant Sci 2008;2:118–23.

[191] Datta KS, Sharma KD. Effect of chloride and sulphate types of salinity on characteristics of chlorophyll content, photosynthesis and respiration chickpea (*Cicer arietinum* L.). Biol Plant 1990;32:391–5.

[192] Murumkar CV, Chavan PD. Alterations in photosynthetic carbon metabolism of chickpea (*Cicer arietinum* L.) due to imposed NaCl salinity. Agrochimica 1993;37:26–32.

[193] Katerji N, van Hoorn JW, Hamdy A, Mastrorilli M, Oweis T, Malhotra RS. Response to soil salinity of two chickpea varieties differing in drought tolerance. Agric Water Manage 2001;50:8396.

[194] Kukreja S, Nandwal AS, Kumar N, Sharma SK, Sharma SK, Unvi V, et al. Plant water status, H_2O_2 scavenging enzymes, ethylene evolution and membrane integrity of *Cicer arietinum* roots as affected by salinity. Biol Plant 2005;49:305–8.

[195] Maliro MF, McNeil D, Redden B, Kollmorgen JF, Pittock C. Sampling strategies and screening of chickpea (*Cicer arietinum* L.) germplasm for salt tolerance. Genet Resour Crop Evol 2008;55(1):53–63.

[196] Gupta B, Huang B. Mechanism of salinity tolerance in plants: physiological, biochemical, and molecular characterization. Int J Genom 2014;.

[197] Liang W, Ma X, Wan P, Liu L. Plant salt-tolerance mechanism: a review. Biochem Biophys Res Commun 2018;495:286–91.

[198] Roy SJ, Negrão S, Tester M. Salt resistant crop plants. Curr Opin Biotechnol 2014;26:115–24.

[199] Foolad MR. Recent advances in genetics of salt tolerance in tomato. Plant Cell Tiss Organ Cult 2004;76(2):101–19.

[200] Samineni S, Siddique KHM, Gaur PM, Colmer TD. Salt sensitivity of the vegetative and reproductive stages in chickpea (*Cicer arietinum* L.): podding is a particularly sensitive stage. Environ Exp Bot 2011;71:260–8.

[201] Hiremath PJ, Farmer A, Cannon SB, Woodward J, Kudapa H, Tuteja R, et al. Large-scale transcriptome analysis in chickpea (*Cicer arietinum* L.), an orphan legume crop of the semi-arid tropics of Asia and Africa. Plant Biotechnol J 2011;9:922–31.

[202] Jain M, Misra G, Patel RK, Priya P, Jhanwar S, Khan AW, et al. A draft genome sequence of the pulse crop chickpea (*Cicer arietinum* L.). Plant J 2013;74:715–29.

[203] Kudapa H, Azam S, Sharpe AG, Taran B, Li R, Deonovic B, et al. Comprehensive transcriptome assembly of chickpea (*Cicer arietinum* L.) using Sanger and next generation sequencing platforms: development and applications. PloS ONE 2014;9(1):e86039.

[204] Varshney RK, Hoisington DA, Upadhyaya HD, Gaur PM, Nigam SN, Saxena K, et al. Molecular genetics and breeding of grain legume crops for the semi-arid tropics. In: Varshney RK, Tuberosa R, editors. Genomics-assisted crop improvementDordrecht: Springer; 2007. p. 207–41.

[205] Vadez V, Krishnamurthy L, Thudi M, Anuradha C, Colmer TD, Turner NC, et al. Assessment of ICCV 2 × JG 62 chickpea progenies shows sensitivity of reproduction to salt stress and reveals QTL for seed yield and yield components. Mol Breed 2012;30:9–21.

[206] Pushpavalli R, Krishnamurthy L, Thudi M, Gaur PM, Rao MV, Siddique KH, et al. Two key genomic regions harbour QTLs for salinity tolerance in ICCV 2× JG 11 derived chickpea (*Cicer arietinum* L.) recombinant inbred lines. BMC Plant Biol 2015;15(1):124.

[207] Mantri NL, Ford R, Coram TE, Pang EC. Evidence of unique and shared responses to major biotic and abiotic stresses in chickpea. Environ Exp Bot 2010;69(3):286–92.

[208] Molina C, Zaman-Allah M, Khan F, Fatnassi N, Horres R, Rotter B, et al. The salt-responsive transcriptome of chickpea roots and nodules via deep Super SAGE. BMC Plant Biol 2011;11:31.

[209] Varshney RK, Hiremath PJ, Lekha P, Kashiwagi J, Balaji J, Deokar AA, et al. A comprehensive resource of drought-and salinity-responsive ESTs for gene discovery and marker development in chickpea (*Cicer arietinum* L.). BMC Genom 2009;10(1):523.

[210] Johansen C, Saxena NP, Chauhan YS, Subba Rao GV, Pundir RPS, Kumar Rao JVDK, et al. Genotypic variation in salinity response of chickpea and pigeonpea. In: Sinha SK, Sane PV, Bhargava SC, Agrawal PK, editors. Proceedings of the international congress of plant physiology, vol. 1. New Delhi: Indian Society for Plant Physiology and Biochemistry, Indian Agriculture Research Institute; 1990. p. 977–83.

[211] Saxena NP. Chickpea. In: Goldworthy F, Fisher NM, editors. The physiology of tropical field cropsChichester: John Wiley & Sons Ltd; 1984. p. 419–52.

[212] Vadez V, Krishnamurthy L, Serraj R, Gaur PM, Upadhyaya HD, Hoisington DA, et al. Large variation in salinity tolerance in chickpea is explained by differences in sensitivity at the reproductive stage. Field Crops Res 2007;104:123–9.

Chapter 16

Mungbean Wild Genetic Resource: A Potential Source of Genetic Improvement for Biotic and Abiotic Stress Tolerance

Muhammad Jawad Asghar[a], Amjad Hameed[a], Muhammad Shahid[a], Muhammad Rizwan[b] and Farrukh Azeem[c]

[a]Nuclear Institute for Agriculture and Biology, Faisalabad, Pakistan; [b]Nuclear Institute of Agriculture, Tandojam, Sindh, Pakistan; [c]Department of Bioinformatics and Biotechnology, Government College University, Faisalabad, Pakistan

16.1 Introduction

Mungbean [*Vigna radiate* (L.) Wilczek] is the most important summer legume. It is grown throughout the tropics and subtropics [1]. About 90% of global production is in South Asia. It is a short duration crop and can be grown in wide range of soils and environments [2]. The seeds and sprouts of mungbean are used as an excellent source of antioxidants in China, India, Bangladesh, and South-east Asia. Mungbean sprouts are also used as salad vegetable by many countries [1,3]. It contains vitamin A (94 mg), iron (7.3 mg), zinc (3 mg), calcium (124 mg), and folate (549 mg) per 100 g dry seed. Usually, it is used in split form (Dhal) and in many other food products [4]. It is also rich in digestible protein (24%), due to which it is used in cereal-based diets [5].

Mungbean productivity is constrained by biotic and abiotic factors. The major diseases of this crop are yellow mosaic, anthracnose, powdery mildew, cercospora leaf spot, dry root rot, halo blight, bacterial leaf spot, and tan spot. Bruchids, whitefly, thrips, army worm, spinola bug, stem fly, aphids, and pod borers are the major insect-pests of mungbean [6,7]. Drought, water logging, salinity, and heat stress are major abiotic factors [2,8]. The biotic and abiotic stresses not only decrease yield but also quality of existing varieties.

Mungbean wild relatives would be important sources to develop varieties with broad genetic base and adaptability [9,10]. So far, genetic improvement of legume crops lags behind when compared with cereal crops. The legume breeders have under-utilized the wild relatives, because of limited information

Wild Germplasm for Genetic Improvement in Crop Plants. http://dx.doi.org/10.1016/B978-0-12-822137-2.00016-3

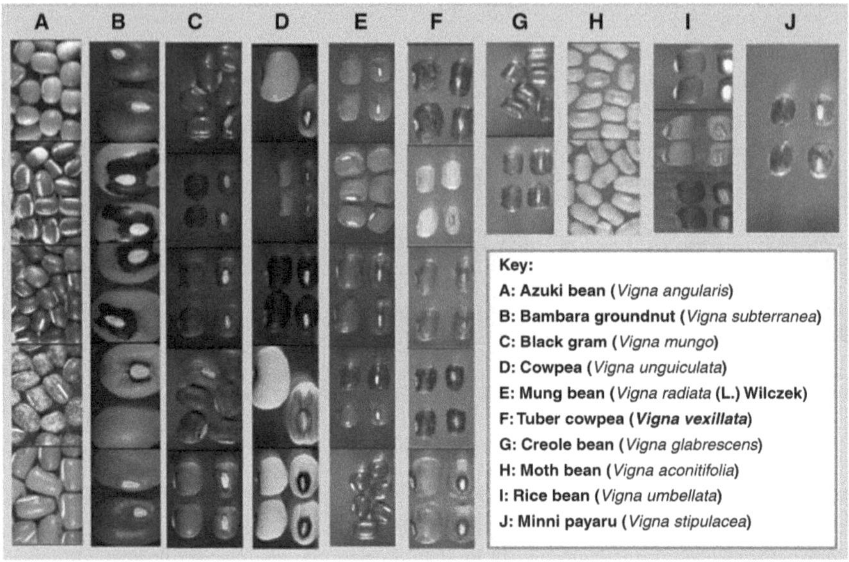

FIGURE 16.1 **Wild sources of resistance against biotic stresses including some cultivated species.**

of traits of wild species, crossing barriers, and linkage of undesirable traits [11]. Wide hybridization enables the interspecific gene transfer which may lead to additional source of variation for desirable characters. *Vigna* genus contains 88 taxa including important crops. Sources of pest and disease resistance are present in 75% taxa while sources of abiotic stress are present in 30% taxa of *Vigna*. These taxa along with cultivated *Vigna* species can be used as a source of resistance against biotic and abiotic stresses [12]. Some of the important *Vigna* species are displayed in Fig. 16.1.

16.2 Biotic stresses

Any disturbance in the metabolic processes caused by various biotic and abiotic stresses faced by the plant may reduce the actual yield. The severity of various stresses is largely due to varying weather conditions that prevail year after year, thus lowering pulses yield at farmer's field and potential yield. The low yielding cultivars are susceptibility to diseases, that is, mungbean yellow mosaic disease (MYMD), Cercospora leaf spot (CLS), and urdbean leaf crinkle disease (ULCD) which are the major diseases of mungbean.

16.2.1 Mungbean yellow mosaic disease (MYMD)

MYMD is the most serious limiting factor in mungbean cultivation [13]. Yield losses due to this disease vary from 5% to 100% depending upon disease severity, susceptibility of cultivars, and population of whitefly [14,15]. This disease is transmitted by whitefly (vector) and has a wide range of hosts. MYMD is characterized by small irregular yellow specks and spots along the veins which enlarge until leaves completely become yellow. Affected plants show stunted growth and produce a fewer flowers and pods with smaller and shriveled seeds [16,17]. Breakdown of chlorophyll contents is caused by the mungbean yellow mosaic disease which results in yellowing of leaves. In addition to yellowing of leaves, destruction of chlorophyll content is responsible for reduced photo assimilation and subsequently plant yield reduces. It can ultimately cause up to 100% yield losses in severe epidemics [17,18].

Disease incidence ranges between 30% and 100% causing yield losses that range between 60% and 80%, depending upon the disease severity and the crop stage at which plants get infected. However, in naturally infected susceptible cultivars it varies with the time of infection and yield losses may reach up to 100% (complete crop failure). Apart from Pakistan, MYMD is the major threat to mungbean production in India, Sri Lanka, Bangladesh, Papua New Guinea, Philippines, and Thailand [19–22] and inflicts heavy yields losses annually. MYMD is caused by mungbean yellow mosaic begomoviruses belonging to family *Geminiviridae*. Like other begomoviruses its viral particles are isometric and greminate having 18–30 nm in size with two single stranded DNA molecules (DNA A and DNA B) of 2726 and 2775 nucleotides, respectively [23–26]. This virus is transmitted by whitefly (*Bemisia tabaci* Genn) and through grafting but not through seed, sap, and soil [27,28]. Mungbean yellow mosaic virus (MYMV) may render a resistant variety to become susceptible to some extent after a certain period when new viral strains are produced. Thus, varieties generally appear to be resistant or tolerant at one location may show susceptible behavior at another location.

Genetic resistance against diseases is most economical and viable solution for crop improvement. Different levels of disease resistance are found in mungbean germplasm. Disease resistance may be dominant or recessive. Mostly dominant genes controlled resistance against bacterial and fungal pathogens and recessive genes against virus in plants [29,30]. Nature of disease resistance is very important for determination of mechanism underlying the resistance and resistance development. The number of genes controlling resistance is also very important for plant breeding perspectives [31]. Different scientists have different opinions about the number of genes controlling MYMD resistance in plants. According to some scientists, resistance against MYMD is controlled by single recessive gene. While some other scientists claimed that resistance is controlled by two or three recessive genes or a single dominant gene. Recently, a scientist claimed that the resistance is controlled by two genes showing additive gene action. This conflict regarding the resistance controlled by number of genes is due to disease severity index and number of classes used by the scientists [31].

Most of the scientists believe that resistance against MYMD is qualitative in nature but some believe it as quantitative. It is quantitative because there are no clear cut classes of disease reaction. Plants show continuous change in disease

TABLE 16.1 Resistant genotypes of mungbean against MYMD.

Genotype(S)	Resistant level	References
NM-10-12-01, NM-2, VC-3960 (A-88), 98-CMH-016, VC-3960 (A-89), BRM-195, 014043, 014133, 014249, 014250, ML 1265, ML 1229, SML 1815, MH 421, BPMR-145, Vaibhav, Phule M-2003-3, TARM-18, Phule M-2002-13, Phule M-2001-3, Phule M-2002-17, Phule M-2001, EC300072, K141, VC-6960-88, VC-6773 (B-G), VC-3960-89, ACC-12840014, VC-1089 A, NCM-15-11, AZRI-1, AZRI-06, NCM-21, 14063, NCM-11-8, VRMG(g)1, LM 235 (GY), K 851, T 44, Nelambur, Sona Moong, AVRDC 1785/5, LM 150, Madura moong, TNAU 26, WBM 202 (GY), KM 2, TARM 22, HUM 1, LGG 429/1, TARM2/2, TARM2/1, NM 94, Bari mung 2, ML267, LGG407, ML267, LGG407, ML-5, ML 405, ML 408, ML 337, MUM 2, VGG3 45, Pusa 8773, ML-818, IPM 2-14, PDM139, Ganga-14, HUM-I, PDM-262, HUM-8	R	[33] [34] [35] [36] [37] [38] [39] [40] [41] [42] [43] [44] [45] [46]
ML-9	MR	[47]
LGG424B, LM108B, GG-89 and GG-39, R: TM-98-50, TM-97-55, Co-5	I	[48]
NM 94	T- Odisha and Andhra Pradesh MR- Tirunelveli	[49]
ML1628	T	
HUM 1, HUM 12, DMS 03-17-2, Pant Mung 4, Pusa 9531, HUM 16, Meha, RMG 62, TMB 37	MR	
ML-881, UPM-98,	HR	[50]

HR, Highly resistant; I, immune; MR, moderately resistant; R, Resistant; T, Tolerant.

severity which means no discrete resistance is present. One explanation to this continuous variation of disease severity is that there are some modifying genes along with major genes. These modifying genes change the degree of expression of major genes [32].

Use of resistant varieties against MYMD is the practical approach for the management of the disease [13]. Resistant varieties can be developed through identification of resistant sources against disease. A massive work is done to search out MYMD resistant sources in Pakistan, India, Bangladesh, Thailand, and in other countries. The resistant sources of MYMD (Table 16.1) can be used as potential donors for the development of the varieties.

Improvement work on mungbean at Nuclear Institute for Agriculture and Biology (NIAB), Faisalabad (Pakistan) yielded the evolution of 12 varieties of mungbean. Out of these, five are early maturing, short duration, small seeded along with high seed yield. Varieties, NM 28, NM 13-1, NM 19-19, NM 20-21, and NM 121-25 developed through irradiation in local genetic background, whereas other six varieties namely; NM 51, NM 54, NM 92, NM 98, NM 2006, and NM 2011 have bold seed size evolved through introgression of large seed size into well adapted germplasm. Area under cultivation of mungbean in the country has increased 3 times through the introduction of these short duration and high yielding varieties. These are not only constituting a substantial economic share in the country but also improving the economy of the growers of neighboring countries like India, Nepal, Bangladesh, Burma, Sri Lanka, etc., where NIAB varieties were either released for general cultivation and/or used in mungbean breeding programmes for the introgression of desirable traits especially MYMV resistance and earliness. But the yield potential and disease resistance is lost with the passage of time due to the development of new races/types of pathogens; hence continuous and substantial efforts are required to fill the gap by the evolution of cultivars with better plant type, having high yield potential, and resistance to abiotic and biotic stresses.

For the development of short duration varieties; NM-92 is a good source of earliness [51] and can be used in hybridization program to incorporate this trait in high yielding and disease resistant cultivars. World Vegetable Centre (World Veg) also utilized NM 92 and NM 94 in international mungbean breeding program. Presently, these varieties have lost resistance and become susceptible to MYMD during June-July. Hence, finding new resistant lines has become a laborious work. So, the introgression of wild relative genes to mungbean could broaden the genetic base and resistance would be long term. In a study on mungbean, 24 accessions of *V. umbellata* proved to be highly resistant to MYMD. These accessions of *V. umbellata* can be used for introduction of MYMD resistant gene into *V. radiata* [52].

16.2.2 Fungal diseases

Fungal pathogens can infect mungbean plants at different stages such as during emergence, seedlings, vegetative and reproductive stages, and cause substantial damage leading to yield loss or complete failure of production [53]. Fungal diseases are responsible for reducing yield up to 40%–60% in mungbean [54]. Species of the genera *Fusarium* (wilt), *Rhizoctonia* (wet root rot), and *Macrophomina* (dry root rot) infect mungbean plants during seed/seedlings stages (seed borne or soil borne), while species of the genera *Colletotrichum* (anthracnose), Alternaria and Cercospora (leaf spot), Erysiphe/Podospheare (Sphaerotheca) (powdery mildew) affect plants during vegetative and reproductive stages [53].

CLS, a fungal disease caused by *Cercospora canesens* is distributed by spores of infected leaves. Characteristic symptoms include leaf spots with brown to grayish center and reddish brown border. This is an important disease of mungbean and is usually occurs in severe form, causing heavy yield losses. CLS causes severe yield losses to mungbean; upto 23% [55] and is a serious disease in mungbean growing areas of Pakistan especially where high humidity prevails. During rainy season, hot and humid conditions are favorable to encounter this disease. It starts appearing after 35 days of planting, spreads rapidly depending on the humidity during growing season and causes premature defoliation of pods and grains. Severe leaf spotting and defoliation occurs at the time of flowering and pod formation under conditions favorable for this disease [56].

Leaf crinkle disease caused by ULCV is a very serious, widely distributed disease. The virus has been reported to be transmitted by Aphid species (*Aphis craccivora* and *A. gossypii*), whitefly (*Bemisia tabaci*), and leaf hoppers [57]. The disease starts appearing within 5 weeks after sowing and is characterized by crinkling, curling, and puckering of leaves often coupled with stunting and malformation of floral organs. Enlargement in size followed by crinkle surface of laminae are the characteristics symptoms on affected trifoliate leaves. Pollen production, fertility, and subsequent pod formation is severely reduced thus affecting seed weight and size of seeds in infected plants leading to decrease in yield [13].

Use of host plant resistance is an effective, economical, and eco-friendly method for managing mungbean fungal diseases [7]. Sources of resistance against powdery mildew, CLS, anthracnose, Macrophomina blight and dry root rot have been identified (Table 16.2). The majority of studies targeted resistance to CLS and powdery mildews and were conducted in the field. There have been fewer studies to identify root rot and anthracnose resistance sources, and these were conducted in both lab/glasshouse and field experiments. Most of the identified resistant materials were derived from cultivars/recombinant lines/breeding lines/land races; however, some were from wild relatives and mutant lines [58,59]. Since screening trials for resistance against Alternaria leaf spot, anthracnose, and root rot diseases are limited, more attention is required on these. These resistant lines from difference sources can be utilized as donors for developing resistant varieties.

16.2.3 Bacterial diseases

Bacterial leaf spot caused by *Xanthomonas campestris* pv. *Vigna radiata* is the most prevalent disease in India. While halo blight (*Pseudomonas savastanoi* pv. phaseolicola) and tan spot (*Curtobacterium flaccumfaciens* pv. flaccumfaciens) are economically significant diseases of mungbean in Australia [53,73]. All the diseases together can cause significant yield loss of up to 10%–100%. All bacterial pathogens are seed-borne and can persist in stubbles [74]. Resistant varieties are the best solution for disease management. Little work has been done on the screening of mungbean genotypes against bacterial diseases. The resistant sources of bacterial leaf spot (Table 16.3) can be used as potential donors for the development of the varieties.

TABLE 16.2 Resistant genotypes of mungbean against fungal diseases.

Disease	Genotype(S)	Resistant level	References
Powdery Mildew	V2159, V4189, V4207, V4574, V4668, V4990, M5-10, M5-25, BPMR-145, Vaibhav, TARM-18, Phule M-2002-13, Phule M-2001-3, Phule M-2003-3, Phule M-2002-17, and Phule M-2001-5, TARM-18, LGG-460, KMP – 2,3, 5,19,20,24,30,34,38,42,47,52, MLGG-8 591	R	[60] [59] [38] [61] [62] [63]
	V1104, V4631, V4658, V4662, V4717, V4883, V3912, V4186, BL 849, LM1668, BL 865, AKM 8803, PBM, PMB 63, KGS 83, MH 96-1, Pusa 572, GS 33-5, AKM 99-4, GS 21-5, COGG 936, ML 1299, TMB 47, HUM 1, MH 429, MH 429, MH 530, KMP-36,39 and 41, F4: C1-34-23, F5: C1-15-10, C1-15A-11, C1-21A-17, C1-25-19, C1-28-20, C1-32-22, C1-37-23, C1-38-27, C1-41-28, C1-44-31, C1-175-111, C1-236-152, C1-246-159, C1-275-177	HR	[60] [64] [65] [63] [66]
Cercospora leaf spot (CLS)	V1471, V2757, V2773, V4718, V5036, M5-22, M5-25, NCM 255-2, NCM 257-6, ML-267, NCM 251-1, NCM 259-2, NCM 251-13, NCM 257-2, NM-92, NCM 251-12, VC-3960-A88, NCM 257-10, NCM-209, Mung-6 C1/94-4-19, VC 3960-A89, ML5, 443, 453, 515, 610, 611, 613, 682, 688, 713, 728, 735,746, 759, 769, PANT M103, PANT M3, PUSA 105, ML 613, PANT M2, ML 173, ML 347, ML 561, PANT M4, PDM 11, GM-02-08, GM-02-13, GM-03-03, ML-5, ML-4, HUM-9, HUM-4, HUM-1, SM-9-124, LGG-450, SM-9-107, AKM 9910, IPM 02-5, ML 1299, SML 668	R	[60] [59] [67] [68] ([58] [62] [69] [65]
	BRM-188, NM-98, C2/94-4-42, 98-cmg-003, NM-2, NM-1, 98cmg-018, Basanti, CO-3, PDM-11, VC3960-88, BARI-Mung-2, LGG-460, 1224-52, 12404	HR	[67] [62] [70]
	KMP-13	MR	[63]
	Vigna aconitifolia, V. glabrascence, V. sublobata, V. umbellata and a mutant PBM.	No infection	[58]

(*Continued*)

TABLE 16.2 Resistant genotypes of mungbean against fungal diseases. (*Cont.*)

Disease	Genotype(S)	Resistant level	References
CLS, anthracnose, Macrophomina blight	ML1486, ML1464, ML1194, ML1349	R	[54]
Dry root rot	MSJ 118, KM 4-44, KM 4-59, 40504, NCM 257-5, 40457, NCM 251-4, 6368-64-72	R	[71,72]
	NCM 252-10 and 40536	HR	[72]

TABLE 16.3 Resistant genotypes of mungbean against bacterial leaf spot disease.

Genotype(S)	Resistant level	References
Jalgaon781, P646, P475, PLM501 NCM 11-8, NCM 15-1, AZRI-1 and 14063	R	[75] [41]
NCM 258-10, NCM-21, NCM 11-6, AZRI-06 and NCM 11-3	MR	[41]

R, Resistant; MR, moderately resistant.

16.2.4 Molecular approach to combat diseases

Molecular tools are increasingly important in modern plant research. Evaluation of disease resistance is not always straight forward as assays are often influenced by environmental factors. Inaccuracies of the field selections can be overcome by an alternative approach, that is, marker assisted selection (MAS). The use of molecular markers for resistance genes is particularly powerful as it removes the delays in breeding programs associated with the phenotypic analysis. Now a day, this is possible due to availability of many kinds of markers namely, RAPD, AFLP, and SSR that are used for genotyping purposes. RAPD is one among the various molecular markers that are being used to screen resistant genes. This RAPD technique is reliable, faster, and easier for exploiting genetic polymorphism within and among species and populations. RAPD markers have been already used successfully in many crops for mapping genes, as RAPD markers are more rapidly and easily detectable than RFLP markers. The RAPD marker had also been used effectively for tagging MYMV resistant genes in mungbean [76]. Distantly linked RAPD primer OPAJ20 of the MYMV resistance has also been reported [77]. Two MYMV resistance marker loci, YR4 and CYR1, were identified in mungbean and urdbean using R gene homolog sequences [78]. In molecular MAS, once a gene of interest is tagged with molecular marker, selection for that gene can be made based on the marker rather than the phenotype of individual which embodies the gene. Use of molecular markers may help in locating important genes and facilitate their transfer to cultivars with greater production potential. Validating the DNA markers for MYMV in mungbean germplasm breeding lines with different genetic backgrounds, using a high-throughput genotyping system and evaluating the usefulness of the markers for those genes in MAS and fingerprinting of parents and selected mutant lines generates useful information that aids breeders in selecting markers for use in MAS and gene pyramiding to enhance durability of MYMV resistance. Use of molecular markers may help in locating important genes and facilitate their transfer to cultivars with greater production potential.

16.2.5 Major insect pests and economic aspect

Different insects damage mungbean crop at different growth stages. The available losses due to pest complex on different varieties of mungbean ranged from 27.03%–38.06% [79]. The crop is the most vulnerable to the attacks of different insect pests from sowing to harvest and thus pests cause huge production losses (up to 70% of grain yield). In some cases, 100% yield loss is also reported if crop is left unattended against injuries by different insect pests (particularly late season bugs including spinola). The insect pest complex of mungbean and urdbean has undergone a tremendous change during the last 2 decades due to change in climate, cropping pattern, insecticide application pattern, and introduction of high yielding varieties [80]. The key plant bugs like *Riptortus* spp., *Nezara* spp. and the pod bug, *Clavigralla* spp (spinola in particular) have serious plant injuries and their damages result in greater yield losses.

Insect-pests attack mungbean at all growing stages of crop from sowing to storage and cause a heavy loss to yield. Some insect-pests directly damage the crop, while others act as vectors of diseases. The economically important insect-pests in mungbean include stem fly, thrips, aphids, whitefly, pod borer complex, pod bugs, and bruchids [81]. Stem fly (bean fly), *Ophiomyia phaseoli* (Tryon), is one of the major pests of mungbean [82]. This pest infests the crop within a week after germination and under epidemic conditions, it can cause total crop loss [83]. Whitefly, *B. tabaci* is a serious pest in mungbean and damages the crop either directly or indirectly by transmitting MYMD. Whitefly causes yield loss from 17% to 71% in mungbean [84,85].

Thrips infest mungbean both in the seedling and in flowering stages. During the seedling stage, thrips infest the seedling's growing point while flowering thrips cause heavy damage and attack during flowering and pod formation. They feed on the pedicles and stigma of flowers. Under severe infestation, flowers drop and no pod formation takes place. Spotted pod borer, *Maruca vitrata* (Fab.) is a major insect-pest of mungbean in the tropics and subtropics. The pest causes a yield loss of 2%–84% in mungbean amounting the US $30 million [86]. The larvae damage all the stages of the crop including flowers, stems, peduncles, and pods; however, heavy damage occurs at the flowering stage where the larvae form webs combining flowers and leaves [87]. Cowpea aphid, *Aphis craccivora* Koch., sucks plant sap that causes loss of plant vigor and may lead to yellowing, stunting, or distortion of plant parts. Further, aphids secrete honeydew (unused sap) that leads to the development of sooty mould on plant parts. Cowpea aphid also acts as a vector of bean common mosaic virus. Bruchids are the most important stored pests of legume seeds worldwide. They infest seeds both in field and in the storage, however, major damage is caused in storage. Bruchid damage can cause up to 100% losses within 3–6 months, if not controlled [88,89]. Twenty species of bruchids have been reported infesting different pulse crops [90]. Of these, the Azuki bean weevil (*Callosobruchus chinensis* L.) and cowpea weevil (*Callosobruchus maculatus* Fab.) are the most serious pests of mungbean. The behavior of bruchids where the grubs feed inside the legume seeds makes it easy to spread them through international trade.

Among pod sucking bugs, *Clavigrella gibbosa* (Spinola) is prominent in Eastern Uttar Pradesh (India) and the total grain loss worked out to be about 50,000 tons annually for that region alone. In addition to that our farmers are mostly illiterate and it is very difficult for them to control it with the use of chemicals and management practices. Up till now no technique and recommendation is available for effective control of this devastating insect. Farmers of Pakistan are facing complete losses of their crops in summer season due to this insect. Therefore, it is very essential to combat these serious problems either through the development of tolerant genotypes and through management practices to sustain mungbean and to curtail the import bill of the country. Spinola bug (*Clavigrella gibbosa)* seriously affecting mungbean crop by feeding on pods from 4–5 years and damaged them even after harvesting and causes complete crop failure in summer, the main growing season of Pakistan. Spinola bug observed to be a major pest of mungbean [79]. Feeding of this bug causes premature shedding of flower-buds, flowers and pods, deformation of pods, and shriveling of grains resulting in substantial losses to pigeonpea crops [91]. Therefore, it is very essential to combat these serious problems either through the development of tolerant genotypes and through management practices to sustain mungbean and to curtail the import bill of the country.

16.2.6 Breeding for insect resistance

Identification and introgression of resistance gene(s) from other accessions or wild-relatives into mungbean breeding germplasm is dire need of time.

Identification of sources of resistance is a pre-requisite for the introgression of resistance into cultivars through breeding. The primary gene pool forms the first choice for the breeder for source of resistance. The secondary and tertiary gene pools provide further choices of variation to be incorporated into the crop. Host plant resistance plays an important role in crop protection against insect pests. The identification of new insect resistance sources provides breeders with avenues to breed for resistance to insect pests. The variability of primary gene-pool available with the breeders could serve an important source for various traits including insect resistance. Evaluation and successful monitoring of the introgressed disease and insect resistance genes into superior breeding lines through use of closely linked molecular markers would be useful tools for selection of desirable material. Extensive screening studies have been carried out under controlled and natural conditions to identify insect resistance sources in mungbean. These resistant sources are given in Table 16.4).

TABLE 16.4 Resistant genotypes of mungbean against insects/pests.

Insect/Pest	Genotype(S)	Resistant level	References
Stem fly (*Ophiomyia* spp.)	V2396, V3495, V4281, G05253, G05776, 02005, G02472, Co 3, V3726, BM 4 and Vaibhav	R	[82] [92] [93] [94] [95]
	Chai Nat 72 (CN72)	MR	[96]
Whitefly (Bemisia tabaci)	ML 1265, ML 1229, TMB-36, RMG-1004, PKV Green Gold, Bari Mung-6, MDGVV-16	R	[80] [69] [95] [97] [98]
	ML 1, ML 6, ML 7, P 290, P 292, P 131, P 293, P 325, P 364, 11,148, NM 92, NM 98, 99.CMG-059, NM 2003-06, NM. 2003-24, NM. 2003-26, NCM. 258, PDM-54, VBN 2, CO 8, VGG10-002, KM 200, NM 04-2-38, NM 10-12-1, NM 46-5-2-21, NM 013, NM 0183, NM 04-1-11, NM 15-11, MH 3153, NM-92, NM-2006, Azri 2006, NM-121, CO 3, CO 4 and CO 5	MR	[99] [100] [101] [102] [43] [103] [104] [105]

(Continued)

TABLE 16.4 Resistant genotypes of mungbean against insects/pests. (*Cont.*)

Insect/Pest	Genotype(S)	Resistant level	References
Thrips (Megalurothrips spp., Thrips palmi)	NM-92	R	[100]
	MGG 362, MGG 365	MR	[106]
Spotted pod borer (Maruca spp.)	PKV Green Gold, PDM-54-146, ML 131, ML 372, JRUM1, JRUM11, JRUM33, DP1703, LAM 14-2, UPM-83-6, UPM 83-10, LGG 505, ML 267, LGG 502, LGG 407, LGG 460, LGG 485, CGG 08-007, CGG 08-028, ML 337, ML 5, MH 85-61, ML 325 and PM 10-18	R	[95] [107] [108] [109] [110] [66]
	LU-3, LU-15, LU-33, LU-173, LU-190, LU-196, LU-397, LU-426, LU-434, J-1, LM-11, P-527, P-536, ML-65, B-101, B-103, KM-9-128, KM-9-136, RMG-492, LGG-527, LGG-538, MGG-336, KM-8-655, MGG-335 and RVSm-11-9	MR	[111] [105] [112] [113,114] [69]
Cowpea aphid (Aphis craccivora)	Bari Mung-6, Phule M702-1	R	[97] [95]
Bruchid (Callosobruchus spp.)	V2709, V2802, V1128, V2817, TC1966, Zhonglv 3, Zhonglv 4, Zhonglv 6, Jangan, VC1535-11-1-B-1-3-B, VC2764-B-7- 2-B, VC2764-B-7-1-B, VC1209-3-B- 1-2-B and VC1482-C-12-2-B	R	[115] [88] [96] [116] [117] [118] [119] [120] [121] [122]

R, Resistant; MR, moderately resistant.

16.3 Abiotic stress

Abiotic stresses negatively affect the plant growth and productivity. They are the main cause of agricultural losses worldwide [123,124]. Crop yield is steadily decreasing due to environmental variations over the decades [125]. Abiotic stresses include heat, cold, frost, water, radiation, soil, and mechanical factors [2]. Crops utilize resources from their immediate environment for their growth.

Water and salt stress are the main environmental limiting abiotic factors. The stresses change the growth pattern of mungbean by affecting the physical and biochemical parameters. They mainly affect root production as compared to shoot production [126].

16.3.1 Drought stress

Drought is a global problem. It affects the basic metabolism which controls the yield, growth, and quality of a crop. Water deficiency can also change the growth pattern of the grain [127,128]. Mungbean crop is also more sensitive to drought during flowering period. As mungbean is usually grown on marginal lands, it is largely considered as a drought tolerant. However, decrease in soil moisture level can also decrease growth and productivity of the plants. If water stress is imposed on vegetative phase then mungbean yield can be reduced to 30%. And if the stress is imposed on flowering phase then yield can be decreased to 50%–60% [129]. Many mungbean drought tolerant genotypes were screened for the development of drought tolerant varieties [47]. But these genotypes become susceptible to drought after short time. So, wild relatives of *Vigna* species would be a better choice to increase the genetic variability and sustainability for drought resistant genes in domesticated mungbean via hybridization. Among many of the wild *Vigna* species *V. aridicola* showed highest level of tolerance of dehydration, while other wild tolerant species are *V.* trilobata, *V.* vexillata var. ovate, and *V. monantha* [12].

16.3.2 Salinity stress

Salt is also one of the most predominant factors affecting the growth of the plants. Salinity is not the alone cause of decreasing the quality of the crop. There are other factors, which increase the salt stress in the environment. These factors either can be natural or induced by human. Naturally, salinity occurs by the breakdown of the rocks containing high content of calcium, chlorine, sodium, magnesium, and sulfate Industry, agriculture, transportation, mining, construction, and deforestation are the human induced activities contributing in salinity [130,131]. Salinity not only delays plants growth but also decreases flowering and yield [132]. It may affect pollination and thus decrease seed set and grain yield [133]. Reduction in yield due to salinity has been studied in many crops namely, wheat, barley, bean, rice, and cotton [134]. Mungbean is an important legume crop of the dry land agriculture. Salinity stress is restricting the productivity of mungbean in arid and semiarid regions. Mungbean production is also stagnant from the last few decades due to susceptibility toward various abiotic stresses. Wild relatives of mungbean (Table 16.5) would be good option to increase the genetic variability in germplasm and to incorporate salt tolerant genes into mungbean. High level of salinity tolerance is reported in wild relatives of mungbean namely, *V. luteola, V. marina, V. nakashimae,*

TABLE 16.5 Resistant genotypes of mungbean against insects/pests.

Abiotic stress	Wild relatives of mungbean	Tolerance level	References
Drought	*V. aridicola, V. trilobata, V. vexillata var. ovate, and V. monantha*	T	[12]
Salinity	*V. luteola, V. marina, V. nakashimae, V. riukiuensis, and V. vexillata var. macrosperma.* *V. trilobata, V. vexillata var. ovate, and V. monantha.*	HT T	[12]

HT, highly tolerant; T, tolerant

V. riukiuensis, and *V. vexillata* var. macrosperma. Tolerance to salinity was also shown by *V. trilobata, V. vexillata* var. ovate, and *V. monantha* [12].

16.4 Future outlook

Biotic and abiotic stresses greatly reduce the yield of the mungbean. The best and economical solution of the problem is to develop resistant varieties. Mungbean germplasm has been extensively screened against biotic and abiotic stresses. Many varieties have also been developed by using the germplasm. But these varieties have narrow genetic base, which rendered the varieties susceptible after a short time. Wild relatives of the mungbean have been limitedly utilized for incorporation of resistant/tolerant genes into cultivated varieties. There is a large scope for the identification of wild relatives having desirable resistant/tolerant genes. These wild relatives would be used in breeding program and will broaden the genetic vase of the varieties.

References

[1] Li L, Dong Y, Ren H, Xue Y, Meng H, Li M. Increased antioxidant activity and polyphenol metabolites in methyl jasmonate treated mung bean (*Vigna radiata*) sprouts. Food Sci Technol 2017;37(3):411–7.

[2] HanumanthaRao B, Nair RM, Nayyar H. Salinity and high temperature tolerance in mungbean [Vigna radiata (L.) Wilczek] from a physiological perspective. Front Plant Sci 2016;7:957.

[3] Xue Z, Wang C, Zhai L, Yu W, Chang H, Kou X, et al. Bioactive compounds and antioxidant activity of mung bean (*Vigna radiata* L.), soybean (*Glycine max* L.) and black bean (Phaseolus vulgaris L.) during the germination process. Czech J Food Sci 2016;34(1):68–78.

[4] Rasul F, Cheema M, Sattar A, Saleem M, Wahid M. Evaluating the performance of three mungbean varieties grown under varying inter-row spacing. J Anim Plant Sci 2012;22(4):1030–5.

[5] Khattak GS, Ashraf M, Elahi T, Abbas G. Selection for large seed size at the seedling stage in mungbean (*Vigna radiata* (L.) Wilczek). Breed Sci 2003;53(2):141–3.

[6] War AR, Murugesan S, Boddepalli VN, Srinivasan R, Nair RM. Mechanism of resistance in mungbean [*Vigna radiata* (L.) R. Wilczek var. radiata] to bruchids, *Callosobruchus* spp. (Coleoptera: *Bruchidae*). Fron Plant Sci 2017;8:1031.
[7] Pandey AK, Burlakoti RR, Kenyon L, Nair RM. Perspectives and challenges for sustainable management of fungal diseases of mungbean [*Vigna radiata* (L.) R. Wilczek var. radiata]: A review. Front Environ Sci 2018;6(53):1–15.
[8] Singh D, Singh B. Breeding for tolerance to abiotic stresses in mungbean. J Food Legumes 2011;24(2):83–90.
[9] Dwivedi SL, Ceccarelli S, Blair MW, Upadhyaya HD, Are AK, Ortiz R. Landrace germplasm for improving yield and abiotic stress adaptation. Trends Plant Sci 2016;21(1):31–42.
[10] Sharma S, Upadhyaya HD, Varshney RK, Gowda C. Pre-breeding for diversification of primary gene pool and genetic enhancement of grain legumes. Front Plant Sci 2013;4:309.
[11] Prohens J, Gramazio P, Plazas M, Dempewolf H, Kilian B, Díez MJ, et al. Introgressiomics: a new approach for using crop wild relatives in breeding for adaptation to climate change. Euphytica 2017;213(7):158.
[12] van Zonneveld M, Rakha M, yee Tan S, Chou Y-Y, Chang C-H, Yen J-Y, et al. Mapping patterns of abiotic and biotic stress resilience uncovers conservation gaps and breeding potential of *Vigna* wild relatives. Sci Rep 2020;10(1):1–11.
[13] Singh J, Rai JP, Mishra K. Viral diseases of mungbean and their management. In: Indian agriculture and farmers. BHU, Varanasi: Poddar Publication; 2016. p. 197–205.
[14] In: Rathi Y, editor. Epidemiology, yield losses and management of major diseases of Kharif pulses in IndiaPlant Pathology and Asian Congress of Mycology and Plant Pathology; 2002.
[15] Karthikeyan A, Shobhana V, Sudha M, Raveendran M, Senthil N, Pandiyan M, et al. Mungbean yellow mosaic virus (MYMV): a threat to green gram (*Vigna radiata*) production in Asia. Int J Pest Manage 2014;60(4):314–24.
[16] Habib S, Shad N, Javaid A, Iqbal U. Screening of mungbean germplasm for resistance/tolerance against yellow mosaic disease. Mycopath 2007;5(2):89–94.
[17] Sudha M, Karthikeyan A, Anusuya P, Ganesh N, Pandiyan M, Senthil N, et al. Inheritance of resistance to mungbean yellow mosaic virus (MYMV) in inter and intra specific crosses of mungbean (*Vigna radiata*). Am J Plant Sci 2013;4(10):1924.
[18] Mohan S, Sheeba A, Murugan E, Ibrahim S. Screening of mungbean germplasm for resistance to mungbean yellow mosaic virus under natural condition. Indian J Sci Technol 2014;7(7):891.
[19] Honda Y, Iwaki M, Saito Y, Thongmeearkom P, Kittisak K, Deema N. Mechanical transmission, purification, and some properties of white-fly-borne mung bean yellow mosaic virus in Thailand. Plant Dis 1983;67:801–804.
[20] Chenulu V, Verma A. Virus and virus-like diseases of pulse crops commonly grown in India. Pulse cropsNew Delhi: Oxford and IBH; 1988. p. 338–70.
[21] Malik B, Bashir M, editors. Major diseases of food legume crops of Islamic countries. Proceedings of COMSTECH NIAB international workshop of agro climatology pests and disease and their control by Jamil F F and SHM Naqvi; 1992.
[22] Jones DR. Plant viruses transmitted by whiteflies. Eur J Plant Pathol 2003;109(3):195–219.
[23] Morinaga T, Ikegami M, Miura K. Physical mapping and molecular cloning of mung bean yellow mosaic virus DNA. Intervirology 1990;31(1):50–6.
[24] Morinaga T, Ikegami M, Miura KI. The nucleotide sequence and genome structure of mung bean yellow mosaic geminivirus. Microbiol Immunol 1993;37(6):471–6.
[25] Bos L. Plant viruses, unique and intriguing pathogens: a textbook of plant virology. Backhuys Publishers; 1999.
[26] Matthews REF, Hull R. Matthews' plant virology. Gulf Professional Publishing; 2002.

[27] Nariani T. Yellow mosaic of mung (*Phaseolus aureus* L.). Indian Phytopathol 1960;13(1):24–9.
[28] Nene Y. A survey of viral diseases of pulse crops in Uttar Pradesh. UP: G.B. Pant University, Agriculture and Technology; 1972. p. 191.
[29] Soosaar JL, Burch-Smith TM, Dinesh-Kumar SP. Mechanisms of plant resistance to viruses. Nat Rev Microbiol 2005;3(10):789–98.
[30] Truniger V, Aranda MA. Recessive resistance to plant viruses. Advances in virus research, 75. Elsevier; 2009. p. 119–231.
[31] Akbar W, Akhtar MA, Murtaza G, Hussain A, Javed HM, Arshad M, et al. Mungbean yellow mosaic disease and its management, vol. 4. 2019. p. 13.
[32] Khattak G, Haq M, Rana S, Abass G, Irfag M. Effect of Mungbean yellow mosaic virus (MYMV) on yield and yield components of Mungbean (*Vigna radiata* (L.) Wilczek). Agricult Nat Resour 2000;34(1):12–6.
[33] Akhtar KP, Kitsanachandee R, Srinives P, Abbas G, Asghar MJ, Shah TM, et al. Field evaluation of mungbean recombinant inbred lines against mungbean yellow mosaic disease using new disease scale in Thailand. Plant Pathol J 2009;25(4):422–8.
[34] Bashir M, Ahmad Z, Ghafoor A. Sources of genetic resistance in mungbean and blackgram against urdbean leaf crinkle virus (ULCV). Pak J Bot 2005;37(1):47–51.
[35] Iqbal U, Iqbal SM, Afzal R, Jamal A, Farooq MA, Zahid A. Screening of mungbean germplasm against Mungbean yellow mosaic virus (MYMV) under field conditions. Pak J Phytopathol 2011;23(1):48–51.
[36] Kooner BS, Cheema HK. Screening of mungbean germplasm against whitefly, Bemisia tabaci and MYMV. Acta Hortic 2007;752:307–10.
[37] Mahalingam A, Satya VK, Manivannan N, Narayanan SL, Sathya P. Inheritance of mungbean yellow mosaic virus disease resistance in greengram [Vigna radiata (L.) Wilczek]. Int J Curr Microbiol App Sci 2018;7(1):880–5.
[38] Mandhare VK, Suryawanshi AV. Dual resistant against powdery mildew and yellow mosaic virus in greengram. Agric Sci Digest 2008;28(1):39–41.
[39] Manivannan N, Sethuraman K, Natarajan S. Screening of Greengram (Vigna radiata (L.) Wilczek) germplasm for yellow mosaic resistance. Leg Res 2001;24(4):268–71.
[40] Mondol MEA, Rahman H, Rashid MH, Hossain MA, Islam MM. Screening of mungbean germplasm for resistance to mungbean yellow mosaic virus. Int J Sustain Crop Prod 2013;8:11–5.
[41] Munawar MH, Iqbal SM, Malik SR, Chattha MR, Ali A. Identification of resistant sources in mungbean to bacterial leaf spot disease. Mycopath 2011;9(2):71–2.
[42] Pandiyan M, Subbalakshmi B, Alice D, Marimuthu R. Screening of Mungbean [Vigna radiata (L.) Wilczek] germplasm for mungbean yellow mosaic virus. Plant Arch 2007;7(1):375–6.
[43] Panduranga GS, Vijayalakshmi K, Loka RK, Rajashekara H. Evaluation of mungbean germplasm for resistance against whitefly (Bemisia Tabaci Genn.) and mungbean yellow mosaic virus (MYMV) disease. Ind J Entomol 2011;73(4):338–42.
[44] Patel MB, Srivastava KP. Field screening of some high yielding genotypes of mungbean, Vigna radiata (Linnaeus) Wilczek to whitefly Bemisia Tabaci (Gennadius) and yellow mosaic virus (YMV). Ind J Entomol 1990;52(4):547–51.
[45] Paul PC, Biswas MK, Mandal D, Pal P. Studies on host resistance of mungbean against mungbean yellow mosaic virus in the agro-ecological condition of lateritic zone of West Bengal. Bioscan 2013;8(2):583–7.
[46] Suman S, Sharma SK, Kumar H, Shahi VK. Screening of mungbean [Vigna radiata (L.) Wilczek] genotypes for resistance to mungbean yellow mosaic virus (MYMV). Environ Ecol 2015;33:855–9.

[47] Nair RM, Pandey AK, War AR, Hanumantharao B, Shwe T, Alam A, et al. Biotic and abiotic constraints in mungbean production—Progress in genetic improvement. Front Plant Sci 2019;10:1340.

[48] Salam SA, Patil MS, Salimath PM. Evaluation of mungbean cultures against MYMV in Karnataka under natural conditions. Leg Res 2009;32(4):286–9.

[49] Nair RM, Götz M, Winter S, Giri RR, Boddepalli VN, Sirari A, et al. Identification of mungbean lines with tolerance or resistance to yellow mosaic in fields in India where different begomovirus species and different Bemisia tabaci cryptic species predominate. Eur J Plant Path 2017;149(2):349–65.

[50] Yadav GS, Dahiya B. Performance of mungbean genotypes against whitefly and yellow mosaic. Ann. Biol. 2004;20:57–9.

[51] ur Rehman A, Ali MA, Saleem M, Tadesse W. Study of heritable variation and genetics of earliness in mungbean (*Vigna radiata* L Wilczek). Euphytica 2010;176(3):331–9.

[52] Sudha M, Karthikeyan A, Shobhana V, Nagarajan P, Raveendran M, Senthil N, et al. Search for *Vigna* species conferring resistance to mungbean yellow mosaic virus in mungbean. Plant Genet Resour 2015;13(2):162.

[53] Ryley M, Tatnell J, editors. Management of the major foliar diseases of mungbeans and peanuts in Australia. In: Proceedings of the fourth Asian conference on plant pathology and the eighteenth Biennial Australasian Plant Pathology Society Conference (ACPP/APPS 2011). Australasian Plant Pathology Society; 2011.

[54] Kaur L, Singh P, Sirari A. Biplot analysis for locating multiple disease resistance diversity in mungbean germplasm. Plant Dis Res 2011;26(1):55–60.

[55] Quebral F, Cagampang I. Influence of *Cercospora* leaf spot control on yield of mungbean. Agriculture at Los Banos; 1970. p. 7–12.

[56] Bashir M, Zubair M. Survey report of kharif pulses in Islamabad, Rawalpindi and Sialkot districts during 1985. Pulses Programme, NARC, Pakistan Agricultural Research Council; 1985.

[57] Narayanasamy P, Jaganathan T. Vector transmission of black gram leaf crinkle virus. Madras Agric J 1973;60(7):651–2.

[58] Marappa N. Screening of mungbean genotypes and its wild relatives for resistant sources to *Cercospora* leaf spot disease. Asian J Bio Sci 2008;3:324–6.

[59] Wongpiyasatid A, Chotechuen S, Hormchan P, Srihuttagum M. Evaluation of yield and resistance to powdery mildew, *Cercospora* leaf spot and cowpea weevil in mungbean mutant lines. Agric Nat Resour 1999;33(2):204–15.

[60] Hartman GL, Wang TC, Kim D. Field evaluation of mungbeans for resistance to Cercospora leaf spot and powdery mildew. Int J Pest Manag 1993;39(4):418–21.

[61] Sujatha K, Kajjidoni ST, Patil PV, Somashekhar G. Heterosis for productivity related traits involving diverse parents for powdery mildew reaction in mungbean. J Food Leg 2011;24(2):101–5.

[62] Yadav DL, Jaisani P, Pandey RN. Identification of sources of resistant in mungbean genotypes and influence of fungicidal application to powdery mildew epidemics. Int J Curr Microbiol Appl Sci 2014;3(2):513–9.

[63] Bhaskar AV. Genotypes against major diseases in green gram and black gram under natural field conditions. Int J Curr Microbiol App Sci 2017;6(6):832–43.

[64] Ramakrishnan CKD, Savithramma DL. Screening of mungbean germplasm for powdery mildew disease Resistant. Int J Agron Agric Res 2014;4(6):16–21.

[65] Akhtar J, Lal HC, Kumar Y, Singh PK, Ghosh J, Khan Z, et al. Multiple disease resistant in greengram and blackgram germplasm and management through chemicals under rain-fed conditions. Leg Res 2014;37(1):101–9.

[66] Kumar R, Singh PS. Screening of certain mungbean, Vigna radiata (L.) Wilczek genotypes against spotted pod borer and pod bugs. J Exp Zool Ind 2017;20(1):595–7.
[67] Iqbal SM, Zubair M, Haqqani AM. Resistant in Mungbean to Cercospora leaf spot disease. Int J Agric Biol 2004;06(5):792–3.
[68] Singh G, Sharma YR, Shanmugasundaram S, Shih SL, Green S K. *Improving income and nutrition by incorporating mungbean in cereal fallows in the Indo-Gangetic Plains of South Asia DFID Mungbean Project for 2002-2004*, in Proceedings of the Final Workshop and Planning Meeting, Status of Mung Bean Yellow Mosaic Virus Resistance Breeding. Ludhiana: Punjab Agricultural University; 2004. p. 204–13.
[69] Singh SP, Singh SK. Sources of resistant in mungbean for Cercospora leaf spot diseases. Ann Agric Biosci Res 2014;19(2):280–1.
[70] Zhimo VY, Panja BN, Saha J, Nath R. Evaluation of mungbean genotypes for resistance against Cercospora leaf spot and Yellow Mosaic disease under field condition. J Mycopathol Res 2013;51(2):273–8.
[71] Choudhary S, Choudhary AK, Sharma OP. Screening of mungbean (Vigna radiata) genotypes to identify source of resistant to dry root rot. J Food Leg 2011;24(2):117–9.
[72] Khan KSH, Shuaib M. Identification of sources of resistant in mungbean (Vigna radiata L.) against charcoal rot Macrophomina phaseolina (Tassi) Goid. Afr. Crop Sci. Conf. Proc. 2007;8:2101–2.
[73] Munawar MH, Iqbal SM, Malik SR, Chattha MR, Ali A. Identification of resistant sources in mungbean to bacterial leaf spot disease. Mycopath 2011;9(2):71–2.
[74] Noble TJ, Young AJ, Douglas CA, Williams B, Mundree S. Diagnosis and management of halo blight in Australian mungbeans: a review. Crop Pasture Sci 2019;70(3):195–203.
[75] Patel PN, Jindal JK. Bacterial leaf spot and halo blight disease of mungbean and other legume in India. Ind Phytopath 1972;25:526–9.
[76] Selvi R, Muthiah A, Manivannan N, Raveendran T, Manickam A, Samiyappan R. Tagging of RAPD marker for MYMV resistance in mungbean (*Vigna radiata* (L.) Wilczek). Asian J Plant Sci 2006;5(2):277–80.
[77] Lambrides C, Lawn R, Godwin I, Manners J, Imrie B. Two genetic linkage maps of mungbean using RFLP and RAPD markers. Austr J Agric Res 2000;51(4):415–25.
[78] Maiti S, Basak J, Kundagrami S, Kundu A, Pal A. Molecular marker-assisted genotyping of mungbean yellow mosaic India virus resistant germplasms of mungbean and urdbean. Mol Biotechnol 2011;47(2):95–104.
[79] Duraimurugan P, Tyagi K. Pest spectra, succession and its yield losses in mungbean and urdbean under changing climatic scenario. Legume Res 2014;37(2):212.
[80] Kooner B, editor. Screening of mungbean germplasm against whitefly (*Bemisia tabaci* genn.) and mungbean yellow mosaic virus. I International conference on indigenous vegetables and legumes prospectus for fighting poverty. Hunger and Malnutrition; 2006. p. 752.
[81] Swaminathan R, Singh K, Nepalia V. Insect pests of greengram *Vigna radiata* (L.) Wilczek and their management. Agric Sci 2012;10:197–222.
[82] Talekar N. Agromyzid flies of food legumes in the tropics. Wiley Eastern; 1990.
[83] Chiang H, Talekar N. Identification of sources of resistance to the beanfly and two other agromyzid flies in soybean and mungbean. J Econom Entomol 1980;73(2):197–9.
[84] Marimuthu G, Rajan S, Chandrashekaran M. Social entrainment of the circadian rhythm in the flight activity of the microchiropteran bat Hipposideros speoris. Behav Ecol Sociobiol 1981;8(2):147–50.
[85] Chhabra K, Kooner B. Insect pest management in mungbean and black gram: Status and strategies. IPM Syst Agric 1998;4:233–310.

[86] Zahid M, Islam M, Begum M. Determination of economic injury levels of *Maruca vitrata* in mungbean. J Agric Rural Dev 2008;6(1):91–7.
[87] Sharma HC, Saxena K, Bhagwat V. The legume pod borer, *Maruca vitrata*: bionomics and management. Patancheru, India: ICRISAT; 1999.
[88] Tomooka N, Lairungreang C, Nakeeraks P, Egawa Y, Thavarasook C. Development of bruchid-resistant mungbean line using wild mungbean germplasm in Thailand. Plant Breed 1992;109(1):60–6.
[89] Somta P, Ammaranan C, Ooi PA-C, Srinives P. Inheritance of seed resistance to bruchids in cultivated mungbean (*Vigna radiata*, L. Wilczek). Euphytica 2007;155(1–2):47–55.
[90] Southgate B. Biology of the Bruchidae. Ann Rev Entomol 1979;24(1):449–73.
[91] Narasimhamurthy G, Keval R. Field evaluation of some insecticides and bio-pesticide against tur pod bug, *Clavigralla gibbosa* (Spinola) in long duration pigeonpea. Afr J Agric Res 2013;8(38):4876–81.
[92] Abate T, Girma A, Ayalew G. Progress in host plant resistance research against bean stem maggots. Afr Crop Sci Conf Proc 1995;2:167–73.
[93] Devasthali S, Joshi M. Infestation and varietal preference of insect-pests in green gram. Ind. Agric 1994;38(4):263–72.
[94] Thi TO, Aung K, Myint T. Natural incidence of the bean stem fly Ophiomyia phaseoli (Tryon) (Diptera: Agromyzidae) in different plant growth stages of greengram in different growing seasons. Proceedings of the Fourth Agricultural Research Conference in Myanmar 2005;135–43.
[95] Bhople SK, Dhandge SR, Aravindarajan G, Patange NR. Varietal screening of mungbean genotypes for their resistance against pest complex of mungbean. AGRES 2017;6(1):123–8.
[96] Watanasit A, Pichitporn S. Improvement of mungbeanfor resistance to bruchids. In: Srinives P, Kitbamroong Miyazaki CS, editors. Mungbean germplasm: collection, evaluation and utilization for breeding program. Tsukuba, Japan: Japan International Research Center for Agricultural Sciences; 1996. p. 67–71.
[97] Abdullah-Al-Rahad M, Rahman MS, Akter T, Akter J, Rahman MA, Aziz SMS. Varietal screening of mungbean against whitefly and aphid. J Biosci Agric Res 2018;18:1478–87.
[98] Chauhan R, Singh AK, Sharma KR, Ali A. Screening of mungbean (Vigna radiata L.) germplasm against major sucking pest. J Pharm Phytochem 2018;7:1784–7.
[99] Kooner BS, Chhabra KS, Arora BS. Resistant sources in mungbean to manage whitefly, jassids and yellow mosaic virus. In: Proceedings of third agricultural science congress. India: PAU Ludhiana; 1997. p. 12–5.
[100] Khattak MK, Shafqat A, Chisti JI. Varietal resistance of mungbean (Vigna radiata L.) against whitefly (Bemisia tabaci Genn.), jassid (Amrasca devastans Dist.), and thrips (Thrips tabaci Lind.). Pak Entomol 2004;26(1):9–12.
[101] Shad N, Mughal SM, Farooq K, Bashir M. Evaluation of mungbean germplasm for resistance against mungbean yellow mosaic begomovirus. Pak J Bot 2006;38(2):449–57.
[102] Sekar S, Nalini R. Varietal Screening of Mungbean Genotypes against Whitefly (Bemisia tabaci Genn.), Mungbean Yellow Mosaic Virus (MYMV) and Cercospora leaf Spot. Int J Curr Microbial App Sci 2017;6(3):1278–85.
[103] Akhtar KP, Sarwar G, Abbas. G, Asghar MJ, Sarwar N, Shah TM. Screening of mungbean germplasm against mungbean yellow mosaic India virus and its vector Bemisia tabaci. Crop Prot 2011;30(9):1202–9.
[104] Nadeem S, Hamed M, Asghar MJ, Abbas G, Saeed NA. Screening of mungbean (Vigna radiata (L.) Wilczek) genotypes against sucking insect-pests under natural field conditions. Pak J Zool 2014;46(3):863–6.

[105] Lal SS. Insect-pests of mungbean, urd, cowpea, and pea and their management. In: Rao VM, Sithanantham S, editors. *Plant protection in field crops*. Hyderabad: Plant Protection Association of India; 1987. p. 185–202.
[106] Sandhya Rani C, Eshwari KB, Sudarshanam A: Field screening of greengram (Vigna radiata L.) entries against thrips (Thrips palmi) and spotted pod borer (Maruca vitrata), J Res ANGRAU 36(2/3):17–22, 2008.
[107] Sahoo BK, Sontakhe BK, Ruth LK. Varietal susceptibility of different greengram and blackgram cultivars to the leaf beetles and pod borer complex. Environ Ecol 1989;7(2):345–7.
[108] Sahoo BK, Hota AK. Field screening of greengram germplasm against insect pest and disease complex. Madras Agric J 1991;78(1–4):84–6.
[109] Swarnalatha P. Germplasm screening and insecticidal management of pest complex in greengram (Vigna radiata (L.) Wilczek). M.Sc.(Ag.) Thesis. Rajendranagar, Hyderabad: Acharya NG Ranga Agricultural University; 2007.
[110] Soundararajan RP, Chitra N, Ramasamy M. Host Plant Resistance to insect-pests of urdbean and mungbean. In: National workshop on paradigm shifts in research on crop resistance to pests. Annamalai University, Annamalai Nagar 2010;57–8.
[111] Chhabra, K.S., Kooner, B.S., Sharma, A.K., Saxena, A.K., Shanmugasundaran, S., Mungbean Proceedings of Second International Symposium. Bangkok, Thailand 1988:16–22.
[112] Gangwar B, Ahmed R. Performance of mungbean varieties under andaman and nicobar island condition. Ind J Pulse Res 1991;4(1):115–6.
[113] Sandhya Rani C, Rao GR, Chalam MSV, Kumar PA, Rao VS. Field screening of greengram genotypes against Maruca vitrata in Summer. J Agri Crop Sci 2014;1:18–25.
[114] Sandhya Rani C, Rao GR, Chalam MSV, Kumar PA, Rao VS. Estimation of avoidable losses in mungbean genotypes evaluated under field conditions during summer against Maruca vitrata. Int Res J Biol Sci 2015;4:47–54.
[115] Somta C, Somta P, Tomooka N, Ooi PAC, Vaughan DA, Srinives P. Characterization of new sources of mungbean (Vigna radiata (L.) Wilczek) resistance to bruchids, Callosobruchus spp. (Coleoptera: Bruchidae). J Stored Prod Res 2008;44(4):316–21.
[116] Fujii K, Miyazaki S. Infestation resistance of wild legumes (Vigna sublobata) to azuki bean weevil, Callosobruchus chinensis (L.) (Coleoptera: Bruchidae) and its relationship with cytogenetic classification. Appl Entomol Zool 1987;22(2):319–22.
[117] Kitamura K, Ishimoto M, Sawa M. Inheritance of resistance to infestation with azuki bean weevil in Vigna sublobata and successful incorporation to V. radiata. Jpn J Breed 1988;38(4):459–64.
[118] Fujii K, Ishimoto M, Kitamura K. Patterns of resistance to bean weevils (Bruchidae) in Vigna radiata-mungo sublobata complex inform the breeding of new resistant variety. Appl Ent Zool 1989;24(1):126–32.
[119] Talekar NS, Lin YH. Two sources with differing modes of resistance to Callosobruchus chinensis in mungbean. J Econ Entomol 1981;74(5):639–42.
[120] Talekar NS, Lin YH. Characterization of Callosobruchus chinensis resistance in mungbean. J Econ Entomol 1992;85(4):1150–53.
[121] Yao Y, Cheng X, Ren G. A 90-day study of three bruchid-resistant mungbean cultivars in Sprague–Dawley rats. Food Chem Toxicol 2015;76:80–5.
[122] Hong MG, Kim KH, Ku JH, Jeong JK, Seo MJ, Park CH, et al. Inheritance and quantitative trait loci analysis of resistance genes to bruchid and bean bug in mungbean (Vigna radiata L. Wilczek). Plant Breed Biotechnol 2015;3:39–46.
[123] Shanker A, Venkateswarlu B. Abiotic stress in plants: mechanisms and adaptations: BoD–Books on Demand. IntechOpen; 2011.

[124] Ye H, Liu S, Tang B, Chen J, Xie Z, Nolan TM, et al. RD26 mediates crosstalk between drought and brassinosteroid signalling pathways. Nat Commun 2017;8(1):1–13.

[125] Boyer J, Byrne P, Cassman K, Cooper M, Delmer D, Greene T, et al. The US drought of 2012 in perspective: A call to action. Global Food Secur 2013;2(3):139–43.

[126] Sharma A, Dhanda S. Abiotic stress response in *Vigna radiata* L(Mungbean). Int J Life Sci Biotechnol Pharma Res 2014;3(4):14.

[127] Hsiao TC, Acevedo E. Plant responses to water deficits, water-use efficiency, and drought resistance. Developments in agricultural and managed forest ecology, vol. 1. Elsevier; 1975. p. 59–84.

[128] Begg JE, Turner NC. Crop water deficits. Advances in agronomy, vol. 28. Elsevier; 1976. p. 161–217.

[129] El-Nakhlawy FS, Ismail SM, Basahi JM. Optimizing mungbean productivity and irrigation water use efficiency through the use of low water-consumption during plant growth stages. Legume Res Int J 2018;41(1):108–13.

[130] Munns R. Genes and salt tolerance: bringing them together. New Phytol 2005;167(3):645–63.

[131] Manchanda G, Garg N. Salinity and its effects on the functional biology of legumes. Acta Physiol Plant 2008;30(5):595–618.

[132] Gill K. Effect of soil salinity on grain filling and grain development in barley. Biol Plant 1979;21(4):241–4.

[133] Maas E. Salt tolerance of plants. Apple Agric. Res. 1, 12-26. World J Agric Sci 1986;4(3):351–8.

[134] Keating B, Fisher M. Comparative tolerance of tropical grain legumes to salinity. Austr J Agric Res 1985;36(3):373–83.

Chapter 17

Lentil Wild Genetic Resource: A Potential Source of Genetic Improvement for Biotic and Abiotic Stress Tolerance

Muhammad Jawad Asghar[a], Amjad Hameed[a], Muhammad Rizwan[b], Muhammad Shahid[a] and Rana Muhammad Atif[c,d]
[a]Nuclear Institute for Agriculture and Biology, Faisalabad, Pakistan; [b]Nuclear Institute of Agriculture, Tandojam, Sindh, Pakistan; [c]Centre for Advanced Studies in Agriculture and Food Security, University of Agriculture, Faisalabad, Pakistan; [d]Department of Plant Breeding and Genetics, University of Agriculture, Faisalabad, Pakistan

17.1 Introduction

Domesticated lentil (*Lens culinaris* Medik.) is an important pulse crop grown in Mediterranean and semi-arid climates. It is a rich source of dietary protein (24%–28%) along with essential amino acids (lysine and tryptophan), minerals (Fe, Zn, and K), carbohydrates, and vitamins, which makes it a balancing human diet in supplementation with cereals [1,2]. In 2018, world production was about 6.33 million tons from an area of 6.10 million ha with largest production share of 47.3% from Asia, 39.6% by Americas, and 5.2% by Europe. Canada is the biggest lentil producing country in the world, followed by India, United States of America, Turkey, and Australia which collectively produce 74% of total world lentil production [3]. Despite the highest production share, average lentil yield in Asia (817 kg/ha) is significantly below than the world average of 926 kg/ha [4]. Lentil records poor yield because of its general cultivation on marginal lands under rainfed conditions [5]. Further this crop is affected by several biotic (fungal, viral, bacterial diseases, and insect pest) and abiotic stresses (drought, high temperature, salinity, winter hardiness, waterlogging, and herbicide) which leads to severe yield losses [6–8]. Despite of its higher economic significance, this crop has remained under-researched until recently. Only a small number of local landraces have exploited in the development of modern lentil cultivars through intraspecific hybridization, pure-line and mass selection. This has narrowed the genetic base of these varieties and made them vulnerable to disease epidemics, pests, and unpredictable environmental factors

Wild Germplasm for Genetic Improvement in Crop Plants. http://dx.doi.org/10.1016/B978-0-12-822137-2.00017-5

[9]. Therefore, there is an immediate need to broaden the genetic base of lentil cultivars by accumulating favorable genes and alleles from distantly related wild *Lens* species in elite lentil backgrounds [4,10].

Recently, wild species of lentil have attained considerable attention for being a potential genetic resource for breaking yield barriers and enhancing tolerance to biotic and abiotic stresses. In past, efforts have been made to characterize and evaluate global wild accessions in search for genes imparting resistance/tolerance to biotic and abiotic stresses, for which substantial variation were recorded. Success in introgression of desirable genes from wild species to cultivated lentil has been achieved for some diseases, insect pests, and environmental influences [11]. Although, there are some constraints in utilizing these wild genetic resources but advances in molecular breeding techniques have opened the ways for successful exploitation of these resources and improving stress resistance in cultivated lentil. This chapter summarizes the progress, achievements, and impact of lentil wild genetic resources in improving biotic and abiotic stress resistance in lentil.

17.2 Wild genetic resources

The genus *Lens* is comprised of six main species *L. culinaris* ssp. *orientalis*, *L. culinaris* ssp. *tomentosus*, *L. odemensis*, *L. lamottei*, *L. ervoides*, and *Lens nigricans* [12]. On the basis of crossability with cultivated lentil, subsequent cytogenetic and recent sequence analysis, the species of lentil have been classified into primary (*L. orientalis and L. tomentosus*), secondary (*L. odemensis and L. lamottei*), tertiary (*L. ervoides*), and quaternary (*L. nigricans*) gene pools [13–15]. These wild species show large genetic diversity for plant morphological and seed traits (Fig. 17.1). *L. orientalis* represents the wild lentil with lanceolate stipules and small adjunct populations consisting a small number of plants in sparse stands. *L. tomentosus*, morphologically resembles to *L. orientalis* but distinguished as having hairy pods. Both have flattened seeds of 3.85–4.17 mm in size [13,16–17]. *L. odemensis* is charcterized by semi-hastate stipules forming horizontal positions on the stem [13] whereas *L. lamottei* possesses horizontal, less dentate stipules [18]. *L. odemensis* is also characterized by flattened seeds with up to 3.62 mm diameter whereas *L. lamottei*'s seed is small (1.92 mm) and round in shape [17]. *L. ervoides* is a wild relative with semi-hastate or lanceolate stipules. It has smaller leaves, calyx teeth, pods, and seeds [13,19]. *L. nigricans* has stipules that are considerably semi-hastate and dentate at their base. It consists of both types of accessions with upright and horizontal stipules [13].

The wild lentil species are widely dispersed in the Mediterranean region [20]. The subspecies *L. orientalis* and *L. tomentosus* have an eastern and southeastern spread from Turkey [21]. *L. odemensis* has distribution in the east, spreading from Turkey to Syria and Palestine [22]. *L. lamottei* has wide distribution in Morocco [23], *L. ervoides* from Spain to Ukraine and *L. nigricans* from Spain to Turkey [20,24]. These genetic resources which act as reservoir of useful genes

FIGURE 17.1 Lentil wild genetic resources; variation in plant and seed morphology. *(Images of wild lentil seed and plant are sourced from https://knowpulse.usask.ca; http://herbier.sesa-aude.fr/cgi-sys and different research articles cited earlier.)*

are being eroded quickly. Therefore, conservation of wild genetic resources is essential to meet the current and future needs of lentil breeding programs. The wild species of lentil have been collected from their diversity rich areas and further utilized in lentil improvement programs [25]. At present, 1295 accessions of wild lentil species are conserved in global genebanks (Table 17.1). About 474 accessions belong to the wild progenitor *L. culinaris* ssp. *orientalis* (Boiss.) Ponert. Wild lentil accessions which are available for distribution represent 52% of global wild germplasm of this genus. International Centre for Agriculture Research in Dry Areas (ICARDA) holds maximum 651 accessions of wild lentil species followed by Western Regional Plant Introduction Station, USDA-ARS, United States (197 accessions), and Australian Grains Genebanks, Department of Economic Development Jobs Transport and Resources, Australia which holds 133 accessions (Table 17.2). The use of this diverse germplasm is still inadequate as only a few common lines have repeatedly been used in hybridization with cultivated lentil resulting in a narrow genetic base. Only 15 wild accessions out of 1295 are advanced as breeding material and improved cultivar.

However, these wild germplasm resources hold a wealth of alleles controlling the important traits that can help break the yield barriers and enhance tolerance to biotic and abiotic stresses [26]. Genetic diversity present in wild germplasm is potentially an important resource for the genetic improvement of cultivated lentil [27]. Gupta and Sharma [28] observed wide variation for

TABLE 17.1 Lentil wild genetic resources; collection, biological status, and availability.

Wild species	Total accessions	Biological status						No. of accessions available for distribution
		Wild	Natural	Semi-natural	Breeding material	Other	Not specified	
L. culinaris ssp. Orientalis	474	398	47	5	7	1	16	286
L. culinaris ssp. Tomentosus	22	22	0	0	0	0	0	21
L. odemensis	88	77	4	0	0	5	2	70
L. lamottei	40	26	5	0	0	4	5	13
L. ervoides	421	273	31	9	7	6	95	206
L. nigricans	250	163	28	2	1	2	54	75

Source: Genesys: a global portal for plant genetic resources for food and agriculture (PGRFA) conserved in genebanks worldwide.

TABLE 17.2 Genebanks and institutes in the world holding lentil wild accessions.

Gene Bank/Institute/Country holding accessions	*L. culinaris* ssp. orientalis	*L. culinaris* ssp. tomentosus	*L. odemensis*	*L. lamottei*	*L. ervoides*	*L. nigricans*
International Centre for Agricultural Research in Dry Areas, Lebanon	286	21	68	10	193	73
Western Regional Plant Introduction Station, USDA-ARS, United States	90	—	8	—	62	37
Millennium Seed Bank Project, Seed Conservation Department, Royal Botanic Gardens, United Kingdom	48	—	3	9	38	32
Institute of Botany, Armenia	14	—	—	—	2	—
Laboratory of Plants Gene Pool and Breeding, Armenia	9	—	—	—	—	—
Junta de Comunidades de Castilla-La Mancha. Consejería de Agricultura. Centro de Investigación Agraria de Albaladejito., Spain	8	—	—	4	—	31
Agricultural Research Service, Jordan	8	—	—	—	21	—
Israel Gene Bank for Agricultural Crops, Agricultural Research Organisation, Israel	4	—	6	—	8	—

(Continued)

TABLE 17.2 Genebanks and institutes in the world holding lentil wild accessions. (*Cont.*)

Gene Bank/Institute/Country holding accessions	*L. culinaris* ssp. orientalis	*L. culinaris* ssp. tomentosus	*L. odemensis*	*L. lamottei*	*L. ervoides*	*L. nigricans*
Plant Genetic Resources Department, Turkey	3	—	—	—	—	2
Genebank, Leibniz Institute of Plant Genetics and Crop Plant Research, Germany	1	—	—	—	4	7
Institute of Plant Production n.a. V.Y. Yurjev of UAAS, Ukraine	—	1	1	1	—	—
Institute for Agrobotany, Hungary	—	—	2	—	—	2
Centro Nacional de Recursos Fitogenéticos, Spain	—	—	—	11	1	20
Australian Grains Genebanks, Department of Economic Development Jobs Transport and Resources, Australia	—	—	—	5	84	44
Genetic Resources Institute, Azerbaijan	—	—	—	—	5	—
Institute for Plant Genetic Resources 'K.Malkov', Bulgaria	—	—	—	—	—	1

Source: Genesys: a global portal for plant genetic resources for food and agriculture (PGRFA) conserved in genebanks worldwide.

agro-morphological traits in 70 accessions of wild lentil species. Similarly, a wide range of genetic variation for resistance against biotic and abiotic stresses has been observed in wild accessions of *Lens* gene pool [8,29].

17.3 Wild genetic resources for utilization in breeding for resistance to biotic stresses

Evaluation of wild relatives of lentil has resulted in identification of resistance or tolerance for key biotic stresses including Pea seed-borne mosaic virus, Pea enation mosaic virus, Anthracnose, Stemphylium blight, *Ascochyta* blight, Rust, Powdery mildew, *Fusarium* wilt, Botrytis grey mould, Root-knot nematodes, Aphid, Sitona weevils, Bruchid weevils, and Orobanche (Table 17.3).

17.3.1 Viral diseases

Worldwide, lentils are susceptible to more than 25 different virus species [30] causing little to complete crop failure depending on the environmental conditions during the cropping season [31]. Among them, the most important viruses that infect lentil crop includes Pea seed-borne mosaic virus (PSbMV), Pea enation mosaic virus (PEMV), Bean leafroll virus (BLV), Cucumber mosaic virus (CMV), Beet western yellow virus (BWYV), Faba bean necrotic yellow virus (FBNYV), Bean yellow mosaic virus (BYMV), and Pea streak virus (PeSV). These viruses tend to be transmitted by aphids and through seed infection thus controlling the insect vector and planting healthy seed could control these viral diseases. Another way is the use of resistant cultivars but unfortunately not much attention has given in the past to explore resistant sources against viral diseases among wild *Lens* species. Only few sources of resistance to different viruses have been identified. Hadad and coworkers [32] have identified immunity source against PSbMV controlled by recessive gene. Hussain and coworkers [33] also reported an accession (66013-6) with high level of resistance to PSbMV. In a study to identify sources of resistance, all nine accessions from three wild *Lens* species, *L. orientalis*, *L. nigricans*, and *L. ervoides* were found susceptible against PEMV, inoculated either mechanically or by aphids as vector [34].

17.3.2 Fungal diseases

Lentil production is often constrained by several fungal diseases which causes significant yield loss and down-grading of seed quality. Intensive selection pressure, narrow genetic base, and co-evolution of fungal pathogens have accelerated susceptibility to different fungal diseases in cultivated lentil [35]. Lentil anthracnose caused by *Colletotrichum lentis* is a major disease of lentil especially in Canada which accounts for 70%–78% crop losses under high disease pressure [36]. In cultivated lentil, partial resistance to *C. lentis* has been reported;

TABLE 17.3 Wild sources of resistance in lentil against biotic stresses.

Useful traits	Wild species	Wild accessions	References
Anthracnose resistance	*Lens ervoides, Lens nigricans*	IG 72815, PI 72847, L-01-827A, PI 320937	[37,38,53,54]
Stemphylium blight resistance	*Lens ervoides*	IG 72803, IG 116033, IG 72815, L-01-827A	[35,40,41]
	Lens nigricans	IG 72548, IG 72553, IG 72713, IG 136645, IG 72843	
	Lens orientalis	IG 72829	
	Lens tomentosus	IG 72643	
	Lens odemensis	IG 72606	
	Lens lamottei	IG 110809, IG 72537	
Ascochyta blight resistance	*Lens orientalis*	ILWL 7, ILWL 180	[42,43]
	Lens nigricans	PI 572348, PI 572359, PI 615677, ILWL 37, PI 572351	
	Lens ervoides	IG 72815, L01-827A	
Rust resistance	*Lens tomentosus*	ILWL 90, 198, 195, 480	[28,44]
	Lens orientalis	ILWL 230, 349, 476	
	Lens odemensis	ILWL 203, 235, 357	
	Lens ervoides	ILWL 60, 204, 292, 414, 442	
	Lens nigricans	ILWL 16, 18, 37, 460, EC718266	
Powdery mildew resistance	*Lens orientalis*	ILWL 230, 476	[44,46,47]
	Lens odemensis	ILWL 39, 203, 198, IG 136788	
	Lens tomentosus	ILWL 480	
	Lens nigricans	ILWL 14	
Fusarium wilt resistance	*Lens orientalis*	ILWL 7, 96, 117, 227, 246, 344, 359	[28,44,45]
	Lens odemensis	ILWL 15, 39	
	Lens tomentosus	ILWL 199, 308	
	Lens ervoides	ILWL 58, 398, 414, IG 136626	
	Lens nigricans	ILWL 19, 23, 34, 38, 474	
	Lens lamottei	ILWL 20, 430	
Root-knot nematodes	*Lens odemensis*	ILWL 18	[46]
	Lens nigricans	ILWL 14, ILWL 27	

TABLE 17.3 Wild sources of resistance in lentil against biotic stresses. (*Cont.*)

Useful traits	Wild species	Wild accessions	References
Broomrape resistance	*Lens ervoides*	UJWL 3, UJWL 5, L01-827A	[52,55]
	Lens orientalis	UJWL 6, UJWL 22	
Bruchid weevils resistance	*Lens orientalis*	BCU001418, 21, 22, 23	[51]
	Lens lamottei	BCU001899	
	Lens nigricans	BCU001413, 14, 15, 26, 30, 31 BCU001524, 25, 51, 77 BCU001890, 92, 93, 95, 96	
Sitona weevils resistance	*Lens orientalis*	ILWL 245	[50]
	Lens ervoides	ILWL 136, ILWL 258	
	Lens odemensis	ILWL 166, ILWL 203, ILWL 207	
	Lens nigricans	ILWL 110	

however, wild *lens* accessions IG 72815 and PI 320937 belonging to *L. ervoides* and *L. nigricans,* respectively, have shown complete resistance against the pathogen [37,38]. Similarly, Stemphylium blight, caused by *Stemphylium botryosum* causes susceptibility and significant yield loss to cultivated lentil in India and Bangladesh [39]. But some wild accessions, L01-827A, IG 116033 of *L. ervoides*, IG 72548, IG 72843 of *L. nigricans*, IG 72829 of *L. orientalis*, IG 72643 of *L. tomentosus*, IG 72606 of *L. odemensis*, IG 110809 of *L. lamottei* have been identified as resistant sources against *S. botryosum* [40,41]. Podder and coworkers [35] evaluated a set of 70 accessions from 7 species of the genus *Lens* for resistance against highly aggressive isolate of *S. botryosum*. Most of the cultivated lentil accessions were susceptible, whereas more than 70% of wild accessions were found resistant to *S. botryosum*. *L. ervoides* and *L. nigricans* showed the highest frequency of resistance to Stemphylium blight disease. The germplasm accessions, which showed resistance to Ascochyta blight included PI 572348, PI 572359, PI 615677 of *L. nigricans*, and IG 72815, L-01-827A of *L. ervoides* [42]. Some other potential wild accessions, namely; ILWL 7, ILWL 180 of *L. culinaris* ssp. *orientalis* have been identified against newly evolved highly aggressive isolates of *Ascochyta lentis* [43]. *Uromyces fabae* causing lentil rust is also a serious constraint in areas having mild temperature and humid weather conditions. Gupta and Sharma [28] screened a set of 93 accessions of wild *Lens* taxa and found sources of resistance for rust and Fusarium wilt. The wild accessions which showed resistance to rust were found in almost all wild species except *L. lamottei* [44]. Similarly, 22 accessions from annual wild *Lens* taxa have shown complete resistance against the Fusarium wilt disease [45]. Likewise, for powdery mildew disease, cultivated lentil gene pool has shown partial resistance against the *Erysiphe polygoni DC*. But some wild accessions, ILWL 230, ILWL 476 of *L. orientalis*, IG 136788, ILWL 39 of *L. odemensis*, ILWL 480 of *L. tomentosus*, and ILWL 14 of *L. nigricans* have been identified as resistant sources against the pathogen [46,47]. These promising wild genetic resources can be exploited for the transfer of resistance into elite genetic background of cultivated gene pool.

17.3.3 Insect-pests

Lentil is also damaged by insects. Among them, aphid, bruchid, and weevils are considered to be of significance [26]. Sitona leaf weevil (*Sitona crinitus* Herbst) is the main insect pest of lentil in North Africa and West Asia [48]. Adult insects feed on the leaflets causing damage to young seedlings. Larvae feed on the nodules, thus affecting the ability of the plant to fix atmospheric nitrogen. At high infestation level >90%, the insect causes 14% loss in grain yield [49]. Very limited work on resistance to insect pests in cultivated or wild *Lens* species has been reported. El-Bouhssini and coworkers [50] searched for resistant sources against sitona weevil in wild species of lentil. They screened 315 wild accessions and observed large variation in the percent nodule damage. Eight accessions, ILWL 245 of *L. orientalis*, ILWL 136, ILWL 258 of *L. ervoides*, ILWL 166, ILWL 203, ILWL 207, ILWL 254 of *L. odemensis*, and ILWL 110 of *L. nigricans* were identified as sitona weevil resistant sources.

Seed bruchid (*Bruchus* spp.) is another damaging pest to lentil crop. Laserna-Ruiz and coworkers [51] screened 571 accessions from cultivated and wild lentil species for susceptibility to seed bruchids. Large variation was observed among accessions for number of eggs per pod and infestation. Total 32 accessions including cultivated and wild *Lens* species were selected as genetic sources of resistance to this pest.

17.3.4 Plant parasitic weeds and nematodes

Broomrape (*Orobanche crenata* Forsk.) is a parasitic weed that causes significant yield and quality losses in lentil. Cultivated lentil germplasm are susceptible to *O. crenata*. No proper sources of resistance to *Orobanche* spp. were reported in cultivated lentil. Abu Irmailah and Haddad [52] evaluated a set of 22 wild accessions for resistance against *O. crenata*. They observed a wide range of response, from highly tolerant to susceptible. Accessions of *L. ervoides* (UJWL 3, UJWL 5, and L-01-827A) and *L. orientalis* (UJWL 6, UJWL 22) showed higher levels of tolerance. Root-knot nematode (*Meloidogyne incognita*) resistance was also observed in wild accessions ILWL 18 of *L. odemensis* and ILWL 14, ILWL 27 belonging to *L. nigricans* [46].

17.4 Utilization of wild relative in breeding for tolerance to abiotic stresses

Among abiotic stresses in lentil, the widely distributed are drought, heat, cold, salinity, and alkalinity stress. Of these, drought and heat are considered as important abiotic stresses globally [56,57]. Salinity is another emerging problem and lentil is more salt sensitive as compared to other legumes [58]. Similarly, cold and alkalinity stresses are endemic abiotic constraints affecting lentil production worldwide. Tolerant sources against these stresses in cultivated lentil are rare; however, evaluation of wild relatives of lentil has resulted in the identification of tolerance for key abiotic stresses (Table 17.4).

17.4.1 Drought stress

Wild lentil species are evolved in drought prone areas therefore can be a good source of drought tolerant traits for introgression into cultivated lentils. Identification of drought conferring traits and genotypes is an important researchable issue in view of climate changes [59]. Traits such as early seedling growth, early

TABLE 17.4 Wild sources of resistance in lentil against abiotic stresses.

Useful traits	Wild species	Wild accessions	References
Drought tolerance	*Lens orientalis*	PI 572376, IG72643, ILWL 73, 91, 314	[60–63]
	Lens ervoides	IG 72815, L-01-827A, ILWL 311, 303	
	Lens tomentosus	IG 72805	
	Lens odemensis	IG 72623, ILWL 175, 167, 436	
	Lens lamottei	IG 110813	
	Lens nigricans	ILWL 266, 305	
Cold tolerance	*Lens orientalis*	-	[29,69,70]
Salinity tolerance	*Lens orientalis*	ILWL 203, 95, 297, 368, 371, 417, IG 136670	[71,72]
	Lens ervoides	ILWL 137	
	Lens lamottei	ILWL 428	
	Lens nigricans	ILWL 9	
Alkalinity tolerance	*Lens orientalis*	ILWL 192	[73]
	Lens odemensis	ILWL 20	
	Lens nigricans	ILWL 15	
	Lens lamottei	ILWL 20, 430	

flowering, deep roots, reduced transpiration rate, rapid ground coverage, and early maturity were considered useful for selecting drought tolerant germplasm from wild resources [60–62]. Several drought tolerant accessions from wild *Lens* taxa have been reported. Following main tolerant sources have been identified as PI 572376, IG 72643, ILWL 73 of *L. orientalis*; IG 72815, L-01-827A of *L. ervoides*; IG 72805 of *L. tomentosus*, IG 72623 of *L. odemensis*, IG 110813 of *L. lamottei*, and ILWL 266 of *L. nigricans* [28,61–63]. Gorim and Vandenberg [61] further concluded that wild *Lens* species employed one or more strategies to survive against drought stress. Even same genotype employed different strategies to cope with drought stress. For instance, IG 72623, IG 72815, L-01-827A, PI 572376 responded to drought by developing deep roots with the exception of PI 572376 which also exhibited delayed flowering as another form of drought avoidance. This wide variation among accessions across the genus is important for future development of drought tolerant varieties given that all species can be hybridized with elite cultivated lentil germplasm.

17.4.2 Heat stress

Lentil being a cool season crop is highly sensitive to rising temperature. It needs low temperature at vegetative growth stage and warm temperature at the time of maturity. It grows best in a temperature range of 18°C–30°C. Temperature above 32°C during flowering and pod filling can drastically reduce seed yield [64]. Further it has been observed that winter season are becoming shorter and due to prolonged heat periods, lentil crop is more exposed to heat stress particularly in the reproductive stage [65]. Therefore, heat tolerant germplasms, suited to such environment are needed to sustain lentil production. Wild *Lens* taxa have yet to be evaluated for heat tolerance. Few sources of tolerance including IG 2507, IG 4258, IG 3263, and FLIP 2009 have only been reported in cultivated lentil [66]. With the assumption that crop wild relatives are adapted to their environment of collection [67], sources of heat stress tolerance in *L. orientalis* may occur in Turkmenistan and northern Syria [68].

17.4.3 Cold stress

Susceptibility to low temperature is another serious constraint which limits lentil production in areas of severe winter and in highland regions of the world. The effects of prolonged low temperature on lentil plant included slow growth, flower injury, and pod abortion [69]. Limited germplasm is available with a desirable degree of resistance to cold which allows breeding cold tolerant cultivars. To date, only a single effort has been made to identify genetic resources for tolerance in wild *Lens* species. Hamdi and coworkers [29] reported the highest level of winter hardiness in *L. orientalis* accessions and susceptibility in *L. ervoides*. However, some tolerant lines have been identified in cultivated lentil which can be utilized in breeding cold tolerant cultivars [70].

17.4.4 Salinity stress

Lentil is sensitive to salinity as it adversely affects its nodulation, limits plant growth, and productivity. It is mostly grown in arid and semi-arid regions. Soil and water management approaches for amelioration of soil salinity are expensive; therefore, salt tolerant genotypes can help in stabilizing yield in salt affected area. Wild genetic resources with substantial variation have been reported. Singh and coworkers [71] evaluated 162 accessions of different *Lens* species for salinity tolerance. Impressive variations were observed among accessions across the genus for salinity tolerance. Five wild accessions ILWL 9, ILWL 137, ILWL 203, ILWL 428, and ILWL 95 were reported as tolerant to salinity stress. Some other potential wild accessions namely; ILWL 297, ILWL 417ILWL 371, ILWL 368, IG 136670 of *L. orientalis* were identified as salinity tolerant and promising sources for utilization in breeding salinity tolerant cultivars [72].

17.4.5 Alkalinity stress

Lentil yield is severely affected by alkalinity caused by carbonate and bicarbonate ions. Presence of these ions increases the pH level of the soil and lead to drastic reduction in plant's germination, growth, and development. For alkaline soils, it is paramount to grow lentil varieties that are tolerant to soil alkalinity. Limited information is available on alkalinity tolerant sources in cultivated lentil. However, some potential sources of resistance were found in wild *Lens* germplasm. Singh and coworkers [73] evaluated 285 genotypes involving germplasm collections, breeding lines, landraces, and wild accessions. Almost all genotypes were found susceptible except some wild *Lens* accession ILWL 192, ILWL 20, ILWL 15, thus proving their tolerance to alkalinity stress. These wild genetic resources can be used for introgression of genes to improve alkalinity tolerance of the cultivated one.

17.5 Breeding constraints in developing stress resilient lentil by using wild resources

Gene introgression from wild *Lens* species into cultivated one is a long and tedious process due to lack of chromosomes homology of species involved in distant hybridization. Pre- and post-zygotic crossability barriers including reproductive isolation, embryo abortion, hybrid sterility, and limited genetic recombination are some major obstacles for utilization of wild resources at a larger extent in breeding stress resilient cultivated lentil. Crossability potential of wild *Lens* species with cultivated lentil is shown in Fig. 17.2.

17.5.1 Reproductive isolation

All species in the genus *Lens* have a common chromosome number of $2n = 2x = 14$. However chromosomal rearrangements lead to reproductive

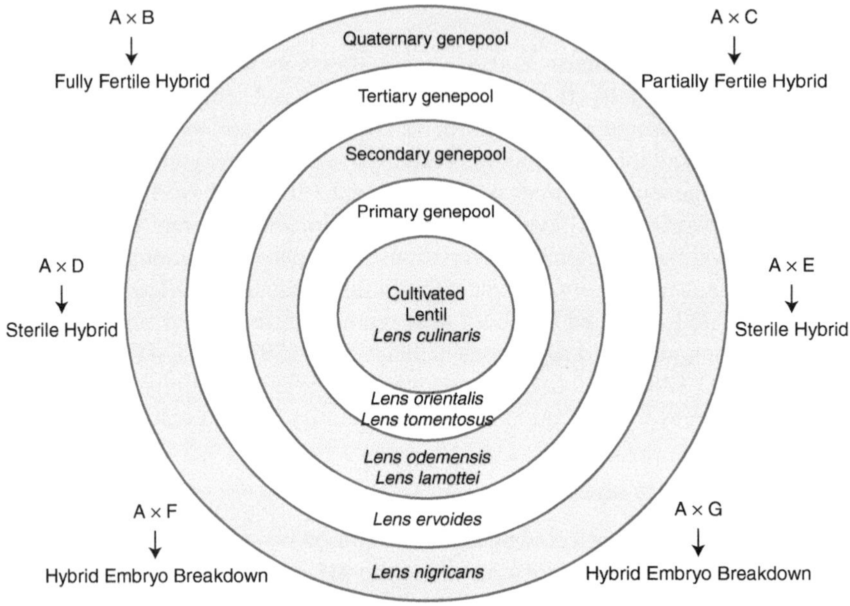

FIGURE 17.2 Crossability potential of wild lens species with cultivated lentil. (A) *Lens culinaris*; (B) *Lens orientalis*; (C) *Lens tomentosus*; (D) *Lens odemensis*; (E) *Lens lamottei*; (F) *Lens ervoides*; (G) *Lens nigricans.*

isolation between the species. Abnormal chromosome pairing has been observed in the crosses of *Lens culinaris* × *Lens tomentosus* [74]. *Lens* species are reproductively isolated from one another by cross-incompatibility [75]. Although closely related, these lentil species are separated from each other by a range of reproductive barriers. Many of the reproductive barriers so far reported in plants have been found in *Lens* species. These include pre-zygotic barriers (asynchronous flowering, reduced cross-fertility) and post-zygotic barriers (hybrid inviability, hybrid sterility).

17.5.2 Embryo abortion

Hybrid embryo abortion is an effective crossing barrier in genus *Lens* and first described by Ladizinsky and coworkers [13]. Lens taxa are free of embryo abortion following self-pollination. Hybrid embryos abort after interspecific hybridization and this is apparently due to heterozygous condition, an epistatic condition or both [76]. Interestingly, embryo abortion is limited to a few cross combinations within the genus *Lens* [77]. Another problem is that hybrid embryos cease to grow about 7–14 days after pollination. In some *L. culinaris* × *L. culinaris* ssp. *orientalis* crosses, development of endosperm was observed to be normal, while the hybrid embryo ceased growing. Embryo abortion was also observed in *L. culinaris* × *L. ervoides* and *L. culinaris* × *L. nigricans* crosses [78].

17.5.3 Hybrid inviability and sterility

Crosses are possible between cultivated and wild *Lens* species, but they are characterized by reduced fertility, sterility, and even non-viable hybrids [13]. Hybrids showed varying degrees of fertility usually due to chromosome translocations and subsequent problems with chromosome pairing at meiosis in *L. culinaris* × *L. nigricans* [79]. Fertility is often very low with little viable pollen produced in anthers and varies depending on the accession in *L. culinaris* × *L. culinaris* ssp. *orientalis* crosses from 2% to 69% [13]. These problems can occur in the F_1 and also persist in later generations causing partial or complete sterility.

17.6 Strategies to overcome breeding constraints

17.6.1 Use of cross compatibility

During past years, efforts have been made to identify the cross-compatible wild species with the cultivated lentil. The wild progenitors of cultivated *L. culinaris* spp. *orientalis* and *L. culinaris* spp. *tomentosus* are in primary gene pools, so they can be crossed easily without any fertility problem [80]. Ladizinsky [81] also reported the cross-compatibility of *L. orientalis* and *L. odemensis* with cultivated lentil and ranked them in the common gene pool. Gupta and Sharma [82] crossed *L. culinaris* with *L. orientalis* and *L. odemensis*; normal were produced in the former whereas varying degree of pollen fertility was observed in the later. Suvorova [16] successfully produced F_2-F_7 generation of *L. culinaris* × *L. tomentosus* and was able to select lines from the segregation populations. Table 17.5 enlisted the success in overcoming crossability barriers using different techniques.

17.6.2 Use of bridge species

Large genetic distances between cultivated lentil and quaternary gene pool (*L. nigricans*) indicate that direct hybridization between accessions belonging to these gene pools may be challenging. To cover this genetic distance, the use of an intermediary specie as a bridge may be helpful in indirect transfer of genes into the desired cultivated background [83]. As gene flow from *L. nigricans* to *L. culinaris* is limited, so attempts have been made to transfer genes from *L. nigricans* (quaternary gene pool) to *L. culinaris* using *L. ervoides* (tertiary gene pool) as bridge species [5].

17.6.3 Using embryo rescue technique and growth hormones

Almost no success is achieved in obtaining fertile hybrids from crossing between cultivated and tertiary and quaternary gene pools of lentil. Goshen and coworkers [79] found irregularities in meiotic analysis of the hybrids arising

TABLE 17.5 Overcoming cross-ability barriers using different techniques.

Techniques	Crosses attempted successfully	Hybrid status	References
Use of cross-compatibility	*L. culinaris* × *L. orientalis*	Viable hybrids with no fertility problem	[80,82]
	L. culinaris × *L. tomentosus*	Successful production of F_2-F_7 generations	[16]
	L. culinaris× *L. odemensis*	Viable hybrid with varying degree of pollen fertility in later generations	[82]
Use of bridge species	*L. culinaris* × *L. nigrican* *using L. ervoides as bridge specie*	*L. ervoides* readily crossed with *L. nigricans*. *L. culinaris* × *L. ervoides* hybrid rescued using embryo rescue.	[5,87,88]
Embryo rescue	*L. culinaris* × *L. orientalis* *L. culinaris* × *L. tomentosus* *L. culinaris* × *L. odemensis* *L. culinaris* × *L. lamottei* *L. culinaris* × *L. ervoides* *L. culinaris* × *L. nigricans*	3%–90% of hybrids were recovered	[80,89–92]
Growth hormones	*L. culinaris* × *L. odemensis* *L. culinaris* × *L. ervoides* *L. culinaris* × *L. nigricans*	Viable hybrids were produced with artificial supplementation of GA_3	[87]

from such crosses. Some other researchers too had little success when crossing *L. culinaris* with *L. ervoides* and *L. nigricans* [84,85] likely due to embryo failure [86] which may be resolved through embryo rescue technique. The embryo culture coupled with in vivo hormonal treatments (gibberellic acid, benzyl adenine, and kinetin) has improved the scope of wild hybridization in lentil [87,88] by overcoming post-fertilization barriers. Fratini and Ruiz [80] reported the use of embryo-ovule rescue technique which could be able to rescue the hybrid ovules in crosses between cultivated lentil with *L. odemensis* and *L. nigricans*.

17.7 Conclusion and future prospects

The lentil wild genetic resources are important sources of alien genes controlling desirable traits for making genetic improvement of cultivated lentil. These traits can be helpful to develop improved cultivars with high productivity due to tolerance to various biotic and abiotic stresses. Many of the useful alleles/genes are expected to be different from those of the cultivated species and are thus useful in broadening the base of resistance to various stresses. Wild gene pool of lentil showed tolerance against drought and cold stresses. Some of the wild accessions with multiple disease resistance have also been identified. Crossability potential of these wild resources with cultivated germplasm has also been revealed in many studies. Although, gene introgression from wild *Lens* species into cultivated one is a long and tedious process and involve different pre- and post-zygotic crossability barriers. However, better understanding of the crossability barriers and invention of different strategies to break the barriers have improved the possibilities of genetic improvement of cultivated lentil. Therefore there is a need to make embryo rescue technique much more routine and also to study the wild species extensively to search for high-crossability genes.

References

[1] Erskine W, Rihawe S, Capper BS. Variation in lentil straw quality. Anim Food Sci Technol 1990;28:61–9.

[2] Kumar S, Chaudhary AK, Rana KS, Sarker A, Singh M. Biofortification potential of global wild annual lentil core collection. PLoS ONE 2018;13:e0191122.

[3] FAO. FAOSTAT: Food and Agriculture Organization of the United Nations. Rome; 2018.

[4] Singh M, editor. Lentils: Potential resources for enhancing genetic gains. Academic Press; 2018.

[5] Kumar J, Srivastava E, Singh M, Mahto D, Pratap A, Kumar S, et al. Alien gene transfer in crop plants, vol. 2. New York, NY: Springer; 2014. p. 191–205.

[6] Clement SL, Cristofaro M, Cowgill SE, Weigand S. Germplasm resources, insect resistance, and grain legume improvement. In: Clement SL, Quisenberry SS, editors. Global plant genetic resources for insect resistant cropsBoca Raton, FL: CRC Press; 1999. p. 131–48.

[7] Dita MA, Rispail N, Prats E, Rubiales D, Singh KB. Biotechnology approaches to overcome biotic and abiotic stress constraints in legumes. Euphytica 2006;147:1–24.

[8] Stoddard FL, Balko C, Erskine W, Khan HR, Link W, Sarker A. Screening techniques and sources of resistance to abiotic stresses in cool-season food legumes. Euphytica 2006;147:167–86.

[9] Duran Y, Fratini R, Garcia P, Perez de la Vega M. An inter-sub-specific genetic map of *Lens*. Theor Appl Genet 2004;108:1265–73.

[10] Kumar S, Imtiaz M, Gupta S, Pratap A. Distant hybridization and alien gene introgression. In: Pratap A, Kumar J, editors. Biology and breeding of food legumesCABI, Oxfordshire; 2011. p. 81–110.

[11] Hajjar R, Hodgkin T. The use of wild relatives in crop improvement, a survey of developments over the last 20 years. Euphytica 2007;156:1–13.

[12] Cubero JI, Perez de la Vega M, Fratini R. Origin, phylogeny, domestication and spread. In: Erskine W, Muehlbauer FJ, Sarker BS, editors. The lentil: botany, production and usesCABI, Oxfordshire; 2009. p. 13–33.

[13] Ladizinsky G, Braun D, Goshen D, Muehlbauer FJ. The biological species of the genus *Lens* L. Bot Gazette 1984;145(2):253–61.

[14] Muehlbauer FJ, McPhee KE. Lentil *Lens culinaris* Medik. In: Singh RJ, Jauhar PP, editors. Genetic resources, chromosome engineering and crop improvement, grain legumesBoca Raton, FL: Taylor & Francis; 2005. p. 219–30.

[15] Wong MM, Gujaria-Verma N, Ramsay L, Yuan HY, Caron C, Diapari M, et al. Classification and characterization of species within the genus *Lens* using genotyping-by-sequencing (GBS). PLoS ONE 2015;10(3):e0122025.

[16] Suvorova G. Hybridization of cultivated lentil *Lens culinaris* Medik. and wild species *Lens tomentosus* Ladizinsky. Czech J Genet Plant Breed 2014;50(2):130–4.

[17] Jha TB, Halder M. Searching chromosomal landmarks in Indian lentils through EMA-based Giemsa staining method. Protoplasma 2016;253(5):1223–31.

[18] Ferguson ME, Maxted N, Van Slageren M, Robertson LD. A re-assessment of the taxonomy of *Lens* Mill. (*Leguminosae, Papilionoideae Vicieae*). Bot J Linnean Soc 2000;133(1):41–59.

[19] Chen LA. Assessing impacts of crop wild introgression in lentil using interspecific *Lens* species recombinant inbred line populations. PhD dissertation, University of Saskatchewan; 2018.

[20] Ferguson M, Acikgoz N, Ismail A, Cinsoy A. An eco-geographic survey of wild *Lens* species in Aegean and south West Turkey. Anadolu 1996;6:159–66.

[21] Ladizinsky G. A new species of *Lens* from south-East Turkey. Bot J Linn Soc 1997;123:257–60.

[22] Ferguson ME, Ford-Llyod B, Robertson LD, Maxted N, Newbury HJ. Mapping the geographical distribution of genetic variation in the genus *Lens* for the enhanced conservation of plant genetic diversity. Mol Ecol 1998;7:1743–55.

[23] Van Oss H, Aron Y, Ladizinsky G. Chloroplast DNA variation and evolution in the genus *Lens* Mill. Theor Appl Genet 1997;94:452–7.

[24] Zohary D. Monophyletic vs. polyphyletic origin of the crops on which agriculture was founded in the Near East. Genet Resour Crop Evol 1999;46:133–42.

[25] Fikiru E, Tesfaye K, Bekele E. Genetic diversity and population structure of Ethiopian lentil (*Lens culinaris* Medikus) landraces as revealed by ISSR marker. Afr J Biotechnol 2007;6:1460–8.

[26] Singh M, Sardana S, Sharma SK. Genetic resources of lentil and its utilization in India. Plant Genet Resour 2011;9(1):30–7.

[27] Ferguson ME, Robertson LD. Morphological and phenological variation in the wild relatives of lentil. Genet Resour Crop Evol 1999;46(1):3–12.

[28] Gupta D, Sharma SK. Evaluation of wild Lens taxa for agro-morphological traits, fungal diseases and moisture stress in North Western Indian Hills. Genet Resour Crop Evol 2006;53:1233–41.

[29] Hamdi A, Küsmenoglu I, Erskine W. Sources of winter hardiness in wild lentil. Genet Res Crop Evol 1996;43:63–7.

[30] Abraham AD, Menzel W, Lesemann DE, Varrelmann M, Vetten HJ. Chickpea chlorotic stunt virus: a new *Polerovirus* infecting cool-season food legumes in Ethiopia. Phytopathology 2006;96:437–46.

[31] Kumari SG, Larsen R, Makkouk KM, Muhammad B. Virus diseases and their control. The lentil: Botany, production and usesCABI; 2009. p. 306–25.

[32] Haddad NI, Muehlbauer FJ, Hampton RO. Inheritance of resistance to pea seed-borne mosaic virus in lentils 1. Crop Sci 1978;18(4):613–5.

[33] Hussain S, Ali A, Zahid MA, Iqbal SM. Multiple disease resistance in lentil germplasm from Pakistan. Pak J Bot 2008;40(2):897–903.

[34] Aydin H, Muehlbauer FJ, Kaiser WJ. Pea enation mosaic virus resistance in lentil (*Lens culinaris*). Plant Dis 1987;71(7):635–8.

[35] Podder R, Banniza S, Vandenberg A. Screening of wild and cultivated of lentil germplasm for resistance to Stemphylium blight. Plant Gen Res Char Util 2013;11:26–35.

[36] Chongo G, Bernier CC, Buchwaldt L. Control of anthracnose in lentil using partial resistance and fungicide applications. Can J Plant Pathol 1999;21:16–22.

[37] Tullu A, Buchwaldt L, Warkentin T, Taran B, Vandenberg A. Genetics of resistance to anthracnose and identification of AFLP and RAPD markers linked to the resistance gene in PI 320937 germplasm of lentil (*Lens culinaris* Medikus). Theor Appl Genet 2003;106:428–34.

[38] Fiala JV, Tullu A, Banniza S, Seguin-Swartz G, Vandenberg A. Interspecies transfer of resistance to anthracnose in lentil (*Lens culinaris* Medic.). Crop Sci 2009;49:825–30.

[39] Bakr MA, Ahmed F. Development of Stemphylium blight of lentil and its chemical control. Bangladesh J Plant Pathol 1992;8:39–40.

[40] Bhadauria V, Ramsay L, Bett KE, Banniza S. QTL mapping reveals genetic determinants of fungal disease resistance in the wild lentil species *Lens ervoides*. Sci Rep 2017;7:3231.

[41] Kant P, Materne M, Rodda MS, Slater AT. Screening lentil germplasm for Stemphylium blight resistance. Austr Plant Pathol 2017;46:129–36.

[42] Tullu A, Banniza S, Tar'an B, Warkentin T, Vandenberg A. Sources of resistance to ascochyta blight in wild species of lentil (*Lens culinaris* Medik.). Genet Resour Crop Evol 2010;57:1053–63.

[43] Dadu RH, Ford R, Sambasivam P, Gupta D. A novel *Lens orientalis* resistance source to the recently evolved highly aggressive Australian Ascochyta lentis isolates. Front Plant Sci 2017;8:1038.

[44] Singh M, Kumar S, Basandrai AK, Basandrai D, Malhotra N, Saxena DR, et al. Evaluation and identification of wild lentil accessions for enhancing genetic gains of cultivated varieties. PloS ONE 2020;15(3):e0229554.

[45] Singh M, Rana JC, Singh B, Kumar S, Saxena DR, Saxena A, et al. Comparative agronomic performance and reaction to Fusarium wilt of *Lens culinaris* × *L. orientalis* and *L. culinaris* × *L. ervoides* derivatives. Front Plant Sci 2017;8:1162.

[46] Gautam NK, Singh M, Khan Z, Roy A, Akhtar J, Ram B. Assessment of lentil germplasm with respect to agronomic performance and major biotic stress. Legum Res 2013;36:214–9.

[47] Singh M, Bisht IS, Dutta M, Kumar K, Kumar S, Bansal KC. Genetic studies on morpho-phenological traits in lentil (*Lens culinaris* Medikus) wide crosses. J Genet 2014;93:561–6.

[48] Solh MB, Itani HM, Kawar NS. The effect of sowing date on the growth and yield of lentils and the implication of certain control measures. Lebanese Sci Bul 1986;2:17–27.

[49] Weigand S, Pala M, Saxena MC. Effect of sowing date, fertilizer and insecticide on nodule damage by *Sitona crinitus* Herbst (Coleoptera: *Curculionidae*) and yield of lentil (*Lens culinaris* Medik) in northern Syria. J Plant Dis Prot 1991;99(2):174–81.

[50] El-Bouhssini M, Sarker A, Eriskine W, Joubi A. First sources of resistance to Sitona weevil (Sitona crinitus Herbst) in wild *Lens* species. Genet Resour Crop Evol 2008;55:1–4.

[51] Laserna-Ruiz I, De-Los-Mozos-Pascual M, Santana-Meridas O, Sanchez-Vioque R, Rodriguez- Conde MF. Screening and selection of lentil (*Lens* Miller) germplasm resistant to seed bruchids (Bruchus spp). Euphytica 2012;73:1–10.

[52] Abu Irmailah B, Haddad N. Indication of tolerance to broomrapes (*Orobanche crenata* Forsk) in wild lentil collected from Jordan. Dirasat Shar'ia Law Sci 2010;37(2):101–6.

[53] Tullu A, Buchwaldt L, Lulsdorf M, Banniza S, Barlow B, Slinkard AE, et al. Sources of resistance to anthracnose (*Colletotrichum truncatum*) in wild *Lens* species. Genet Resour Crop Evol 2006;53(1):111–9.

[54] Tullu A, Bett K, Banniza S, Vail S, Vandenberg A. Widening the genetic base of cultivated lentil through hybridization of *Lens culinaris* "Eston" and *L. ervoides* accession IG72815. Can J Plant Sci 2013;93:1037–47.

[55] Bucak B, Bett KE, Banniza S, Vandenberg A. Transfer of resistance to broomrape (*Orobanche crenata*) from *Lens ervoides* to cultivated lentil. In: Sixth international food legume research conference. Saskatoon, Canada; 2014.
[56] Turner NC, Wright GC, Siddique KHM. Adaptation of grain legumes (pulses) to water-limited environments. Adv Agron 2001;71:193–231.
[57] Yau SK, Erskine W. Diversity of boron toxicity tolerance in lentil growth and yield. Genet Resour Crop Evol 2000;47:55–61.
[58] Katerji N, Van Hoorn JW, Hamdy A, Mastrorilli M. Salinity effect on crop development and yield, analysis of salt tolerance according to several classification methods. Agric Water Manage 2003;62(1):37–66.
[59] Malhotra RS, Sarker A, Saxena MC. Drought tolerance in chickpea and lentil: Present status and future strategies. In: Rao SC, Ryan J, editors. Challenges and strategies for dry-land agricultureMadison, WI: Crop Science Society of America (CSSA), Special Publication 32, CSSA; 2004. p. 257–74.
[60] Hamdi A, Erskine W. Reaction of wild species of the genus *Lens* to drought. Euphytica 1996;91(2):173–9.
[61] Gorim LY, Vandenberg A. Evaluation of wild lentil species as genetic resources to improve drought tolerance in cultivated lentil. Front Plant Sci 2017;8:1129.
[62] Gorim LY, Vandenberg A. Can wild lentil genotypes help improve water use and transpiration efficiency in cultivated lentil? Plant Genet Resour 2018;16(5):459–68.
[63] Singh D, Singh CK, Tomar RSS, Taunk J, Singh R, Maurya S, et al. Molecular assortment of Lens species with different adaptations to drought conditions using SSR markers. PLoS ONE 2016;11(1):e0147213.
[64] Delahunty A, Nuttall J, Nicolas M, Brand J. Genotypic heat tolerance in lentil. In: Proceedings of the seventeenth ASA conference (Hobart). 2015. p. 20–24.
[65] Hasanuzzaman M, Nahar K, Alam MM, Roychowdhury R, Fujita M. Physiological, biochemical, and molecular mechanisms of heat stress tolerance in plants. Int J Mol Sci 2013;14:9643–968.
[66] Sita K, Sehgal A, Kumar J, Kumar S, Singh S, Siddique KH, et al. Identification of high-temperature tolerant lentil (*Lens culinaris* Medik.) genotypes through leaf and pollen traits. Front Plant Sci 2017;8:744.
[67] Baute GJ, Dempewolf H, Reisenberg LH. Using genomic approaches to unlock the potential of CWR for crop adaptation to climate change. In: Redden R, Yadav SS, Maxted M, Dulloo E, Guarino L, Smith P, editors. Crop wild relatives and climate changeUSA: Wiley-Blackwell; 2015. p. 268–80.
[68] Coyne CJ, Kumar S, von Wettberg EJ, Marques E, Berger JD, Redden RJ, et al. Potential and limits of exploitation of crop wild relatives for pea, lentil, and chickpea improvement. Legume Sci 2020;2(2):e36.
[69] Solanki RK, Kumar J, Islam AU, Srivastava E. Response of early flowering lentil genotypes to low temperature under field conditions. J Food Leg 2011;24(2):142–4.
[70] Kahraman A, Kusmenoglu I, Aydin N, Aydogan A, Erskine W, Muehlbauer FJ. QTL mapping of winter hardiness genes in lentil. Crop Sci 2004;44:13–22.
[71] Singh D, Singh CK, Kumari S, Tomar RSS, Karwa S, Singh R, et al. Discerning morpho-anatomical, physiological and molecular multiformity in cultivated and wild genotypes of lentil with reconciliation to salinity stress. PloS ONE 2017;12(12):e0190462.
[72] DAC-ICAR-ICARDA, Annual progress report. International centre for agricultural research in dry areas, South Asia and China Regional Programme. New Delhi; 2014. p. 145.

[73] Singh D, Singh CK, Singh YP, Singh V, Singh R, Tomar RSS, et al. Evaluation of cultivated and wild genotypes of *Lens* species under alkalinity stress and their molecular collocation using microsatellite markers. PloS ONE 2018;13(8):e0199933.

[74] Ladizinsky G. The origin of lentil and its wild gene pool. Euphytica 1979;28:179–87.

[75] Ladizinsky G, Abbo S. The *Lens* genus. The search for wild relatives of cool season legumesSpringer Briefs in Plant Science. Springer, Cham; 2015.

[76] Abbo S, Ladizinsky G. Genetical aspects of hybrid embryo abortion in the genus Lens L. Heredity 1994;72(2):193–200.

[77] Brown DI. Anatomical and genetic characterization of embryo breakdown in the genus *Lens*. M. Sc Thesis, Faculty of Agriculture. The Hebrew University of Jerusalem (Hebrew); 1985.

[78] Abbo S, Ladizinsky G. Anatomical aspects of hybrid embryo abortion in the genus *Lens* L. Bot Gazette 1991;152(3):316–20.

[79] Goshen D, Ladizinsky G, Muehlbauer FJ. Restoration of meiotic regularity and fertility among derivatives of *Lens culinaris* × *L. nigricans* hybrids. Euphytica 1982;31:795–9.

[80] Fratini R, Ruiz ML. Interspecific hybridization in the genus *Lens* applying in vitro embryo rescue. Euphytica 2006;150(1):271–80.

[81] Ladizinsky G. Wild lentils. Crit Rev Plant Sci 1993;12:169–84.

[82] Gupta D, Sharma SK. Widening the gene pool of cultivated lentils through introgression of alien chromatin from wild *Lens* subspecies. Plant Breed 2007;126(1):58–61.

[83] Dissanayake R, Braich S, Cogan NO, Smith K, Kaur S. Characterization of genetic and allelic diversity amongst cultivated and wild lentil accessions for germplasm enhancement. Front Genet 2020;11:546.

[84] Hamdi A, Erskine W. Germplasm program legumes–Annual report. Aleppo, Syria: ICARDA; 1994.

[85] Gupta D. Genetic and cytogenetic studies in populations derived from crosses between wild *Lens* Species and cultivated lentils (Doctoral dissertation). Palampur, India: CSK Himachal Pradesh, Krishi Vishvavidyalaya; 2003.

[86] Gupta D, Sharma SK. Embryo-ovule rescue technique for overcoming post-fertilization barriers in interspecific crosses of *Lens*. J Lentil Res 2005;2:27–30.

[87] Ahmad MAGF, Fautrier AG, McNeil DL, Burritt DJ, Hill GD. Attempts to overcome post fertilization barrier in interspecific crosses of the genus *Lens*. Plant Breed 1995;114(6):558–60.

[88] Ye G, McNiel DL, Hill GD. Breeding for resistance to lentil Ascochyta blight. Plant Breed 2002;121:185–91.

[89] Ladizinsky G, Cohen D, Muehlbauer FJ. Hybridization in the genus *Lens* by means of embryo culture. Theor Appl Genet 1985;70:97–101.

[90] Ladizinsky G, Abbo S. Cryptic speciation in *Lens culinaris*. Genet Res Crop Evol 1993;40:1–5.

[91] Fiala JV. Transferring resistance to *Colletotrichum truncatum* from wild lentil species to cultivated lentil species (*Lens culinaris* subsp culinaris). MSc thesis, Canada: University of Saskatchewan, Saskatoon; 2006. p. 131.

[92] Cohen D, Ladizinsky G, Ziv M, Muehlbauer FJ. Rescue of interspecific *Lens* hybrids by means of embryo culture. Plant Cell Tissue Organ Cult 1984;3:343–7.

Chapter 18

Wild Relatives and Modern Plant Breeding Technologies

Qandeel-e-Arsh[a], Tehreem Jabbar[a], Shahzad Khalid[b], Rana Muhammad Atif[c,d], Hafiz Mamoon Rehman[e] and Iqrar Ahmad Rana[a,c]
[a]Centre for Agricultural Biochemistry and Biotechnology, University of Agriculture, Faisalabad, Pakistan; [b]Forestry Research Institute, Gatwala, Faisalabad, Pakistan; [c]Centre for Advanced Studies in Agriculture and Food Security, University of Agriculture, Faisalabad, Pakistan; [d]Department of Plant Breeding and Genetics, University of Agriculture, Faisalabad, Pakistan; [e]Centre for Soybean Research of the State Key Laboratory of Agrobiotechnology and School of Life Sciences, The Chinese University of Hong Kong, Shatin, Hong Kong, China

18.1 Introduction

The earliest mankind depended largely on nature for its survival by feeding on plants and hunting animals for food and shelter. This hunter's society transitioned to an agricultural society wherein man introduced deliberate changes to the natural breeding procedures of plants. Healthy plants yielding benefits were chosen and then grown under observation and somewhat controlled conditions, a practice that introduced selection in plants. Unbeknownst to mankind, such agricultural practices brought in changes to the genetic makeup of the plants resulting in decreased allelic diversity compared to their wild progenitors. As a result of these genetic changes, desired traits have been developed, and these traits were fixed by preserving the seeds for cultivation in next growing season. For instance, Teosinte, a wild parent of modern maize, used to shed its seeds upon ripening as a means of survival. This trait is considered undesirable if seed is to be stored for longer use. Selection of mutants with least shedding ability in Teosinte resulted in the kernels remaining attached to the cob. Such variants with desired trait for human consumption is thus selected and grown by farmers. As time passed, agriculture centered upon producing increased yield in response to ever-rising demand for food and to overcome food insecurity. The Green Revolution of the 1960s and 70s saw a focus on increasing crop yield by growing dwarf varieties leading to improved grain quality and quantity such as the semi-dwarf wheat [1] and rice [2] varieties that lead to major yield increases in the 1960s. Additionally, the period saw a significant rise in the use of pesticides [3], insecticides [4], and fertilizers [5]. Agriculture, in the contemporary

Wild Germplasm for Genetic Improvement in Crop Plants. http://dx.doi.org/10.1016/B978-0-12-822137-2.00018-7

era, has taken a new turn with the introduction of sustainable means of breeding cultivars. Farmers currently desire plant varieties capable of efficient production within limited resources and changing climatic conditions. Techniques for genetic modifications are being used to increase crop production by improving farming efficiency, yields, and resistance to disease [6,7].

During previous century cross breeding was practiced to develop high yielding genotypes, and most of the land races developed by natural selection of the farmer were either eliminated or were dropped from crossing schemes in the later years. Currently, cross breeding means crossing a group of genotypes developed during last couple of decades. This resulted in the loss of certain essential genes required for various stresses and deficiencies. Hence conventional plant breeding techniques seem unable to cope up with the ever-increasing demands of food especially considering changing environmental patterns and abiotic and biotic stresses. As the genetic makeup of the genotypes being crossed is less divergent, therefore, crossbreeding leads to inbreeding depressions and plant vigor is observed to reduce by continuous selfing. There is a lack of an interdisciplinary approach for successful plant breeding. Much of the data gathered through emerging Omics technologies and bioinformatics tools [8] is too vast and diverse to be completely utilized and benefitted from. Omics needs to be applied to wild species for comparisons to draw and identify sources of traits development in modern genotypes.

Modern plant breeding practices incorporate tools from biotechnology [9] to make them cost and time effective such as use of molecular markers, marker-assisted selection, genomic selection, genetic engineering, gene editing, crop modeling, and virtual breeding. Despite the apparent benefits of domestication of crops, the wild varieties too render many benefits including the provision of genetic variation which is otherwise unavailable in breeding germplasm and these in turn need efficient characterization and conservation [10]. By crossing domesticated varieties with the wild varieties, breeders can reintroduce desired wild traits that are usually associated with resistance to harsh environments and pests. Similarly, gene editing in wild relatives may turn the wild relative to a cultivated crop species such as in wheat and potato, focus is upon the tolerance to abiotic stress while in sunflower, improvement revolves around biotic stresses and fertility traits [11]. During the 20th century, an estimated 30% of increased crop yields were a result of incorporating crop wild relatives (CWR) in plant breeding programs, making up a global value of about US $100 billion. Thus, the conservation of CWR resources carries immense significance [12]. In this review article, an overview of the past achievements and future prospects with the availability of gene editing technologies will be discussed.

18.2 Plant breeding and food security

Plant breeding has been practiced since the ancient plant breeders would hunt for wild species with desirable traits. They would intuitively try to integrate traits of such wild plants with those lacking these important traits to propagate the genetic variants [13]. This was truly long before mankind learnt of genetics and cytogenetics, majorly recognizing the importance in the last decade. The presence of vast amounts of molecular information including cloned and characterized genes and sequencing model genomes makes it difficult to integrate genetics in plant breeding practices. Implementation of well thought policies and methodologies [14] would help in bridging the interface between model systems and crops, to ultimately fulfill the increasing food demand around the globe.

As defined by FAO, food security refers to economic and physical potential of mankind, to access safe and nutritionally sufficient food that fulfills all dietary needs required for an active and healthy life, round the year [15]. Food security is said to be achieved when its four dimensions are fulfilled simultaneously [16], which include availability of healthy and nutritive food from local, and international market; physical and economic access to food that helps maintain good health; safe and healthy food consumption; and stability of food system that comprise access to food without being affected by unprecedented events such as economic or climatic crisis. According to a report published by FAO, more than 820 million people were undernourished in 2019 [17]. Malnourishment is a major indication of widespread food insecurity which occurs upon a lack of fulfillment of one or more of the dimensions of food security.

The fundamentals of plant breeding comprises of deliberately introducing selected genetic traits in breeding lines by intercrossing chosen plant individuals with complementing traits; followed by continual selection of superior plant cultivars over generations to achieve genetic uniformity, increased adaptation, and agronomic stability [18,19]. The breeding objectives and the mode of crop generation determine the breeding methodology to be used. With the use of new biotechnologies, plant breeding techniques can be improved to overcome food security issues. Indirectly yields may be allowed to increase such as in Pakistan where operation timings can be changed [20] such as early maturity and harvesting of cotton followed by early sowing of wheat generate higher yields. Modern plant breeding techniques require highly trained personnel, skill, and advanced infrastructure facilities and resources. Developing countries are unable to maintain these inputs to carry out efficient modern breeding techniques. Breeding for abiotic stress tolerance including the stresses such as heat stress, cold stress, salinity stress, and biotic stresses such as pests, makes it further challenging thus complicating the accuracy of phenotyping. However, if properly exploited, it carries great potential in terms of genetic gains especially in developing countries [21]. Defects within an elite variety can be improved without any loss of agronomic traits.

18.2.1 Plant Genetic Resources and Breeding

It was during the 1950s that the importance of wild species for improvement of crops was recognized using cytogenetic techniques. Ernest Sears took a gene segment from *Aegilopsumbellulata*, having resistance to leaf rust caused by

Puccinia triticina and translocated onto wheat chromosome 6B [13]. This was a pioneering work that aimed to utilize wild species genome through chromosome manipulation thus helping in the improvement of various important crops. Thereafter, much focus has been given to intergeneric and interspecific hybridization to develop cultivars with increased pest tolerance, yields, and various other agronomic traits.

To prevent the loss of valuable genetic diversity, development of genebanks and preservation of germplasm carry immense significance. Genebanks protect natural genomic composition by collecting and conserving alive genepool in the form of seeds of individual genotypes of plant species and their variants. Genepools also maintain characterization of accessions and do evaluations to make them available for breeding community. Ex-situ measures ensure that important genetic makeup of the wild species can also be accessed even if they are eliminated from natural environment. In-situ measures involve the preservation of wild, un-cultivated, and at times endangered species in natural parks and reserves. They are provided with natural conditions to help them thrive, ultimately preserving varying allele combinations. Genebanks maintain cultivated, cross-compatible, and cross-incompatible wild type. Plant breeding methodologies, in the current era, revolve around introducing desirable traits in commercially important crops thus pooling up a uniform genomic composition of limited variability. The availability of diverse genetic resources means important traits from wild species can be obtained when needed [12,22,23]. To facilitate the rapid use of the genetic potential present in the reserves, developing stable and long-lasting seed resources with introgressed lines would prove beneficial. Employing genetic markers and high-density marker maps help determine monogenic traits along with quantitative trait loci (QTL) mapping which greatly reduces the time taken up by the initial analysis of the wild germplasm [24]. For the identification of QTL in agriculture, an exotic library is suitable which has set of lines. Each of these lines carries a defined donor chromosome which differs from an elite variety by a defined segment of chromosome. The plants thus developed do not have sterility issues, are like cultivars and can be analyzed for yield-related traits.

Biotechnology has contributed in preserving valuable plant samples through various in-vitro techniques. Temporary conservation includes growing plantlets on nutrient gels with recurrent sub-culturing under conditions that encourage slow growth, while long-term preservation methods include cryopreservation wherein explants are kept in liquid nitrogen [25]. Physical or osmotic dehydration of plant tissues can also be employed for preservation especially for sensitive explants such as shoot apices and embryos. Vitrification is the process that prevents the formation of ice crystals that may damage the cellular contents. Dehydration is followed by ultra-rapid freezing leading to formation of amorphous glassy structures but no ice crystals occur. Vitrification can be easily carried out in basic labs without the need for any programmable freezer. An increased number of species have been preserved over time due to advanced vitrification procedures (such as encapsulation-dehydration, desiccation, pre-growth-desiccation, vitrification, encapsulation-vitrification, and droplet freezing). Vitrification is especially successful in oil palm and potato [26,27].

18.3 Domestication of wild relative species

Plant breeders have always focused on domesticating wild species by propagation of selected genetic determinants that appeal to the consumption needs of the masses. Domestication has, however, led to the integration of traits in cultivated crops that distinguish them from their wild progenitors. Maize has evolved such that it consists of one or two short branches with the cob filled entirely of kernels and an increased apical dominance compared to its wild parents, hinting at varying gene content among them [28]. Fruit dehiscence is another agronomic trait that carries significance to prevent seed dispersal that has developed overtime in cultivated crop varieties such as soybean [29] and genetic diversity is observed to decrease by domestication such as in cultivated rice varieties [30].With an increase in genetic uniformity, cultivated crops have an increased susceptibility to various pests and diseases [31]. A notable example is the damage to corn in 1970 by fungus *Helminthosporium maydis* causing corn leaf blight epidemic, due to the presence of Texas male-sterile cytoplasm in the cultivated corn hybrids. Similarly, in 1999, wheat rust caused by Ug99 strain of *Puccinia graminis* was transmitted throughout the globe [32], a consequence of genetic uniformity leading to destruction of wheat crops in Kenya, Yemen, Sudan, and even Iran.

A wild crop hybrid requires more than 20 generations to alter their phenotypes suitable for domestication of cultivars. Both the wild and domesticated variants are used as reservoir to multiply desirable determinants. CWR contain desirable traits that are polygenic in nature. These traits are usually related to abiotic and biotic resistance and are deficient in domesticated cultivars. Because these determinants are diffused throughout the entire genome and require the simultaneous functioning of more than one gene, this complicates the integration of wild traits into domesticated cultivars. On the contrary, development of cultivated crop plants usually focus upon monogenic traits linked to yield due to ease of integration in wild species—a phenomenon referred to as de novo domestication [33,34]. For instance, whole genome sequencing of soybean reveals the presence of salt tolerance in wild varieties [35] which has been lost over years of domestication. Fig. 18.1 shows a flowchart diagram describing the use of various techniques being used in domestication of wild relative species.

18.3.1 Mutation breeding and wild relative species

Mutagenesis is brought about by damaging DNA and altering its replication machinery through various mutagens. Physical mutagens include radiations while chemical mutagens make use of chemicals to allow for mutations [36].

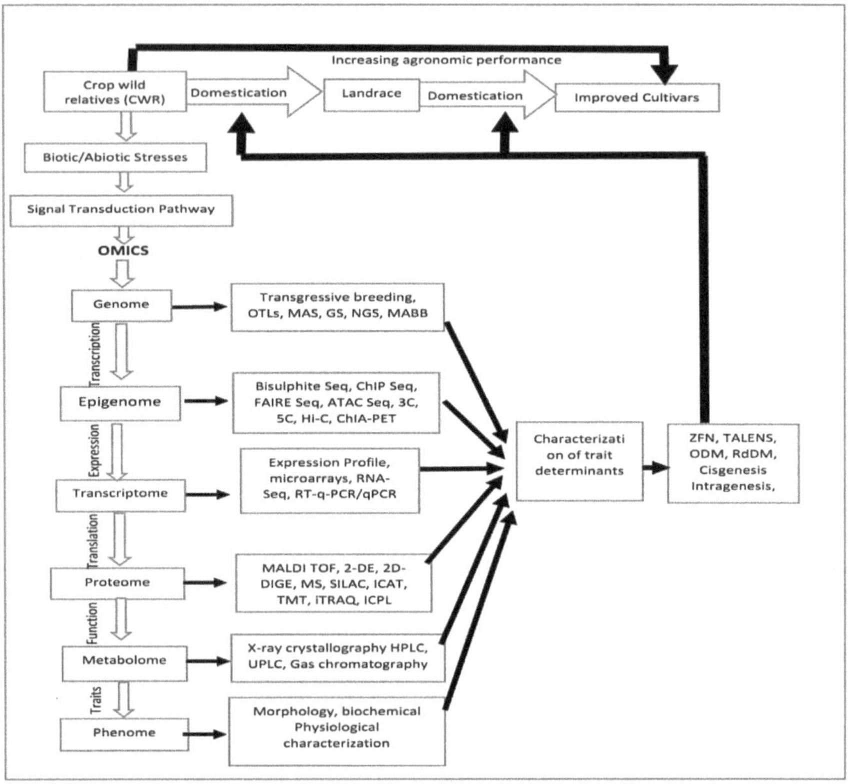

FIGURE 18.1 Modern strategies to accelerate process of domestication and cultivar development.

Back in 1969 mutagens earned an important status in plant breeding when radiations were considered and applied by breeders to allow for induction of mutations in plants [37]. Physical mutagens involve the use of ionizing radiations such as gamma rays to induce mutations. The gamma rays are applied usually in a contained area enclosed by lead or concrete that prevents the escape of gamma rays in the environment by absorption. Radiations result in the production of certain radicals and hydrogen peroxide both of which react with biological molecules, ultimately leading to mutations [38]. Radiations are routinely used in crops [39], for instance, to produce two color flower mutants in *L. Depressa* (Verbenaceae), and various other economically desired mutations in *T. anguina* fruits, *L. acutangula, C. maxima*, and increased herbage yield in *M. citrate*. Compared to conventional breeding that takes around 10–15 years, mutation breeding takes less time spanning 7–9 years. Once the traits are introduced, time span for backcrossing can be reduced by using pots and utilizing immature embryos in the presence of optimum conditions, thus generating as much as eight generations annually. Genetic markers further confirm selection of transgenic individuals [40].

Applications of mutation breeding for improvement resulted in improved crop cultivars in past. Seed size and yield were observed to increase [41] in groundnut mutants by the application of 0.1–1 kGy electron beam [41]. Mutants of wheat developed by both electron beam and gamma rays [42] were observed to have a resistance to *Puccinia striiformis* while also exhibiting earliness with spike increase [43]. Mutation is a preferable medium that is employed for seedless crops such as edible bananas and grapes that lack seeds, root, and tuber crops and plants propagated vegetatively [44].

One of the widely used chemical agent to induce mutations is ethyl methane sulfonate (EMS) that is an alkylating agent. Alkylation of bases results in misreading of the bases leading to mismatched base pairs. Additionally, alkylation may also lead to depurination of the genome which upon replication may be filled in with a different base than the one in place originally. [38]. A comparative analysis by Ref. [45] between physical and chemical mutagens demonstrated the increased mutagenicity of EMS for chickpea. Chemical agents are equally effective depending on the crop and tissue type. Rice can be made palatable and with low amylose content using sodium azide (SA) and EMS. Single-nucleotide substitutions occurred at G2708-A2708 on exon 8 and G3029-A3029 and G3018-A3018occurred on exon 9 of *Waxy* gene [46] to develop premium rice grain with good quality cooking. Nutritional value of crops have been improved such as bio-fortification of maize with iron [47]; iodine enriched vegetables [48]; improving selenium deficiency where needed [49]; and nutritional improvement of staple food crop [50]. High yielding cotton varieties have been produced [51] especially NIAB-999 and CIM-707 [52] producing yields even in soils with increased salinity levels. In addition to good yields, gamma irradiation has also been observed to induce increased yield components such as number of bolls and their weight, seed cotton boll weight, and number of sympodia [53]. Much work has been done on barley [54] such as the generation of semi-dwarf mutant called "Golden Promise" which is greatly salt tolerant and imidazoline herbicide resistant cultivars [55]. Based upon the earlier-mentioned examples of traits improvement, we may conclude that mutation breeding is an excellent option to be used for improving wild species, and make the cultivated ones.

18.4 New plant breeding techniques (NPBT)

Biotechnology refers to make use of living organisms, their biological systems or related derivatives, to develop or change products and processes for specified purposes. This implies that agricultural biotechnology includes various technologies to improve food and agriculture: for example, for genetic modification of crops and livestock targeting increased efficiency and yields; diagnosing diseases in plant and animals and development of vaccine; and to conserve and characterize agricultural biodiversity which is essential for sustainable food production, climate resilient farming, and health security [56]. For accurate characterization and conservation of genetic resources of agriculture, molecular markers, cryopreservation, and various other reproductive technologies are useful.

Due to the rising skepticism against genetically modified crops, new plant breeding techniques (NPBT) are being developed that employ certain steps used to develop genetically modified (GM) crops, however, the resultant products have no foreign genes [57]. New biotechnologies are encouraged that either knock out or modify gene functions. Transgenes may be added during the intermediate steps of breeding and once the purpose is served, these transgenes are eliminated by selection for removal during later crossings [9]. Some of the new plant breeding techniques being developed include: ZFN technology, Cisgenesis and intragenesis development, ODM for variation induction, RdDM, grafting for desired traits, reverse breeding, synthetic genomics [9]. Fig. 18.1 summarizes various modern technologies being used to accelerate breeding programs.

18.4.1 ZFN Technology

ZFN are engineered enzymes that help generate double stranded breaks (DSB) in DNA on specified sequences. They are engineered by fusing a DNA binding zinc-finger domain to a nuclease domain. Three nucleotides are recognized by one zinc finger protein, thus 3–4 proteins per ZFN can determine specificity of 18–24 nucleotides. The DSB initiates the repair system thus either allowing for insertion of any template sequence or correction of a mutation through non-homologous end joining (NHEJ). Additionally, INDELS (insertions and deletions) at the point of DSB can lead to a change of frame-shift thus stopping the production of unwanted proteins. Three methods are developed based on the type of alteration required [9]. ZFN-1 allows for INDELS using the natural cell repair machinery, ZFN-2 utilizes a short repair template that is copied to allow for site-specific point mutations, ZFN-3 employs a large stretch of DNA that is inserted at the DSB carrying homologous flanking sites. ZFNs were used to produce herbicide tolerant maize and tobacco by modifying the IPK1 and SuRA/SuRB loci [58]. This can also become an important tool for those wild species where cultivation is not possible due to the presence of single trait. These traits upon identification of the genes controlling them can be used to mutate the genes and hence eliminate trait.

As mentioned earlier, in maize, IPK1 is modified due to monoallelic insertions of PAT at IPK1. This insertion based disruption generated by ZFN lead to the development of herbicide tolerance and changes in the inositol phosphate profile of developing seeds. Additionally, phytate reduction is also imparted to the mutant thus enabling stacked trait. Hence, ZFN are capable of generating targeted and beneficial mutations which may even lead to dual phenotype [59].

18.4.2 Cisgenesis and Intragenesis

Though area under GM crops has been increasing ever since their introduction but there has been serious debate over the fate of selectable markers and GM crops causing certain allergies, etc. In order to answer these critical questions, the concept of cisgenesis and intragenesisis was introduced which refers to the integration of DNA sequences either from the same species or cross-compatible species that are capable of sexual hybridization [60]. Cisgenesis involves the integration of gene of interest inclusive of introns, along with its native promoter and terminator in sense orientation while intragenesis involves the use of regulatory sequences from cross-compatible species, and if RNA interference (RNAi) is intended then the orientation is antisense. Results from cisgenesis are synonymous with the conventional breeding procedures while intragenesis offers more modification options. Both cisgenesis and intragenesis are exempted from the category of GM because the gene pool is identical to that of conventional breeding.

Commercial apples susceptible to scab are made resistant using HcrVf2 gene from scab resistant apples having endogenous regulatory sequences [61]. The apple cultivar "Florina" is resistant to the apple scab due to the presence of the Rvi6 gene formerly known as HcrVf2 gene. Introduction of an excisable cassette containing a marker fusion gene was mediated by *Agrobacterium* transformation resulting in a true cisgenenic genotype [60,62]. Hence these techniques offer an option for the improvement of cultivated species while the genetic modifications were also accepted with no foreign gene getting integrated into cultivated species. Same strategies can be used to either eliminate or add genes into wild species to make them cultivated ones.

18.4.3 Oligonucleotide Directed Mutagenesis (ODM)

Also referred to as chimeraplasty and genoplasty [9], this technology allows for targeted mutagenesis using one or few adjacent oligonucleotides that are 20–100 nucleotide in length and share homology with the targeted sequence. Oligonucleotides may be single stranded DNA templates or chimeric made of DNA and RNA bases. Induced mutations may include replacement of one to few bases, generation of deletion, and reversing of an existing mutation. Gene transfer is mediated by electroporation and use of polyethylene glycol (PEG) or even particle bombardment. Owing to its accurate, rapid, and non-transgenic breeding alternative, it is employed for agronomic improvements. One of the initial applications of ODM included the targeting of aceto hydroxyl acid synthesis (AHAS) gene in plants such as tobacco [63] and maize [64,65]. Chlorsulfuron resistant rice calli were developed using chimeric DNA/RNA oligonucleotides [66]. A single nucleotide polymorphism (SNP) converts blue fluorescent protein to green fluorescent by targeting the BFP transgene in *Arabidopsis thaliana* [67].

The efficiency of the technology can be demonstrated by generation of single point mutations in less complex genomes such as in rice aceto lactate synthase (ALS) gene which exists as a single copy using chimeric oligonucleotides. When the sequence of ALS gene was matched with herbicide-resistant plants, site-specific mutagenesis was observed indicating a highly efficient application of the technology [67]. Single stranded oligonucleotides may also be used along with double stranded breakers such as transcription activator-like effector nucleases (TALENS) and clustered regularly interspaced short palindromic repeats (CRISPR/Cas9) which increases the target frequency. Using this combination, modifications can be made to the 5'-ENOLPYRUVYLSHIKIMATE-3-PHOSPHATE SYNTHASE (EPSPS) genes. Glyphosate is a herbicide that targets EPSPS by carrying out competitive inhibition to phosphoenolpyruvate. Editing EPSPS imparts tolerance against glyphosate thus producing crops resistant to the herbicide glyphosate. A similar work has been carried out by Sauer, Narváez-Vásquez and coworkers [68] in flax.

18.4.4 RNA Dependent DNA Methylation (RdDM)

No introduction of foreign gene takes place yet the modification of gene expression is brought about by epigenetics. Promoter sequences of DNA are methylated resulting in transcriptional gene silencing (TGS). Genes are delivered to plant cells that code for RNAs carrying homology with the promoter sequences. Upon transcription, these genes formed double-stranded RNAs (dsRNAs) that cause methylation of the promoter sequences thus inhibiting the target gene from transcribing. Such epigenetic changes resulting in desired traits can be passed onto future generations. For instance, inactivation of transcription induced by RdDM ensures inhibited gene expression [69,70]. Fruit ripening greatly involves DNA methylation in various crops. The promoters of many fruit ripening genes contain RIPENING- INHIBITOR (RIN) binding sites. Binding of RIN transcription factors to the promoters is negatively regulated with the DNA methylation. In tomatoes, ripening fruits show an increased DNA demethylase expression. Thus, upon ripening, both ripening induced and ripening repressing genes are targeted. Low methylation levels are synonymous with small fruit size [71].

18.4.5 Grafting for Yield Improvement

Grafting involves combining of scion, a detached vegetative part of one plant with the rootstock of another to form a chimera with improved traits. If the leaves, stem, flowers, seeds, and fruit are desired to be genetically modified, then the scion should be retrieved from a GM plant and vice-versa. Grafting is usually practiced for fruit trees. It is based on the possibility of transmission of recombinant DNA from the GM part to the non-GM through the graft. Generally a non-GM scion grafted on GM rootstock is believed to produce non-GM fruits. However, these are open to doubt and a case by case screening is suggested to differentiate GM from non-GM fruits. According to a survey in Japan, Cd concentrations reduced in eggplants when the scion was grafted on *Solanum torvum* [72]. Grafting is a relatively easy technique that can be employed for trait improvement in crops. A common example includes grafting high-yielding scions on strong rootstalks resistant to various diseases which result in producing high crop yields. Rootstalks of certain crops or plants may impart additional properties such as higher shoot growth and instantaneous water use efficiency contributed from luffa rootstalk to cucumber scion. Drought adaptive response by a rootstalk results in improved drought resistance [73,74].

18.4.6 Reverse Breeding

One of the most sought after goals of plant breeding is reverse breeding wherein homozygous parental lines are generated from a heterozygous plant. Meiosis is engineered to prevent chromosomal recombination. This generates haploid microspores which are cultured in vitro to produce doubled haploid plants that are genotypically homozygous. These plants are then selected for non-transgenic individuals. The resultant parental lines generate offspring that have no additionally altered genes. Reverse breeding finds various applications in agriculture especially where a collection of breeding lines needs to be generated [75,76].

Complementary homozygous lines are an alternate fix to complex and uncharacterized heterozygous genomes. This is helpful in generating parental lines. An F_1 hybrid, which is heterozygous yet homogeneous in nature, is developed by crossing two homozygous lines which in turn allowed to self-fertilize producing segregated F_2 generation. Desirable genotype is selected and followed through reverse breeding. By selection of unknown heterozygous individuals, genetic recombination is decreased because no crossover occurs during meiosis. Non-recombinant male and female gametes thus obtained are used to produce homozygous double haploid plants in vitro which in turn can be used for obtaining heterozygous individuals. Maize, wheat, cucumber, pea, sorghum, and onion are crops wherein double haploids are easily practiced [76,77].

18.4.7 Molecular Assisted Breeding

A pragmatic approach to molecular applications in plant breeding is an important pre-requisite in order to regulate the cost benefit ratio. For instance, the use of markers seems worthwhile in identifying traits of interest and their efficient selection compared to phenotype based selection. Also known as marker assisted selection, it carries great significance in breeding strategies, especially during field trials. Additionally, desired traits too may be characterized simultaneously [78] and usually constitute attributes such as abiotic and biotic stress tolerances, water use efficiency, and traits related to low maintenance and sustainable agriculture.

Traditionally, markers were limited to morphological, biochemical, and cytological characters. Overtime, however, markers have evolved depending on types of plant, tissues, and diversity of need. For instance, probe based: restriction fragment length polymorphism (RFLP); amplification based: random amplification of polymorphic DNA (RAPD), Simple sequence repeats

(SSR), inter-simple sequence repeats (ISSRs), sequence characterized amplified regions (SCAR), cleaved amplified polymorphic sequences (CAPS), sequence-tagged site (STS), specific polymorphic locus amplification test (SPLAT), variable number tandem repeats (VNTRs); probe and PCR based: Amplified fragment length polymorphism (AFLP), second internal transcribed spacer rDNA (rDNA-ITS), and new generation markers: single nucleotide polymorphisms (SNP), expressed sequence tag (EST), single strand conformation polymorphism (SSCP). Various characteristics make them a preferable choice of use such as being neutral, ease of mobility, co-dominance, not limited in number, and availability in mitochondrial and chloroplast genome. One of the major contributions of MAS to plant breeding is the construction of linkage maps that help locate individual gene traits and quantitative trait loci (QTL) on chromosomes. Determining the position of desired genes and their distance from one another on chromosomes makes them easier to identify for further use. A major yield deterrent for rice is drought. In order to breed drought resistance lines, one of the initial steps involve the identification of yield-related QTLs which are stable under drought conditions. Prince and coworkers [79] found these traits on chromosomes 1 (RM8085), 4 (I12S), and 6 (RM6836). Integration of these traits in recipient individuals can help generate drought tolerant rice. Another trait that aids in tolerance against drought in rice is deep rooting which is measured as ratio of deep rooting (RDR). Lou and coworkers [80] carried a genome wide study using SNPs and were able to identify RDR QTLs on chromosomes 1,2,4,7, and 10. Similarly, QTLs related to drought tolerant [81], yield enhancement and biomass production [81] and deep-rooting [82] have also been identified in wheat especially from the wild emmer using SNPs.

MAS has additionally, played a role in combating biotic stresses. QTLs for blast disease and brown plant hopper (BPH) have been reported [83,84]. In sorghum, a QTL referred to as SbAGA01 is characterized to be resistant to green bugs while four QTLs on chromosomes SBI-01, SBI-02, SBI-03, and SBI-08 carry rust resistance [85]. Abdelraheem and coworkers [86] reported various QTLs related to biotic stress such as 85 for root-knot nematode resistance (*Meloiodogyne incognita*), 47 for Fusarium wilt resistance (*Fusarium oxysporum* f. sp. *vasinfectum*), and 201 for Verticillium wilt resistance (*Verticillium dahliae*).

18.4.8 Gene Editing by CRISPR/Cas to Speed up Domestication and Plant Breeding

CRISPR/Cas system; an adaptive immune response system in bacteria, is recently being adopted on a wide scale to engineer desired crop plants. It carries a certain level of supremacy to the traditional mutation breeding techniques based on the efficient generation of novel genomic modifications. Conventional breeding methods are feared to reach a standstill due to a decrease in the natural genetic diversity of the wild relative species [87].

The technique makes use of the Cas protein alongside engineered single guide RNA (sgRNA) that can effectively target specific sites in genome. Moreover, multiple sites can be targeted simultaneously [88]. Some of the frequent genomic modifications include, gene knockouts, transcription modulation in GMOs, multiplex editing, gene replacement, and interruption of gene function [87]. For instance, Gemini viruses are lethal to crop plants and greatly affect crop yields. These crop plants may be modified using CRISPR/Cas technique to confer resistance against Gemini viruses. This involves triggering of mutations within the virus by constraining the copy number of the replicating virus [89]. CRISPR/Cas mediated genome editing is usually propagated by the use of *Agrobacterium* which is the preferred method to generate stable transgenics. Even more complicated is the assembling of multiple sgRNA expression cassettes into one binary CRISPR/Cas vector. A single binary vector can be engineered to contain different sequences by regular cloning following sequential rounds [90–92]. Alternatively, up-to three sgRNA cassettes can be inserted into a vector by multiple restriction enzymes that generate sticky palindromic ends [93,94]. A more efficient methodology to incorporate multiple DNA fragments is the Golden Gate cloning which makes use of type II restriction enzymes that gives rise to sticky ends that are compatible with the sequences to be ligated [95]. This is further developed to integrate multiple sgRNA expression cassettes within CRISPR/Cas9 binary constructs in just a single round of cloning [96,97]. Using Golden Gate cloning, binary constructs consisting of various cassettes, up to eight can be prepared. Alternatively, Taq DNA ligase, Phusion DNA, and T5 exonuclease are employed for the Gibson assembly multiplex cloning [98]. Tandem array of tRNA and gRNA from a polycistronic gene allows efficient multiplex genome editing [99]. Additionally, it is helpful in producing mutants by targeting inaccessible genes. It is also debated that genome edited crops using CRISPR/Cas technology should not be considered GMOs thus reducing speculations around crops modified by CRISPR/Cas technology [87].

18.5 Utilizing crop wild relatives

CWR enjoy the designation of being the progenitors of modern crop cultivars. They continue to carry significance for plant breeders due to their role in providing food, forage, fodder, resistance to pests and diseases, nutritional value, flavor and in pharmacology, ornamental purposes including various industrial purposes. This is primarily because of the genomic diversity present in the CWR which have been used for crop improvement since the dawn of farming. Thus, the CWR are believed to be pivotal to food security, economic stability, and environmental sustainability. Many CWRs are being endangered because much of conservation activities are focused on elite landraces and cultivated varieties. Moreover, global warming has led to a significant change in climate thus affecting the viability of CWRs. There is a dire need to protect the wild relatives of the cultivated or domesticated species and to allow for frequent integration of both [6]. Various conservation activities have been suggested to allow for breeding improvements such as inventories, prioritizing, research, threat and gap analysis, ecogeographic studies, and area designation for genetic reservation followed by managing all the Sociopolitical information on an international, national, and regional level [100].

It is important to define a CWR based on its relatedness to the cultivated crop variety. The measure of relation among them is what defines them. There are two approaches to define this relationship—one proposed by Harlan and de Wet and the second one by Ref. [77]. The former proposition revolves around gene pool set namely GP1, 2, 3 with GP1 being closely related while GP3 being remotely related. However, there is a greater complexity with the variety of crops thus an alternative approach was devised based on taxonomic groups. TG1a is the crop itself, TG1b refers to the same species as the crop. These are followed by measuring relation based on TG2- species, TG3 subgenus, TG4-genus, TG5- tribe [101]. Hence, a CWR can be defined as any wild plant variety that is either closely or remotely related to a cultivated crop based on their taxonomic groups or gene pools [102]. Based on their taxonomic relationship with cultivated crops, CWR's are used to improve domesticated crops for better yields. Over years, much of the trait development in cultivated plants using CWR is driven on economic preferences. For instance, sunflower is globally an important crop due to food, oil, and ornamental purposes [103]. The annual contribution of the wild species to the cultivated variety amounts to US$267–384 million alone. During the 1970s, US maize corn blight destroyed more than 50% of maize yields which almost amounted to about US$1000 million. Yields were recovered by the use of Mexican maize CWR on a national priority basis [101]. The significance of CWRs for crop improvement was recognized back in early 20th century when [104] implied upon the utility of using wild relatives for sugarcane breeding. Increased usage of CWRs for a wide range of crops such as maize and tomato [105] further stressed upon exploiting CWRs. Two of the main procedures to use CWR are: (1) "Choose first": wherein desired plants are chosen on the basis of their phenotype. These varieties are then cross-bred and the resultant progeny is analyzed. (2) "Cross-first": desired wild and domesticated varieties are first crossed and the offsprings are then analyzed for any traits that may be beneficial in domesticated varieties.

The first approach is carried out by observation of traits that are directly expressed by a plant using pathology screens that analyze data related to genotype and agronomic performance. Contrarily, in the second approach, the wild and domesticated varieties are first crossed and the resultant progeny, that is, F_1 or further generations are screened for desirable traits. This latter approach is capable of revealing unexpected diversity in domesticated background due to gene flow from one species to another [11]. Table 18.1 shows selected genes those can be used in domestication in different crops.

TABLE 18.1 Some candidate genes for speeding up domestication through gene editing.

Gene	Crop	Phenotypic function	Causative change	Controls phenotype	References
GhNHX1	Cotton	Improving salt tolerance	Na^+ antiporter	Yes	[106]
Pi-ta Gene	Rice	Bacterial blight and blast resistance	R-Gene	Yes	[107]
OrbHLH001	Rice	Overexpression enhanced *Arabidopsis* freezing and salt tolerance	ICE1 like protein controlling stress tolerance following a pathway other than DREB.	Yes	[108]
bph genes	Rice	Insect tolerance gene cloned from wild rice	Pathway yet to be identified.	Yes	[109]
Pid3-A4	Rice	Rice blast resistance	R-Gene	Candidate	[110]
Pi54rh	Rice	Rice blast resistance	Express constitutively at basal level in leaves, upregulation upon pathogen inoculation	Yes	[111]
Xa21	Rice	Leaf blight	Receptor like Kinase	Yes	[112]
OsSPY	Rice	Negative regulator of GA signaling	Increased leaf and internode elongation upon downregulation	candidate	[113]
GBSS-I and, Waxy gene	Cereals	Accumulate amylose during starch granule development in cereals	Gene sequence in wild species provide better nutritive value	Yes	[114]
Q allele	Wheat	Encodes an AP2 transcription factor	confers a free-threshing character and a partially compact spike inflorescence	Yes	[115]

(*Continued*)

TABLE 18.1 Some candidate genes for speeding up domestication through gene editing. (*Cont.*)

Gene	Crop	Phenotypic function	Causative change	Controls phenotype	References
Wip1, mpi and PR-1 genes from maize (*PR-1*), and *chitinase* gene	Maize	Resistance to fall armyworm (FAW), *Spodoptera frugiperda*	Leaf trichome density	Yes	[116]
ZmWRKY17	Maize	Downregulation prevents salt stress sensitivity and increased ABA sensitivity	Negatively regulates salt stress tolerance	Candidate	[117]
DGAT1-2	Maize	Kernel oil content	Gene isolated from wild maize which controls high oil content	Yes	[118,119]
Brittle Rachis 1(TtBtr1)	Wheat	Controls shattering	Mutations lead to hulled emmer wheat	Yes	[120]
TaGW7	Wheat	Affects grain shape and weight	Mutations increase the grain width and weight	Yes	[93]
rhg1 and Rhg4	Soybean	Soybean cyst nematode (SCN) resistance	Gene from wild soybean causing resistance SCN	Yes	[121]
GmERA1A and *GmERA1B*	Soybean	Negative regulation of ABA signaling to guard cells	least damage under dehydration due to guard cells regulation	Yes	[122]
GsGST	Soybean	High salt and drought resistance	Glutathion S-transferase is pathogenesis related protein involved in biotic stress tolerance	Yes	[123]
GmNAC-2	Soybean	Overexpression leads to hypersensitivity to drought, high salinity, and cold stress	Transcription factor controlling the expression of NAC gene family involved in stress tolerance in soybean	Yes	[124]

GmFAD3	Soybean	Down regulation/silencing increase the stability of the seed oil	The stability of fatty acids is improved by down regulating FAD3 through RNAi	Yes	[125]
LOX H1	Potato	Generates volatile defense and signaling compounds necessary for defense induction	Presence of Lipoxigenase enzyme induces volatile aliphatic C6 aldehydes those induce defense proteins	Yes	[126]
Snakin-1 (SN1)	Potato	Silencing affects cell division, leaf primary metabolism, and cell wall composition	Reduced height, reduction in leaf size, alterations of leaf shape	Yes	[127]
RB/Rpi-blb1	Potato	Resistance to *Phytophthora infestans*	R-Gene	Yes	[128]
Cf-4 and Cf-9	Tomato	Resistance to *Cladosporium fulvum* secreting Avr4 and Avr9 elicitor proteins	R-Gene	Yes	[129]
Aucsia gene	Tomato	Silencing leads to parthenocarpic fruit development	Leaf fusions, and reflexed leaves	Candidate	[130]
LeEIN2	Tomato	Silencing leads to suppression of tomato fruit ripening for longer shelf-life	Transcription factor	Yes	[131]
1RS.1AL	Wheat-rye transloca-tion	Resistance to various fungal and viral pathogens and adaption to marginal environments with good yields	Set of genes found in rye that can be used to breed fungal resistance in wheat	Yes	[132]
Hsdr4	Barley	Water-stress tolerance	Stress responsive protein likely to be substrate for MybMyc and Mtc1 stress responsive transcription factors	Candidate	[133]

(*Continued*)

TABLE 18.1 Some candidate genes for speeding up domestication through gene editing. (*Cont.*)

Gene	Crop	Phenotypic function	Causative change	Controls phenotype	References
BLUFENSIN1 (BLN)	Barley	Negatively impacts basal defense against *Blumeria graminis*	R-Gene	Candidate	[134]
HsCIPKs	Barley	Enhances tolerance to heavy metals in rice	Overexpression results in healthy root growth under metal toxicities	Candidate	[135]
Dehydrin gene (Dhn)	Barley	Drought tolerance	Dehydrin/late embryogenesis abundant protein responsive for dehydration tolerance	Yes	[136]
Sh1	Cereals	Seed shattering	Frameshift mutations lead to prevention of abscission layer production of joint connecting seed hull and pedicel	Yes	[46]
fw2.2 or fas gene	Tomato	Downregulation causes high-locule number thus increased fruit size	Down-regulation due to large insertion in first Intron	Yes	[137]
ZmCCT	Maize	Photoperiod sensitivity	Transposon insertion leads to reduced gene expression thus increasing insensitivity to long days	Yes	[138]
GhTT2_A07	Cotton	Lint color	Sequence inversion imparts brown color and enhanced flame retardancy	Yes	[139]

tgr-1	tomato	Resistance to Gemini viruses	Impairs viral movement	Candidate	[140]
ZmHKT1	Maize	Salt tolerance gene	Root modification are variant dependent	Yes	[141]
Sos1	Maize	Involved in the evolution of paired spikelet	Deletion of a lysine residue	Yes	[142]
Vrs1	Barley	Six-rowed spike	Controls six rowed spikes in barley	Yes	[143]
qSW5	Rice	Related to grain width and size	Deletion within the gene causes increase in rice grain size	Yes	[144]
SD1, SD2	Barley	Moderate seed dormancy	Control moderate seed dormancy which is a desirable trait in barley	Candidate	[145]
ppd-H1	Barley	Circadian clock function	Mutations lead to reduced photoperiod responsiveness advantageous in spring-sown varieties	Yes	[146]

18.6 Challenges and opportunities

Not always do plant breeding programs have sufficient resources to carry out extensive breeding. This is especially true for developing and under-developed countries. Moreover, breeding programs usually focus on priority traits and their incorporation into germplasm with diverse background to be made available in markets. The common priority traits include tolerance to abiotic and biotic stresses, responsiveness to fertilizers, reduced span of maturity, and increased quality of yield, food, and feed. These traits are usually not synonymous to the needs of the farmers and there is a high probability of communication gap among the farmers and the technical plant breeders [147].

Despite its benefits, traditional breeding strategies are associated with various limitations [148] that require highly efficient and accurate techniques. For instance, wild species usually carry genes that impart resistance to harsh conditions and various pests and diseases. However, these may be lost during recurrent backcrossing and inbreeding processes. In addition to the painstaking and laborious activity of domestication, an important aspect to be taken into consideration when formulating any breeding strategy is that domestication may lead to an accumulation of many deleterious allelic variants among the cultivars compared to their wild relatives. Particularly susceptible to accumulating such mutations are the species that have been clonally propagated [33].

18.6.1 Abiotic and Biotic Stresses

Abiotic stress refers to harsh environmental conditions that are detrimental to plants such as salinity, drought, and drastic fluctuations in temperatures. Global crop losses are primarily caused by abiotic factors resulting in a total loss of more than 50% in terms of yield [149]. Biotic stress includes threat from biological living entities such as microorganisms, pests, and insects. For instance, Fusarium head blight in wheat and barley [150,151]; blast in rice [152], late blight in potato [153], and bacterial spot in tomato are all caused by microorganisms [154] promoted by humid conditions.

CWR are major reservoirs of abiotic and biotic stress resistance genes for many cultivars. For instance, in rice (*Oryza sativa*), 22 species are wild, spread around the globe—and only 2 are cultivated, that is, *O. sativa* and *O. glaberrima* [155]. These 22 wild relatives provide germplasm for various traits such as cytoplasmic male sterility to develop hybrid rice; resistance to brown plant hopper imparted by BPH18 gene [156], *hydroperoxide lyase 2* (*OsHPL2*), and its derivative (E) -2- hexenal that helps resist white-backed plant hopper [157]; resistance to bacterial leaf blight caused by *Xanthomonas* sp [158]; adaptation to aerobic soils [159]; tolerance to salinity [160,161] and floods [162]; and viruses such as tungro and grassy stunt [163,164].

Maize fields, especially, those in the United States are severely affected by corn rootworms (Chrysomelidae: *Diabrotica* spp.) causing major economic losses. Eastern gamagrass (*Tripsacum dactyloides*) is a wild relative of maize that is resistant to rootworms. Breeding programs should incorporate gamagrass germplasm [165] to generate maize that is resistant to rootworms. Teosinte is documented to be resistant to 66 insect genera [166]. Z. diploperennis was found to be immune to Striga sp [167,168] and a variety of viruses including maize chlorotic mottle, maize streak viruses, and maize chlorotic dwarf among others [169]. Currently plant breeders are capable of exploiting the resources available in wild relatives through NPBT by gene editing, gene cloning, and developing safe GM crops.

References

[1] Xiong H, Guo H, Xie Y, Zhao L, Gu J, Zhao S, et al. Enhancement of dwarf wheat germplasm with high-yield potential derived from induced mutagenesis. Plant Genet Resour 2018;16(1):74–81.

[2] Fan S, Yao X, Liu J, Dong X, Mao T, Wang J. Characterization and fine mapping of osh15 (t), a novel dwarf mutant gene in rice (*Oryza sativa* L.). Genes Genom 2016;38(9):849–56.

[3] Bag D. Pesticides and health risks. Econom Polit Week 2000;35:3381–3.

[4] Krishnaiah NV. A Global perspective of rice brown planthopper management II-after green revolution era. Mol Entomol 2014;5(1):46–55.

[5] Sebby K. The green revolution of the 1960's and its impact on small farmers in India. An undergraduate thesis by University of Nebraska, USA; 2010.

[6] From plant to crop: The past, present and future of plant breeding. An article published by VIB series of ISAAA. ISAAA; 2016.

[7] Statement on New Breeding techniques released by European Academies Science Advisory Council; 2015.

[8] Hu H, Scheben A, Edwards D. Advances in integrating genomics and bioinformatics in the plant breeding pipeline. Agriculture 2018;8(6):75.

[9] Lusser M, Parisi C, Plan D, Rodriguez-Cerezo E. Deployment of new biotechnologies in plant breeding. Nat Biotechnol 2012;30(3):231.

[10] Castañeda-Álvarez NP, Khoury CK, Achicanoy HA, Bernau V, Dempewolf H, Eastwood RJ, et al. Global conservation priorities for crop wild relatives. Nat Plants 2016;2(4):1–6.

[11] Dempewolf H, Baute G, Anderson J, Kilian B, Smith C, Guarino L. Past and future use of wild relatives in crop breeding. Crop Sci 2017;57(3):1070–82.

[12] Brozynska M, Furtado A, Henry RJ. Genomics of crop wild relatives: expanding the gene pool for crop improvement. Plant Biotechnol J 2016;14(4):1070–85.

[13] Jauhar PP. Modern biotechnology as an integral supplement to conventional plant breeding: the prospects and challenges. Crop Sci 2006;46(4):1841–59.

[14] Snape JW. Challenges of integrating conventional breeding and biotechnology: a personal view. In: Proceedings of the fourth international crop science congress, vol. 26. 2004.

[15] Schmidhuber J, Tubiello FN. Global food security under climate change. Proc Natl Acad Sci 2007;104(50):19703–8.

[16] Ruane J, Sonnino A. Agricultural biotechnologies in developing countries and their possible contribution to food security. J Biotechnol 2011;156(4):356–63.

[17] FAO, IFAD, WFP, WHO, UNICEF. The state of food security and nutrition in the world 2019: safeguarding against economic slowdowns and downturns. Rome, FAO; 2019.

[18] Breseghello F, Coelho ASG. Traditional and modern plant breeding methods with examples in rice (*Oryza sativa* L.). J Agric Food Chem 2013;61(35):8277–86.

[19] Fu YB. Understanding crop genetic diversity under modern plant breeding. Theor Appl Genet 2015;128(11):2131–42.
[20] Ali A, Erenstein O. Assessing farmer use of climate change adaptation practices and impacts on food security and poverty in Pakistan. Clim Risk Manage 2017;16:183–94.
[21] Shao Q, Punt M, Wesseler J. New plant breeding techniques under food security pressure and lobbying. Front Plant Sci 2018;9:1324.
[22] Malmstrom CM, Alexander HM. Effects of crop viruses on wild plants. Curr Opin Virol 2016;19:30–6.
[23] Sharma S, Upadhyaya HD, Varshney RK, Gowda CLL. Pre-breeding for diversification of primary gene pool and genetic enhancement of grain legumes. Front Plant Sci 2013;4:309.
[24] Cobb JN, Biswas PS, Platten JD. Back to the future: revisiting MAS as a tool for modern plant breeding. Theor Appl Genet 2019;132(3):647–67.
[25] Popova E, Shukla M, Kim HH, Saxena PK. Plant cryopreservation for biotechnology and breeding. Advances in plant breeding strategies: breeding, biotechnology and molecular toolsSpringer, Cham; 2015. p. 63–93.
[26] Gantait S, Sinniah UR, Suranthran P, Palanyandy SR, Subramaniam S. Improved cryopreservation of oil palm (Elaeisguineensis Jacq.) polyembryoids using droplet vitrification approach and assessment of genetic fidelity. Protoplasma 2015;252(1):89–101.
[27] Niino T, Arizaga MV. Cryopreservation for preservation of potato genetic resources. Breed Sci 2015;65(1):41–52.
[28] Stitzer MC, Ross-Ibarra J. Maize domestication and gene interaction. New Phytol 2018;220(2):395–408.
[29] Sedivy EJ, Wu F, Hanzawa Y. Soybean domestication: the origin, genetic architecture and molecular bases. New Phytol 2017;214(2):539–53.
[30] Fuchs EJ, Martínez AM, Calvo A, Muñoz M, Arrieta-Espinoza G. Genetic diversity in *Oryza glumaepatula* wild rice populations in Costa Rica and possible gene flow from *O. sativa*. Peer J 2016;4:e1875.
[31] Arora S, Steuernagel B, Gaurav K, Chandramohan S, Long Y, Matny O, et al. Resistance gene cloning from a wild crop relative by sequence capture and association genetics. Nat Biotechnol 2019;37(2):139–43.
[32] Singh RP, Hodson DP, Huerta-Espino J, Jin Y, Bhavani S, Njau P, et al. The emergence of Ug99 races of the stem rust fungus is a threat to world wheat production. Ann Rev Phytopathol 2011;49:465–81.
[33] Fernie AR, Yan J. De novo domestication: an alternative route toward new crops for the future. Mol Plant 2019;12(5):615–31.
[34] Zsögön A, Cermak T, Voytas D, Peres LEP. Genome editing as a tool to achieve the crop ideotype and de novo domestication of wild relatives: case study in tomato. Plant Sci 2017;256:120–30.
[35] Qi X, Li MW, Xie M, Liu X, Ni M, Shao G, et al. Identification of a novel salt tolerance gene in wild soybean by whole-genome sequencing. Nat Commun 2014;5:4340.
[36] Mohan Jain S. Mutation-assisted breeding for improving ornamental plants. In: Twenty-second international eucarpia symposium, section ornamentals, breeding for beauty 714. Genoa, Italy: ISHS; 2006. p. 85–98.
[37] Songstad DD, Petolino JF, Voytas DF, Reichert NA. Genome editing of plants. Crit Rev Plant Sci 2017;36(1):1–23.
[38] Kodym A, Afza R. Physical and chemical mutagenesis. Plant functional genomicsHumana Press; 2003. p. 189–203.

[39] Shu QY. Induced plant mutations in the genomics era. A book published under FAO. ISBN-10: 9251063249. FAO; 2009.
[40] Bado S, Forster BP, Nielen S, Ali AM, Lagoda PJ, Till BJ, et al. Plant mutation breeding: current progress and future assessment. Plant Breed Rev 2015;39:23–88.
[41] Mondal S, Petwal VC, Badigannavar AM, Bhad PG, Verma VP, Goswami SG, et al. Electron beam irradiation revealed genetic differences in radio-sensitivity and generated mutants in groundnut (*Arachis hypogaea* L.). Appl Radiat Isotopes 2017;122:78–83.
[42] Bakshi S, Singh J, Jambhulkar SJ. Isolation and characterisation of yellow rust resistant mutation in wheat (No. IAEA-CN--263). 2018.
[43] Salameh A, Hroub I, Jankuloski L, Basheer-Salimia R, Hamdan Y. Induced mutations in durum wheat (*Triticum durum*) for improve productivity components traits (No. IAEA-CN--263). 2018.
[44] Suprasanna P, Mirajkar SJ, Bhagwat SG. Induced mutations and crop improvement. Plant biology and biotechnologyNew Delhi: Springer; 2015. p. 593–617.
[45] Shah TM, Mirza JI, Haq MA, Atta BM. Induced genetic variability in chickpea (*Cicer arietinum* L.). II. Comparative mutagenic effectiveness and efficiency of physical and chemical mutagens. Pak J Bot 2008;40(2):605–13.
[46] Lin Z, Li X, Shannon LM, Yeh CT, Wang ML, Bai G, et al. Parallel domestication of the Shattering1 genes in cereals. Nat Genet 2012;44(6):720.
[47] Pixley KV, Palacios-Rojas N, Glahn RP. The usefulness of iron bioavailability as a target trait for breeding maize (*Zea mays* L.) with enhanced nutritional value. Field Crops Res 2011;123(2):153–60.
[48] Dai JL, Zhu YG, Zhang M, Huang YZ. Selecting iodine-enriched vegetables and the residual effect of iodate application to soil. Biol Trace Element Res 2004;101(3):265–76.
[49] Hartikainen H. Biogeochemistry of selenium and its impact on food chain quality and human health. J Trace Elements Med Biol 2005;18(4):309–18.
[50] Nestel P, Bouis HE, Meenakshi JV, Pfeiffer W. Biofortification of staple food crops. J Nutr 2006;136(4):1064–7.
[51] Muthusamy A, Vasanth K, Jayabalan N. Induced high yielding mutants in cotton (*Gossypium hirsuthum* L.). Mutat Breed Newslett Rev 2005;1:6–8.
[52] Ali Y, Aslam Z, Hussain F. Genotype and environment interaction effect on yield of cotton under naturally salt stress condition. Int J Environ Sci Technol 2005;2(2):169–73.
[53] Yilmaz A, Boydak E. The effects of Cobalt-60 applications on yield and yield components of cotton (*Gossypium barbadense* L.). Pak J Biol Sci 2006;9(15):2761–9.
[54] Forster BP. Mutation genetics of salt tolerance in barley: an assessment of Golden Promise and other semi-dwarf mutants. Euphytica 2001;120(3):317–28.
[55] Lee H, Rustgi S, Kumar N, Burke I, Yenish JP, et al. Single nucleotide mutation in the barley acetohydroxy acid synthase (AHAS) gene confers resistance to imidazolinone herbicides. Proc Natl Acad Sci 2011;108(21):8909–13.
[56] FAO. Biotechnologies for agricultural development. In: Proceedings of the FAO international technical conference on "Agricultural Biotechnologies in Developing Countries: options and opportunities in crops, forestry, livestock, fisheries and agroindustry to face the challenges of food insecurity and climate change"(ABDC-10). Rome, FAO; 2011. Available from: http://www.fao.org/docrep/014/i2300e/i2300e00.htm.
[57] Schaart JG, van de Wiel CC, Lotz LA, Smulders MJ. Opportunities for products of new plant breeding techniques. Trends Plant Sci 2016;21(5):438–49.
[58] Puchta H, Fauser F. Synthetic nucleases for genome engineering in plants: prospects for a bright future. Plant J 2014;78(5):727–41.

[59] Zhang Y, Massel K, Godwin ID, Gao C. Applications and potential of genome editing in crop improvement. Genome Biol 2018;19(1):210.
[60] Cardi T. Cisgenesis and genome editing: combining concepts and efforts for a smarter use of genetic resources in crop breeding. Plant Breed 2016;135(2):139–47.
[61] Vanblaere T, Szankowski I, Schaart J, Schouten H, Flachowsky H, Broggini GA, et al. The development of a cisgenic apple plant. J Biotechnol 2011;154(4):304–11.
[62] Vanblaere T, Flachowsky H, Gessler C, Broggini GA. Molecular characterization of cisgenic lines of apple 'Gala'carrying the Rvi6 scab resistance gene. Plant Biotechnol J 2014;12(1):2–9.
[63] Kochevenko A, Willmitzer L. Chimeric RNA/DNA oligonucleotide-based site-specific modification of the tobacco acetolactate syntase gene. Plant Physiol 2003;132(1):174–84.
[64] Zhu T, Mettenburg K, Peterson DJ, Tagliani L, Baszczynski CL. Engineering herbicide-resistant maize using chimeric RNA/DNA oligonucleotides. Nat Biotechnol 2000;18(5):555–8.
[65] Zhu T, Peterson DJ, Tagliani L, Clair GS, Baszczynski CL, Bowen B. Targeted manipulation of maize genes in vivo using chimeric RNA/DNA oligonucleotides. Proc Natl Acad Sci 1999;96(15):8768–73.
[66] Okuzaki A, Toriyama K. Chimeric RNA/DNA oligonucleotide-directed gene targeting in rice. Plant Cell Rep 2004;22(7):509–12.
[67] Sauer NJ, Mozoruk J, Miller RB, Warburg ZJ, Walker KA, Beetham PR, et al. Oligonucleotide-directed mutagenesis for precision gene editing. Plant Biotechnol J 2016;14(2):496–502.
[68] Sauer NJ, Narváez-Vásquez J, Mozoruk J, Miller RB, Warburg ZJ, Woodward MJ, et al. Oligonucleotide-mediated genome editing provides precision and function to engineered nucleases and antibiotics in plants. Plant Physiol 2016;170(4):1917–28.
[69] Mette MF, Aufsatz W, Van der Winden J, Matzke MA, Matzke AJM. Transcriptional silencing and promoter methylation triggered by double-stranded RNA. EMBO J 2000;19(19):5194–201.
[70] Sijen T, Vijn I, Rebocho A, van Blokland R, Roelofs D, Mol JN, et al. Transcriptional and posttranscriptional gene silencing are mechanistically related. Curr Biol 2001;11(6):436–40.
[71] Zhang H, Lang Z, Zhu JK. Dynamics and function of DNA methylation in plants. Nat Rev Mol Cell Biol 2018;19(8):489–506.
[72] Arao T, Takeda H, Nishihara E. Reduction of cadmium translocation from roots to shoots in eggplant (*Solanum melongena*) by grafting onto *Solanum torvum* rootstock. Soil Sci Plant Nutr 2008;54(4):555–9.
[73] Kumar P, Rouphael Y, Cardarelli M, Colla G. Vegetable grafting as a tool to improve drought resistance and water use efficiency. Front Plant Sci 2017;8:1130.
[74] Xie F, Du Y, Zhao Q, Li S, Yao X, Zhao M. Shoot/root interactions affect soybean photosynthetic traits and yield formation: a case study of grafting with record-yield cultivars. Front Plant Sci 2019;10:445.
[75] Dirks R, Van Dun K, De Snoo CB, Van Den Berg M, Lelivelt CL, Voermans W, et al. Reverse breeding: a novel breeding approach based on engineered meiosis. Plant Biotechnol J 2009;7(9):837–45.
[76] Mahesh S. The state of art of new transgenic techniques in plant breeding: a review. J Adv Biol Biotechnol 2016;9(4):1–11.
[77] Mandeep K, Avinashe HA, Nidhi D, Kesha R, Simerjeet K, Swapnil K. Reverse Breeding: creating parental line for a heterozygous plant and its complication. Ann Biol 2019;35(1):50–4.
[78] Ahmad F, Akram A, Farman K, Abbas T, Bibi A, Khalid S, et al. Molecular markers and marker assisted plant breeding. Current status and their applications in agricultural development. J Environ Agric Sci 2017;11:35–50.

[79] Prince SJ, Beena R, Gomez SM, Senthivel S, Babu RC. Mapping consistent rice (*Oryza sativa* L.) yield QTLs under drought stress in target rainfed environments. Rice 2015;8(1):25.
[80] Lou Q, Chen L, Mei H, Wei H, Feng F, Wang P, et al. Quantitative trait locus mapping of deep rooting by linkage and association analysis in rice. J Exp Bot 2015;66(15):4749–57.
[81] Merchuk-Ovnat L, Barak V, Fahima T, Ordon F, Lidzbarsky GA, Krugman T, et al. Ancestral QTL alleles from wild emmer wheat improve drought resistance and productivity in modern wheat cultivars. Front Plant Sci 2016;7:452.
[82] Merchuk-Ovnat L, Fahima T, Ephrath JE, Krugman T, Saranga Y. Ancestral QTL alleles from wild emmer wheat enhance root development under drought in modern wheat. Front Plant Sci 2017;8:703.
[83] Jairin J, Kotchasatit U, Saleeto S, Jearakongman S, Srivilai K. Application of marker-assisted breeding to improve biotic stress resistance for rainfed lowland rice in northeastern Thailand. Sabrao J Breed Genet 2017;49(2):168–78.
[84] Korinsak S, Siangliw M, Kotcharerk J, Jairin J, Siangliw JL, Jongdee B, et al. Improvement of the submergence tolerance and the brown planthopper resistance of the Thai jasmine rice cultivar KDML105 by pyramiding Sub1 and Qbph12. Field Crops Res 2016;188:105–12.
[85] Liu G, Han Y, Jiang Y, Wang Y, Lv P, Li H. Genomics approaches to biotic stress resistance. The sorghum genomeSpringer, Cham; 2016. p. 149–67.
[86] Abdelraheem A, Liu F, Song M, Zhang JF. A meta-analysis of quantitative trait loci for abiotic and biotic stress resistance in tetraploid cotton. Mol Genet Genom 2017;292(6):1221–35.
[87] Khatodia S, Bhatotia K, Passricha N, Khurana SMP, Tuteja N. The CRISPR/Cas genome-editing tool: application in improvement of crops. Front Plant Sci 2016;7:506.
[88] Belhaj K, Chaparro-Garcia A, Kamoun S, Nekrasov V. Plant genome editing made easy: targeted mutagenesis in model and crop plants using the CRISPR/Cas system. Plant Methods 2013;9(1):39.
[89] Chaparro-Garcia A, Kamoun S, Nekrasov V. Boosting plant immunity with CRISPR/Cas. Genome Biol 2015;16(1):254.
[90] Li JF, Norville JE, Aach J, McCormack M, Zhang D, Bush J, et al. Multiplex and homologous recombination–mediated genome editing in *Arabidopsis* and *Nicotiana benthamiana* using guide RNA and Cas9. Nature Biotechnol 2013;31(8):688–91.
[91] Yan L, Wei S, Wu Y, Hu R, Li H, Yang W, et al. High-efficiency genome editing in Arabidopsis using YAO promoter-driven CRISPR/Cas9 system. Mol Plant 2015;8(12):1820–3.
[92] Zhou H, Liu B, Weeks DP, Spalding MH, Yang B. Large chromosomal deletions and heritable small genetic changes induced by CRISPR/Cas9 in rice. Nucleic Acids Res 2014;42(17):10903–14.
[93] Wang W, Pan Q, Tian B, He F, Chen Y, Bai G, et al. Gene editing of the wheat homologs of TONNEAU 1-recruiting motif encoding gene affects grain shape and weight in wheat. Plant J 2019;100(2):251–64.
[94] Zhang Z, Mao Y, Ha S, Liu W, Botella JR, Zhu JK. A multiplex CRISPR/Cas9 platform for fast and efficient editing of multiple genes in *Arabidopsis*. Plant Cell Rep 2016;35(7):1519–33.
[95] Engler C, Kandzia R, Marillonnet S. A one pot, one step, precision cloning method with high throughput capability. PloS ONE 2008;3(11):e3647.
[96] Ma X, Zhang Q, Zhu Q, Liu W, Chen Y, Qiu R, et al. A robust CRISPR/Cas9 system for convenient, high-efficiency multiplex genome editing in monocot and dicot plants. Mol Plant 2015;8(8):1274–84.
[97] Xing HL, Dong L, Wang ZP, Zhang HY, Han CY, Liu B, et al. A CRISPR/Cas9 toolkit for multiplex genome editing in plants. BMC Plant Biol 2014;14(1):327.

[98] Gibson DG, Young L, Chuang RY, Venter JC, Hutchison CA, Smith HO. Enzymatic assembly of DNA molecules up to several hundred kilobases. Nat Methods 2009;6(5):343–5.
[99] Xie K, Minkenberg B, Yang Y. Boosting CRISPR/Cas9 multiplex editing capability with the endogenous tRNA-processing system. Proc Natl Acad Sci 2015;112(11):3570–5.
[100] Montenegro de Wit M. Stealing into the wild: conservation science, plant breeding and the makings of new seed enclosures. J Peasant Stud 2017;44(1):169–212.
[101] Maxted N, Kell SP. Establishment of a global network for the in situ conservation of crop wild relatives: status and needsRome, Italy: FAO Commission on Genetic Resources for Food and Agriculture; 2009. p. 266.
[102] Brehm JM, Maxted N, Ford-Lloyd BV, Martins-Louçao MA. National inventories of crop wild relatives and wild harvested plants: case-study for Portugal. Genet Resour Crop Evol 2008;55(6):779–96.
[103] Seiler GJ, Qi LL, Marek LF. Utilization of sunflower crop wild relatives for cultivated sunflower improvement. Crop Sci 2017;57(3):1083–101.
[104] Plucknett D, Smith N, Williams J, Murthi AN. Gene banks and the world's food. Princeton, NJ: Princeton University Press; 1987.
[105] Hoyt E. Conserving the wild relatives of crops. IBPGR, IUCN; 1988.
[106] Long L, Zhao JR, Guo DD, Ma XN, Xu FC, Yang WW, et al. Identification of NHXs in Gossypium species and the positive role of GhNHX1 in salt tolerance. BMC Plant Biol 2020;20:1–13.
[107] Huang CL, Hwang SY, Chiang YC, Lin TP. Molecular evolution of the Pi-ta gene resistant to rice blast in wild rice (*Oryza rufipogon*). Genetics 2008;179(3):1527–38.
[108] Li F, Guo S, Zhao Y, Chen D, Chong K, Xu Y. Overexpression of a homopeptide repeat-containing bHLH protein gene (OrbHLH001) from Dongxiang wild rice confers freezing and salt tolerance in transgenic *Arabidopsis*. Plant Cell Rep 2010;29(9):977–86.
[109] Huang D, Qiu Y, Zhang Y, Huang F, Meng J, Wei S, et al. Fine mapping and characterization of BPH27, a brown planthopper resistance gene from wild rice (*Oryza rufipogon* Griff.). Theor Appl Genet 2013;126(1):219–29.
[110] Lv Q, Xu X, Shang J, Jiang G, Pang Z, Zhou Z, et al. Functional analysis of Pid3-A4, an ortholog of rice blast resistance gene Pid3 revealed by allele mining in common wild rice. Phytopathology 2013;103(6):594–9.
[111] Das A, Soubam D, Singh PK, Thakur S, Singh NK, Sharma TR. A novel blast resistance gene, Pi54rh cloned from wild species of rice, *Oryza ehizomatis* confers broad spectrum resistance to Magnaporthe oryzae. Funct Integr Genom 2012;12(2):215–28.
[112] Qian J, Cheng ZQ, Yang MZ, Liu JM, Wu CJ, Huang XQ. Isolation and sequence analysis of the Xa21 gene Wxon II homologus from different species of wild rice in Yunnan. Yi Chuan Hereditas 2005;27(3). 382-286.
[113] Shimada A, Ueguchi-Tanaka M, Sakamoto T, Fujioka S, Takatsuto S, Yoshida S, et al. The rice SPINDLY gene functions as a negative regulator of gibberellin signaling by controlling the suppressive function of the DELLA protein, SLR1, and modulating brassinosteroid synthesis. Plant J 2006;48(3):390–402.
[114] Shapter FM, Eggler P, Lee LS, Henry RJ. Variation in granule bound starch synthase I (GBSSI) loci amongst Australian wild cereal relatives (*Poaceae*). J Cereal Sci 2009;49(1):4–11.
[115] Greenwood JR, Finnegan EJ, Watanabe N, Trevaskis B, Swain SM. New alleles of the wheat domestication gene Q reveal multiple roles in growth and reproductive development. Development 2017;144(11):1959–65.

[116] Szczepaniec A, Widney SE, Bernal JS, Eubanks MD. Higher expression of induced defenses in teosintes (*Zea* spp.) is correlated with greater resistance to fall armyworm *Spodoptera frugiperda*. Entomol Exp Appl 2013;146(2):242–51.
[117] Cai R, Dai W, Zhang C, Wang Y, Wu M, Zhao Y, et al. The maize WRKY transcription factor ZmWRKY17 negatively regulates salt stress tolerance in transgenic *Arabidopsis* plants. Planta 2017;246(6):1215–31.
[118] Fang H, Fu X, Wang Y, Xu J, Feng H, Li W, et al. Genetic basis of kernel nutritional traits during maize domestication and improvement. Plant J 2020;101(2):278–92.
[119] Zheng PZ, Allen WB, Roesler K, et al. A phenylalanine in DGAT is a key determinant of oil content and composition in maize. Nat Genet 2008;40:367–72.
[120] Avni R, Nave M, Barad O, Baruch K, Twardziok SO, Gundlach H, et al. Wild emmer genome architecture and diversity elucidate wheat evolution and domestication. Science 2017;357(6346):93–7.
[121] Kim M, Hyten DL, Niblack TL, Diers BW. Stacking resistance alleles from wild and domestic soybean sources improves soybean cyst nematode resistance. Crop Sci 2011;51(3):934–43.
[122] Ogata T, Nagatoshi Y, Yamagishi N, Yoshikawa N, Fujita Y. Virus-induced down-regulation of GmERA1A and GmERA1B genes enhances the stomatal response to abscisic acid and drought resistance in soybean. PloS ONE 2017;12(4):e0175650.
[123] Ji W, Zhu Y, Li Y, Yang L, Zhao X, Cai H, et al. Over-expression of a glutathione S-transferase gene, GsGST, from wild soybean (Glycine soja) enhances drought and salt tolerance in transgenic tobacco. Biotechnol Lett 2010;32(8):1173–9.
[124] Jin H, Huang F, Cheng H, Song H, Yu D. Overexpression of the GmNAC2 gene, an NAC transcription factor, reduces abiotic stress tolerance in tobacco. Plant Mol Biol Rep 2013;31(2):435–42.
[125] Flores T, Karpova O, Su X, Zeng P, Bilyeu K, Sleper DA, et al. Silencing of GmFAD3 gene by siRNA leads to low α-linolenic acids (18: 3) of fad3-mutant phenotype in soybean [*Glycine max* (Merr.)]. Transgen Res 2008;17(5):839–50.
[126] León J, Royo J, Vancanneyt G, Sanz C, Silkowski H, Griffiths G, et al. Lipoxygenase H1 gene silencing reveals a specific role in supplying fatty acid hydroperoxides for aliphatic aldehyde production. J Biol Chem 2002;277(1):416–23.
[127] Nahirñak V, Rivarola M, Almasia NI, Baron MPB, Hopp HE, Vile D, et al. Snakin-1 affects reactive oxygen species and ascorbic acid levels and hormone balance in potato. PloS ONE 2019;14(3):e0214165.
[128] Pankin A, Sokolova E, Rogozina E, Kuznetsova M, Deahl K, Jones R, et al. Allele mining in the gene pool of wild *Solanum* species for homologues of late blight resistance gene RB/Rpi-blb1. Plant Genet Resour 2011;9(2):305–8.
[129] Zhao T, Jiang J, Liu G, He S, Zhang H, Chen X, et al. Mapping and candidate gene screening of tomato *Cladosporium fulvum*-resistant gene Cf-19, based on high-throughput sequencing technology. BMC Plant Biol 2016;16(1):51.
[130] Molesini B, Pandolfini T, Rotino GL, Dani V, Spena A. Aucsia gene silencing causes parthenocarpic fruit development in tomato. Plant Physiol 2009;149(1):534–48.
[131] Kou X, Liu C, Han L, Wang S, Xue Z. NAC transcription factors play an important role in ethylene biosynthesis, reception and signaling of tomato fruit ripening. Mol Genet Genom 2016;291(3):1205–17.
[132] Lukaszewski AJ. A set of new 1RS translocations from wheat cv. Amigo in a uniform genetic background. Euphytica 2017;213(9):214.

[133] Suprunova T, Krugman T, Distelfeld A, Fahima T, Nevo E, Korol A. Identification of a novel gene (Hsdr4) involved in water-stress tolerance in wild barley. Plant Mol Biol 2007;64(1–2):17–34.

[134] Meng Y, Moscou MJ, Wise RP. Blufensin1 negatively impacts basal defense in response to barley powdery mildew. Plant Physiol 2009;149(1):271–85.

[135] Pan W, Shen J, Zheng Z, Yan X, Shou J, Wang W, et al. Overexpression of the Tibetan Plateau annual wild barley (*Hordeum spontaneum*) HsCIPKs enhances rice tolerance to heavy metal toxicities and other abiotic stresses. Rice 2018;11(1):1–13.

[136] Abedini R, Mohammadi G, Pishkam FR, Pourabed R, Jafarnezhad E, Shobbar A, et al. Plant dehydrins: shedding light on structure and expression patterns of dehydrin gene family in barley. J Plant Res 2017;130(4):747–63.

[137] Panthee DR, Piotrowski A, Ibrahem R. Mapping quantitative trait loci (QTL) for resistance to Late Blight in tomato. Int J Mol Sci 2018;18:1589.

[138] Yang Q, Li Z, Li W, Ku L, Wang C, Ye J, et al. CACTA-like transposable element in ZmCCT attenuated photoperiod sensitivity and accelerated the postdomestication spread of maize. Proc Natl Acad Sci 2013;110(42):16969–74.

[139] Hinchliffe DJ, Condon BD, Thyssen G, Naoumkina M, Madison CA, Reynolds M, et al. The GhTT2_A07 gene is linked to the brown colour and natural flame retardancy phenotypes of Lc1 cotton (*Gossypium hirsutum* L.) fibres. J Exp Bot 2016;67(18):5461–71.

[140] Bian XY, Thomas MR, Rasheed MS, Saeed M, Hanson P, De Barro PJ, et al. A recessive allele (tgr-1) conditioning tomato resistance to geminivirus infection is associated with impaired viral movement. Phytopathology 2007;97(8):930–7.

[141] Li P, Pan T, Wang H, Wei J, Chen M, Hu X, et al. Natural variation of ZmHKT1 affects root morphology in maize at the seedling stage. Planta 2019;249(3):879–89.

[142] Wu X, Skirpan A, McSteen P. Suppressor of sessile spikelets1 functions in the ramosa pathway controlling meristem determinacy in maize. Plant Physiol 2009;149(1):205–19.

[143] Casas AM, Contreras-Moreira B, Cantalapiedra CP, Sakuma S, Gracia MP, Moralejo M, et al. Resequencing theVrs1 gene in Spanish barley landraces revealed reversion of six-rowed to two-rowed spike. Mol Breed 2018;38(5):51.

[144] Vemireddy LR, Kadambari G, Reddy GE, Kola VSR, Ramireddy E, Puram VRR, et al. Uncovering of natural allelic variants of key yield contributing genes by targeted resequencing in rice (*Oryza sativa* L.). Sci Rep 2019;9(1):1–13.

[145] Gao W, Clancy JA, Han F, Prada D, Kleinhofs A, Ullrich SE. Molecular dissection of a dormancy QTL region near the chromosome 7 (5H) L telomere in barley. Theor Appl Genet 2003;107(3):552–9.

[146] Wiegmann M, Maurer A, Pham A, March TJ, Al-Abdallat A, Thomas WT, et al. Barley yield formation under abiotic stress depends on the interplay between flowering time genes and environmental cues. Sci Rep 2019;9(1):1–16.

[147] Morris ML, Bellon MR. Participatory plant breeding research: opportunities and challenges for the international crop improvement system. Euphytica 2004;136(1):21–35.

[148] In: Jain SM, Brar DS, Ahloowalia BS, editors. Molecular techniques in crop improvement-New York, NY, USA: Springer; 2010.

[149] Qin F, Shinozaki K, Yamaguchi-Shinozaki K. Achievements and challenges in understanding plant abiotic stress responses and tolerance. Plant Cell Physiol 2011;52(9):1569–82.

[150] Buerstmayr H, Ban T, Anderson JA. QTL mapping and marker-assisted selection for Fusarium head blight resistance in wheat: a review. Plant Breed 2009;128(1):1–26.

[151] Massman J, Cooper B, Horsley R, Neate S, Dill-Macky R, Chao S, et al. Genome-wide association mapping of Fusarium head blight resistance in contemporary barley breeding germplasm. Mol Breed 2011;27(4):439–54.

[152] Ballini E, Morel JB, Droc G, Price A, Courtois B, Notteghem JL, et al. A genome-wide meta-analysis of rice blast resistance genes and quantitative trait loci provides new insights into partial and complete resistance. Mol Plant Microbe Interact 2008;21(7):859–68.

[153] Danan S, Veyrieras JB, Lefebvre V. Construction of a potato consensus map and QTL meta-analysis offer new insights into the genetic architecture of late blight resistance and plant maturity traits. BMC Plant Biol 2011;11(1):16.

[154] Hutton SF, Scott JW, Yang W, Sim SC, Francis DM, Jones JB. Identification of QTL associated with resistance to bacterial spot race T4 in tomato. Theor Appl Genet 2010;121(7):1275–87.

[155] Jena KK. The species of the genus *Oryza* and transfer of useful genes from wild species into cultivated rice, *O. sativa*. Breed Sci 2010;60(5):518–23.

[156] Ji H, Kim SR, Kim YH, Suh JP, Park HM, Sreenivasulu N, et al. Map-based cloning and characterization of the BPH18 gene from wild rice conferring resistance to brown plant hopper (BPH) insect pest. Sci Rep 2016;6(1):1–14.

[157] Gomi K, Satoh M, Ozawa R, Shinonaga Y, Sanada S, Sasaki K, et al. Role of hydroperoxide lyase in white-backed planthopper (SogatellafurciferaHorváth)-induced resistance to bacterial blight in rice, *Oryza sativa* L. Plant J 2010;61(1):46–57.

[158] Akhtar MA, Abbasi FM, Ahmad H, Shahzad M, Shah MA, Shah AH. Evaluation of rice germplasm against Xanthomonas oryzae causing bacterial leaf blight. Pak J Bot 2011;43(6):3021–3.

[159] Ismail AM, Johnson DE, Ella ES, Vergara GV, Baltazar AM. Adaptation to flooding during emergence and seedling growth in rice and weeds, and implications for crop establishment. AoB Plants 2012;2012:pls019.

[160] Gregorio GB, Senadhira D, Mendoza RD, Manigbas NL, Roxas JP, Guerta CQ. Progress in breeding for salinity tolerance and associated abiotic stresses in rice. Field Crops Res 2002;76(2–3):91–101.

[161] Mishra S, Singh B, Panda K, Singh BP, Singh N, Misra P, et al. Association of SNP haplotypes of HKT family genes with salt tolerance in Indian wild rice germplasm. Rice 2016;9(1):15.

[162] Niroula RK, Pucciariello C, Ho VT, Novi G, Fukao T, Perata P. SUB1A-dependent and-independent mechanisms are involved in the flooding tolerance of wild rice species. Plant J 2012;72(2):282–93.

[163] Barclay A. Cultivated rice receives help from wild cousins. Appropr Technol 2004;31(4):10.

[164] Khush GS, Angeles E, Virak PS, Brar DS. Breeding rice for resistance to tungro virus at IRRI. SABRAO J Breed Genet 2004;36(2):101–6.

[165] Prischmann DA, Dashiell KE, Schneider DJ, Eubanks MW. Evaluating Tripsacum-introgressed maize germplasm after infestation with western corn rootworms (Coleoptera: *Chrysomelidae*). J Appl Entomol 2009;133(1):10–20.

[166] De la Paz GS, Sánchez GJJ, Ruiz CJA, Ron PJ, Miranda MR, De la Cruz L, et al. Diversidad de especiesinsectilesenmaíz y teocintleen México. Folia Entomol Mexico 2010;48(2):1–6.

[167] Kling JG, Fajemisin JM, Badu-Apraku B, Diallo AO, Menkir A, Melake-Berhan A. Striga resistance breeding in maize. In: Breeding for striga resistance in cereals. Proceedings of a Workshop; Ibadan, Nigeria; 18-20 Aug 1999. Haussmann BIG, Hess DE, Koyama ML, Grivet L, Rattunde HW, Geiger HH, editors. Weikersheim (Germany):Margraf Verlag; C2000 (No. CIS-2475. CIMMYT.).

[168] Rich PJ, Ejeta G. Towards effective resistance to Striga in African maize. Plant Signal Behav 2008;3(9):618–21.

[169] Nault LR, Gordon DT, Damsteegt VD, Iltis HH. Response of annual and perennial teosintes (*Zea*) to six maize viruses. Plant Dis 1982;66(1):61–2.

Index

Note: Page numbers followed by "f" indicate figures, "t" indicate tables.

C

H

I

J

K

L

M

N

O

S

T

FSC
www.fsc.org
MIX
Papier | Fördert
gute Waldnutzung
FSC® C083411

Kolibri 360 GmbH
Ferdinand-Jühlke-Straße 7
99095 Erfurt, Deutschland
produktsicherheit@kolibri360.de